Integrated Design of Multiscale, Multifunctional Materials and Products

Integrated Design of Multiscale, Multifunctional Materials and Products

David L. McDowell
Jitesh H. Panchal
Hae-Jin Choi
Carolyn Conner Seepersad
Janet K. Allen
Farrokh Mistree

AMSTERDAM • BOSTON • HEIDELBERG • LONDON
NEW YORK • OXFORD • PARIS • SAN DIEGO
SAN FRANCISCO • SINGAPORE • SYDNEY • TOKYO

Butterworth-Heinemann is an imprint of Elsevier

Butterworth-Heinemann is an imprint of Elsevier
30 Corporate Drive, Suite 400, Burlington, MA 01803, USA
Linacre House, Jordan Hill, Oxford OX2 8DP, UK

Notices
Knowledge and best practice in this field are constantly changing. As new research and experience broaden our
understanding, changes in research methods, professional practices, or medical treatment may become necessary.
Practitioners and researchers must always rely on their own experience and knowledge in evaluating and using any
information, methods, compounds, or experiments described herein. In using such information or methods they
should be mindful of their own safety and the safety of others, including parties for whom they have a professional
responsibility.
To the fullest extent of the law, neither the Publisher nor the authors, contributors, or editors, assume any liability
for any injury and/or damage to persons or property as a matter of products liability, negligence or otherwise, or
from any use or operation of any methods, products, instructions, or ideas contained in the material herein.

Library of Congress Cataloging-in-Publication Data
Application submitted

British Library Cataloguing-in-Publication Data
A catalogue record for this book is available from the British Library.

ISBN: 978-1-85617-662-0

For information on all Butterworth-Heinemann publications
visit our Web site at www.elsevierdirect.com

Printed in the United States of America
09 10 11 12 13   10 9 8 7 6 5 4 3 2 1

Working together to grow
libraries in developing countries

www.elsevier.com | www.bookaid.org | www.sabre.org

ELSEVIER   BOOK AID International   Sabre Foundation

Contents

Preface

To some extent, design of materials has long been a preoccupation of materials scientists and engineers. However, the historical emphasis on laborious, intuitive, serendipitous materials discovery has only relatively recently been augmented by computer simulations. This speeds up discovery of new material solutions, as well as more rapidly assessing process-structure-property relations upon which modern materials science is based. Fueled by the recent emergence and rapid growth of computational methods, a materials design revolution is underway in which the classical materials selection approach is replaced by the simulation-based design of material microstructure or mesostructure to satisfy multiple performance requirements of the component or system, subject to constraints on certain material properties and other aspects of the system.

Materials typically used in applications today have complex, heterogeneous microstructures with different characteristic length scales, and these microstructures affect processing, manufacture, and in-service performance. In the past fifty years, research in theory of dislocations, phase transformations, and micromechanics of heterogeneous materials enabled explicit consideration of the role of microstructure on the properties of metallic systems and certain classes of composites. Pivotal to progress in the present materials design revolution are the works of Olson (Olson 1997) on combining elements of reductionist, bottom-up modeling with deductionist, top-down systems design of materials. The perspective of a material as a complex hierarchical system is instrumental in drawing analogy to subsystems and components considered in conventional design. Moreover, the contribution of Ashby and coworkers (Ashby 1999) regarding systematic definition and execution of the materials selection problem for various performance requirements is acknowledged as foundational to relating properties to performance.

The core of materials design is the interplay of hierarchical systems-based design of materials and multiscale/multilevel modeling methodologies, embedded within a computational framework that supports coordination of information and human decision making. We add to these developments elements of decision-based robust design of engineering systems. Why? In spite of great advances in material modeling and simulation, from atomic scales upward, these approaches have inherent uncertainty when it comes to predicting the properties of "real" materials; moreover, there are gaps of nearly intractable nature in methods for concurrent bridging of length and time scales in modeling material processing, deformation, and failure. While emerging high-performance computing and related simulation tools provide a more predictive foundation to support materials design, brute force methods based on atomistics or concurrent multiscale modeling are unlikely to have sufficient capabilities, combined with issues such as tractability and uncertainty, to support a broad range of materials design problems. A systems approach is required to address the nonlinear, hierarchical nature of real materials. Such an approach has proven to be beneficial for design of other complex systems (e.g., aircraft, automobiles, circuit boards, etc.).

Systems-based engineering design principles have developed substantially in the last few decades. An October 1998 workshop on materials design science and engineering (McDowell and Story 1998), sponsored by the National Science Foundation and cohosted by Georgia Tech and Morehouse College, was held to discuss interdisciplinary frameworks necessary to facilitate concurrent design of materials and products, and to replace "hit and miss" materials discovery with systematic methods for materials development that draw on combined elements of materials characterization and computational simulation. This is driven by an inexorable technology pull by the marketplace toward rapid development cycles for new and improved materials. Potential benefits include a virtual manufacturing environment that goes well beyond geometric modeling to include many aspects of the physical behavior of materials in simulating system-level response. Moreover, tailoring materials to enhance the performance envelope is an imperative as we consider future requirements for increasing efficiency and environmentally sustainable solutions, for example. This workshop noted that a change of culture is necessary in universities and industries to cultivate new concepts for materials design. The resulting roadmap for materials design focused on the following foundational technologies:

- Principles and approaches for more quantitative materials design.

- Enhanced modeling and simulation tools.

- Validated, reliable, and comprehensive databases.

- Methods for in situ characterization and testing.

In this book, we address the first bullet in the previous list, namely, systems strategies for concurrent robust design of materials and systems, along with elements of distributed modeling and simulation environments. Materials design falls under the general category of simulation-based design, in which computational materials science and multiscale mechanics modeling play key roles in evaluating performance metrics necessary to support materials design. Major findings of the May 2006 Report of the U.S. National Science Foundation Blue

Ribbon Panel on Simulation-Based Engineering Science (SBES) (Oden, Belytschko et al. 2006) can be summarized as follows:

- "SBES is a discipline indispensable to the nation's continued leadership in science and engineering...

- Formidable challenges stand in the way of progress in SBES research. These challenges involve resolving open problems associated with multiscale and multiphysics modeling, real-time integration of simulation methods with measurement systems, model validation and verification, handling large data, and visualization..."

The materials design approach advocated here invokes the notion of robust design, i.e., insensitivity of the desired response to any number of sources of uncertainty or variability having to do with material composition, processing, microstructure, service history, models, and model parameters, coupling of models at different length and/or time scales, chaining of design decisions or simulation outputs/inputs, etc. In many respects, it is a manifestation of SBES. To address robust design of materials, we have developed new concepts that extend existing methods and facilitate top-down design in the presence of complexity and uncertainty that is characteristic of hierarchical material systems.

Figure P.1 encompasses the goals of concurrent material and product design, showing that already established methods of design-for-manufacture of parts, subassemblies, assemblies, and overall systems can be extended to address the multiple scales that control property-performance attributes of materials. Hence, the objective of tailoring the material to specific applications (to the left of the vertical bar in Figure P.1) is patently distinct from traditional materials selection that is common in practice. The problem is that the systems-based design methods used to design parts, components, and assemblies must be extended to consider the nuances of process-structure-property relationships in materials in the presence of significant uncertainty. The hierarchy of scales from quantum to continuum on the left side of Figure P.1 may be viewed as a multiscale modeling problem, but this is a reductionist, bottom-up perspective. The materials design challenge is to develop methods that employ bottom-up modeling and simulation, calibrated and validated by characterization and measurement to the extent possible, yet permit top-down design over the hierarchy of material length scales shown in Figure P.1.

Our intention in this book is to provide a connection between several key primary disciplines or endeavors that have been traditionally distinct but naturally combine to serve as the foundation of modern materials design: (1) systems-based engineering design, (2) computational materials science and engineering, (3) robust systems design, and (4) information technology. It is targeted to serve as a useful reference in emerging methods for concurrent design of materials and products for product designers, materials scientists and

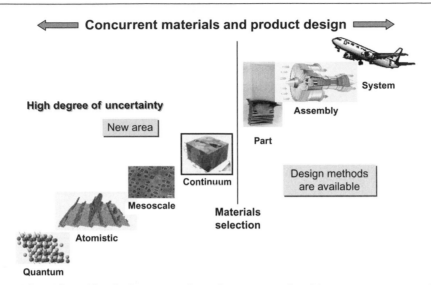

Figure P.1 Hierarchy of levels from atomic scale to system level in concurrent materials and product design. Existing systems design methods focus on levels to the right of the vertical bar, treating the materials design by selecting the material.

engineers, applied mechanics researchers, and other analysts involved in multiscale modeling of material behavior and associated process-structure-property relations.

The reader may find that the premise behind this book—that materials can be designed concurrently with products—is not the usual way of doing business in many organizations that traditionally separate the functions of materials development from systems design along organizational lines, using material properties as the mode of communication/information exchange. It also differs from the way design is taught in most engineering programs, including materials science and engineering. Indeed, the ideas presented here represent our long-term prospectus for how this might be done in the "near tomorrow" as computing power increases, engineering becomes increasingly multidisciplinary, multi-resolution, and multiscale material modeling methods blossom, and systems approaches that emerge from the engineering design, multidisciplinary optimization, and information sciences communities become a part of the engineering lexicon. To this end, we look toward the horizon beyond the current state of the curriculum, university disciplinary structures, or management of design processes in industry. We trust that the initial, embryonic ideas presented here will serve as an impetus for the students of today and technology leaders of tomorrow in various aspects of materials engineering and product development to consider the richness of opportunities that lie ahead in a digital, highly connected world.

In this book, we incorporate ideas and material from the dissertations of three former doctoral students in the Systems Realization Laboratory at Georgia Tech—Carolyn Conner Seepersad,

HaeJin Choi, and Jitesh Panchal. In addition, we have all benefited from the rich interactions among students and faculty in the Systems Realization Laboratory. At the peril of omitting some names, we acknowledge Wei Chen (Northwestern University), Tim Simpson (Penn State University), and Kemper Lewis (University of Buffalo), whose doctoral work at Georgia Tech has had a major impact on the material presented herein. In addition, several graduate students and postdoctoral fellows in mechanics of materials at Georgia Tech have contributed substantially to the formative stages of this work. These include doctoral students Ryan Austin and Jim Shepherd, who provided essential framing of computational mechanics issues and codes that contributed to several materials design scenarios presented here as examples, as well as postdocs Aijun Wang and Rajesh Kumar, who developed valuable analysis tools and methods for use in cellular materials design problems.

Pursuit of many of the concepts outlined in this book received financial support from several sources. Concepts for robust design of materials were initially developed in the context of cellular materials applications with the support of the Defense Advanced Research Projects Agency (DARPA) (N00014-99-1-1016) from 1999 to 2002, monitored by Dr. Leo Christodoulou, and by the Office of Naval Research (ONR) (N0014-99-1-0852), under Dr. Steven Fishman. Support from 2002 to 2007 by the Air Force Office of Scientific Research (AFOSR) Multi-University Research Initiative Grant on Energetic Structural Materials (1606U81, Craig Hartley, monitor) is gratefully acknowledged, as is more recent, ongoing support of the NSF (I/UCRC) Center for Computational Materials Design (CCMD), a joint venture of partner institutions Penn State and Georgia Tech.

References

Ashby, M.F. 1999. Materials Selection in Mechanical Design. Butterworth-Heinemann: Oxford, UK.

McDowell, D.L., Story, T.L., 1998. New directions in materials design science and engineering (MDS&E). Report of a NSF DMR-sponsored workshop on materials design science and engineering, Atlanta, GA, October 19–21.

Oden, J.T., Belytschko, T., Fish, J., Hughes, T.J.R., Johnson, C., Keyes, D., Laub, A., Petzold, L., Srolovitz, D., Yip, S., 2006. Simulation-based engineering science: Revolutionizing engineering science through simulation. In: A Report of the National Science Foundation Blue Ribbon Panel on Simulation-Based Engineering Science. National Science Foundation: Arlington, VA.

Olson, G.B. 1997. Computational design of hierarchically structured materials. Science, 277 (5330), 1237–1242.

Authors

DAVID L. McDOWELL   *Regents' Professor and Carter N. Paden, Jr. Distinguished Chair in Metals Processing*, Woodruff School of Mechanical Engineering, Georgia Institute of Technology, Atlanta, GA 30332
david.mcdowell@me.gatech.edu

JITESH H. PANCHAL *Assistant Professor,* School of Mechanical and Materials Engineering, Washington State University , Pullman, WA 99163
panchal@wsu.edu

HAE-JIN CHOI   *Assistant Professor*, Division of Systems and Engineering Management, School of Mechanical & Aerospace Engineering, College of Engineering, Nanyang Technological University, Singapore
hjchoi@ntu.edu.sg

CAROLYN CONNER SEEPERSAD   *Assistant Professor*, Product, Process, and Materials Design Lab, Mechanical Engineering, The University of Texas at Austin, Austin, TX 78712
ccseepersad@mail.utexas.edu

JANET K. ALLEN   *John and Mary Moore Chair*, School of Industrial Engineering, University of Oklahoma, Norman, OK 73019
janet.allen@ou.edu

FARROKH MISTREE   *L.A. Comp Chair and Director*, School of Aerospace and Mechanical Engineering, University of Oklahoma, Norman, OK 73019
farrokh.mistree@ou.edu

About the Authors

David L. McDowell

Dave McDowell joined Georgia Tech in Fall 1983 after graduating from the University of Illinois at Urbana-Champaign. Regents' Professor and Carter N. Paden, Jr. Distinguished Chair in Metals Processing, McDowell holds a dual appointment in the GWW School of Mechanical Engineering and the School of Materials Science and Engineering. He serves as Director of the Mechanical Properties Research Laboratory and as Chair of the Georgia Tech Materials Council. McDowell's research focuses on the synthesis of experiment and computation to develop physically based constitutive models for nonlinear and time-dependent behavior of materials, with emphasis on wrought and cast metals. Research topics of interest include cyclic plasticity and microstructure-sensitive fatigue modeling of metallic systems, finite strain plasticity and defect field theories, behavior of metallic foams  and honeycomb materials, creep-fatigue-environment interaction, thermo-mechanical fatigue, time-dependent fracture, atomistic simulations with focus on dislocation-interface reactions, phase transformations, and multiscale modeling. McDowell currently serves on the editorial board of the International Journal of Plasticity as well as several others, and is co-Editor of the International Journal of Fatigue. For the past 15 years he has pursued definition and more systematic development of the field of multifunctional design of materials over a broad range of applications, including metal honeycombs for cooling and structural applications, lightweight transportation vehicles, components for hot sections of aircraft gas turbine engines, protective armor systems, and energetic material systems. Author of over 340 technical papers, McDowell serves as Co-Director of the NSF-sponsored Center for Computational Materials Design, a joint Penn State-Georgia Tech I/UCRC.

http://www.me.gatech.edu/faculty/mcdowell.shtml
http://mprl.me.gatech.edu/index.php
e-mail: david.mcdowell@me.gatech.edu

Jitesh H. Panchal

Jitesh Panchal is an Assistant Professor in the School of Mechanical and Materials Engineering at Washington State University. Before joining Washington State University in Fall 2008, he was a Visiting Assistant Professor in the Woodruff School of Mechanical Engineering at the Georgia Institute of Technology. He received his Bachelor of Technology (B.Tech.) degree from Indian Institute of Technology at Guwahati, and MS and PhD in Mechanical Engineering from Georgia Institute of Technology in 2003 and 2005, respectively. Panchal's research interests are in establishing design methodologies and computational tools for integrated product and materials design. His focus is on managing complexity and uncertainty in multilevel design processes. He is interested in developing open computational environments for the integrated design of products and materials. Panchal is excited about developing systematic approaches for leveraging mass-collaboration in engineering design. He is actively involved in developing new pedagogical approaches for fostering collective learning and achieving mass customization in education. He is a member of American Society of Mechanical Engineers (ASME), and American Society for Engineering Education (ASEE).

http://www.mme.wsu.edu/people/faculty/faculty.html?panchal
e-mail: panchal@wsu.edu

Hae-Jin Choi

Hae-Jin Choi is an assistant professor in the School of Mechanical and Aerospace Engineering, Nanyang Technological University (NTU), Singapore. Before joining in NTU, he served as a postdoctoral fellow in the Woodruff School of Mechanical Engineering at Georgia Tech after earning Ph.D. (2005) and M.S. (2001) degrees at Tech. He received B.S. and M.S. degrees from Yonsei University in Seoul, Korea in 1995 and 1997, respectively. His research focuses on management of uncertainty, integrated materials and products design, multiscale simulation-based design, strategic product design, and distributed collaborative product realization. Choi proposed robust design methods for managing variability and uncertainty in complex systems. Applications of the methods include design of multifunctional energetic structural materials, blast resistant panels, and multifunction linear cellular alloys. A wide range of additional design applications, such as bio mass sensors, lightweight tactical bridges, and underwater power generation systems, are currently under development. He is a member of the American Society of Mechanical Engineers and the American Institute of Aeronautics and Astronautics.

http://www3.ntu.edu.sg/home/hjchoi/
e-mail: hjchoi@ntu.edu.sg

Carolyn Conner Seepersad

Carolyn Conner Seepersad is an Assistant Professor of Mechanical Engineering at the University of Texas at Austin. She received a PhD in Mechanical Engineering from Georgia Tech in 2004, an MA/BA in Philosophy, Politics and Economics from Oxford University in 1998, and a BS in Mechanical Engineering from West Virginia University in 1996. She is a former Rhodes Scholar, Hertz Fellow, and NSF Graduate Research Fellow. Dr. Seepersad's research involves the development of methods and computational tools for engineering design, with an emphasis on multilevel design of products and materials. Topics of interest include the computational design and solid freeform (additive) fabrication of customized mesostructure, including multifunctional honeycomb structures for applications such as structural heat exchangers, acoustic energy absorbing devices, and structures with spatially tailored stiffness. Related topics include computational techniques for coordinating design exploration activities on different levels or scales, and predictive process control techniques for incorporating physics-based models into the design of manufacturing processes. Additional research interests include product customization and green design. Dr. Seepersad currently serves as a co-organizer of the annual Solid Freeform Fabrication Symposium, and she is the author of more than 60 journal papers and full-length conference publications. She teaches courses on product design, solid freeform fabrication, and design of complex engineered systems.

http://www.me.utexas.edu/directory/faculty/seepersad/carolyn/121/
http://www.me.utexas.edu/~ppmdlab/
e-mail: ccseepersad@mail.utexas.edu

Janet K. Allen

Janet Allen earned her doctorate from the University of California at Berkeley in 1973 and her S.B. from the Massachusetts Institute of Technology in 1967. Dr. Allen's expertise lies in the area of simulation-based design of engineering systems, especially in the management and mitigation of uncertainty in the early stages of design. There are two conceptually different approaches to accomplishing this. The first is to mitigate uncertainty by accurately modeling uncertainty and/or by reducing it. It is impossible to accomplish this completely. The second approach is to use robust design principles to improve the quality of products and processes by reducing their sensitivity to variations, thereby reducing the effects of uncertainty without removing its sources. During the course of her career,

Dr. Allen has studied both aspects of the problem. Her most significant accomplishments lie in the area of robust design. However, this is by no means the only research thread embodied in Dr. Allen's research portfolio, among other things, she has studied the use of living systems theory in design, design education, modeling design processes, and design for X. Professor Allen is a Fellow of the American Society of Mechanical Engineers, a Senior Member of AIAA, a Member of ASEE and an Honorary Member of the Mechanical Engineering Honor Society, Pi Tau Sigma. She is an Associate Editor of the ASME Journal of Mechanical Design. With her students, she has authored well over 200 technical publications. A Professor Emeritus in the Woodruff School of Mechanical Engineering at the Georgia Institute of Technology, she presently holds the John and Mary Moore Chair in the School of Industrial Engineering at the University of Oklahoma.

http://www.srl.gatech.edu/Members/jallen
e-mail: janet.allen@ou.edu

Farrokh Mistree

Farrokh Mistree earned his doctorate in engineering (with minors in Operations Research and Business Administration) from the University of California, Berkeley in 1974 and earned his first degree in Naval Architecture from the Indian Institute of Technology at Kharagpur. He served at the University of New South Wales, (1974–80), The University of Houston (1981–92)

and joined the Georgia Institute of Technology as Professor of Mechanical Engineering in September 1992. Professor Farrokh Mistree's *design experience* spans mechanical, aeronautical, structural, and industrial engineering. His *teaching experience* spans courses in engineering design, naval architecture, solid mechanics, operations research, and computer science. His current research focuses on learning how to manage design freedom in multiscale design (from molecular scales to reduced order models) to facilitate the integrated design of materials, product and design process chains. He is committed to developing a design pedagogy that is rooted in Decision-Based Design and adaptive action learning. It is in this context that he enjoys experimenting with ways in which design can be learned and taught. He is a Fellow of ASME, an Associate Fellow of the AIAA, a Member of ASEE and the Society of Naval Architects and Marine Engineers. Professor Mistree has co-authored two textbooks and over 350 technical publications. He was recognized for his research by receiving the ASME Design Automation Committee's 1999 Design Automation Award and was honored for outstanding teaching as recipient of the 2001 Jack M. Zeigler Woodruff School Outstanding Educator Award. He served as the National Secretary-Treasurer of Pi Tau Sigma Mechanical Engineering Honor Society (1995–2008) and as a reviewer for ABET (1996–2002). From July 1995 through December 2008, Professor Mistree served as the Associate Chair for the Woodruff School of Mechanical Engineering and the Associate Director of Georgia Tech Savannah. A Professor Emeritus in the Woodruff School of Mechanical Engineering at the Georgia Institute of Technology, he presently holds the L.A. Comp Chair and serves as Director of the School of Aerospace and Mechanical Engineering at the University of Oklahoma.

http://www.srl.gatech.edu/Members/fmistree
e-mail: farrokh.mistree@ou.edu

Photos by Gary Meek and Michelle White

Integrated Material, Product, and Process Design—A New Frontier in Engineering Systems Design

1.1. Motivation for the Integrated Design of Materials and Products

For millennia, the technological capabilities of societies have been so closely linked to available materials that entire eras—the Bronze Age and the Iron Age, for example—have been named for the most advanced engineered materials of the day. Even the modern Information Age owes its name to a revolution in information technology made possible by critical advances in Si-based semiconductors and other materials. The continuing technological advancement of our society is tied closely to our ability to engineer materials that meet the increasingly ambitious requirements of new products. In fact, advancement in materials technology is an enabling element of the exponential rate of technology development. In view of the acceleration of new technology development, it is no longer appropriate to categorize epochs according to material class; we live in an age of synthesis or integration of new and improved materials with electronics, our environment, and even our human bodies via medical implants. The notion of *designing* materials to best suit this integration is a relatively recent trend.

Design has been defined by the U.S. National Science Foundation (NSF) as a process by which products, processes, and systems are created to perform desired functions through specification. The fundamental objective in a design process is "to transform requirements—generally termed 'functions,' which embody the expectations of the purposes of the resulting artifact, into design descriptions" (Gero 1990). Complex new products and systems are currently realized with increasingly sophisticated and effective systems design techniques that have been shown to decrease product development cycle times and increase quality. Like the aircraft illustrated in Figure 1.1, many of these complex systems are realized by concurrently designing the subsystems, components, and parts into which a system is decomposed. However, the design process typically stops at the "part" level—rather than the "material" level—of the system hierarchy illustrated in Figure 1.1. Materials are typically selected—not designed—from a database of available options (Ashby 1999). Accordingly, the performance of many engineered

DOI: 10.1016/B978-1-85617-662-0.00001-6

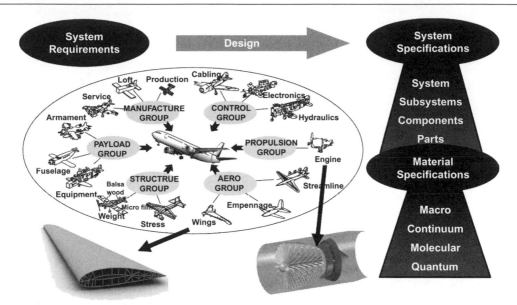

**Figure 1.1 An aircraft as a complex, hierarchical, multiscale system
(Seepersad 2004).**

parts and systems is limited by the properties of available constituent materials. For example, the further increase in efficiency and reduction in emission of aircraft gas turbine engines shown in Figure 1.1 requires high temperature, high strength, structural materials for the engine combustor liners, other hot section components, and gas turbine engine disks and blades. Unfortunately, these combinations of properties are not typically available from materials in current databases. The inherent difficulty with materials selection is the inability to tailor a material for application-specific requirements—such as those of the turbine engine hot section—that may conflict in terms of demands on material structure and properties. On the other hand, lead times for the development of *new* materials have remained relatively constant and unacceptably long relative to the desired product development cycle. The lengthy time frame and expense of new materials development is due in part to the predominantly empirical, trial-and-error approach adopted historically by materials engineers and developers (McDowell and Story 1998, Olson 2000).

A foundational premise for the field of materials design is that systems design techniques offer the potential for tailoring materials—as well as products that employ them—to serve the demands of multifunctional applications. We contend that the concept of materials design is not limited to *selecting* an available material from a database; instead, we actually *tailor* material structure at various levels of hierarchy (atoms, microstructure, etc.) via associated processing paths to achieve properties and performance levels that are customized for a particular application. In the materials science and engineering design communities, momentum is building toward materials design and away from exclusively empirical materials development approaches (McDowell and Story 1998). Materials scientists and

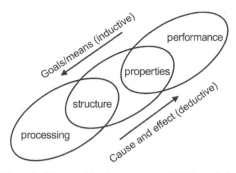

Figure 1.2 Olson's hierarchical concept of "Materials by Design"
(Olson 1997).

engineers facilitate materials design by creating increasingly sophisticated, realistic, and accurate models for material structure and properties that can be used to support a design process to satisfy a ranged set of performance requirements. In addition, product and systems designers recognize the potential technological breakthroughs that can be achieved by concurrently designing *materials and products*, thereby overcoming the property and performance limitations of currently available materials for a host of applications.

At this point, it is useful to formalize a definition for the scope of concurrent design of materials and products, which we will refer to as *materials design*. The term *materials design* may have different meanings to different people and audiences. Accordingly, our definition is *the top-down driven, simulation-supported, decision-based design of material hierarchy to satisfy a ranged set of product-level performance requirements.*

It is quite clear that materials design is a challenging proposition. Materials are complex, multiscale systems with phenomena manifested at a hierarchy of length scales. Time scales range from atomic vibrations (femtoseconds) to relaxation of microstructure (nanoseconds to microseconds) to structural performance (seconds, hours, days, years).

Materials design efforts therefore rely on continuous development and improvement of predictive models and simulations on a hierarchy of length scales, quantitative representations of structure, and effective archiving, management, and visualization of materials-related information and data. Together, these components provide important *deductive* links in the chain of processing-structure-properties-performance, as illustrated in Figure 1.2. Such deductive, bottom-up analytical tools are necessary but not sufficient for materials design. As discussed by Olson (Olson 1997) and illustrated in Figure 1.2, materials design is fundamentally an inductive, goal-oriented **synthesis** activity aimed at identifying material structures and processing paths that deliver required properties and performance. Like engineering systems design, it is a top-down exercise.

While Olson's construct provides a foundation on which to build systems-based materials design, it does not fully illuminate the processes and strategies by which top-down design is to be carried out. Without an accompanying systems-based design strategy, it relegates practical aspects of inductive materials design to the creative will, depth of insight, experience, ability, and knowledge base of the individual designer. Important practical challenges must be addressed:

- Material structure and behavior is uncertain;

- Material models are uncertain and may be incomplete;

- Models are typically applied in bottom-up fashion, both for multiple scales of material structure and within the context of the linear deductive structure shown in Figure 1.2;

- Design goals change with time and adjust to new possibilities, such that the design process must be flexible and adaptable; and

- Materials engineers, analysts, and systems designers are distributed geographically within different organizations and have different backgrounds and expertise.

In view of the prevalent role of uncertainty, robust design methods are desirable. Effective, efficient, *systems-based robust design methods* are needed for modeling and executing complex, hierarchical materials design processes. A broad set of challenges and opportunities for concurrent design of materials and products is outlined in the report by the U.S. National Academy of Engineering National Materials Advisory Board on Integrated Computational Materials Engineering (ICME) (Pollock, Allison et al. 2008). These notions have gained considerable traction in industry. This book is devoted to exploring implications for systems strategies for concurrent design of materials and products, building on the foundations of Olson's ideas embodied in *Materials by Design*® and ICME.

1.2. Systems-based Multilevel Materials Design

Systems-based multilevel design involves accounting for all aspects of systems from lower length and time scales to higher scales, addressing the multiscale nature of physical structure and behavior. The design process is multilevel in the sense that decisions must be made with respect to structure at each level of material and product hierarchies. Multiple levels of models must be integrated with design decisions in concurrent materials/product design. The design process is multiscale in the sense that multiple length and time scales of material structure and responses are addressed by different models. Multilevel design seeks to make risk-informed design decisions at all scales.

Design has traditionally involved selecting a suitable material for a given application (see Figure 1.3). In fact, engineering design textbooks (Norton 2006, Pahl and Beitz 1996, Shigley and Mischke 1989) and undergraduate design courses have typically conveyed the

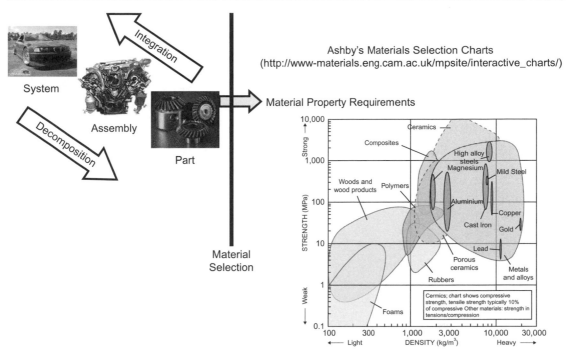

Figure 1.3 Materials selection in systems design.

idea that materials should be selected based on tabulated databases of properties. However, the relatively recent use of tailored materials that did not exist in antiquity, such as graphite epoxy composite materials in sporting goods and aerospace applications or advanced Ni-base superalloys in hot section gas turbine engine components, have promoted reexamination of this concept. The key element in tailoring these materials is a quantitative understanding of the relation of process route to microstructure, structure to properties, and properties to performance. This is not a new realization—ancient samurai swords were processed by experts with extensive empirical knowledge of the interplay of composition and process route on microstructure and resulting properties (Olson 2000). Today, however, we are poised to synthesize knowledge of such relations using computational modeling and simulation.

1.2.1. Mappings in Materials Design

Olson's conceptualization in Figure 1.2 (Olson 1997) has provided a philosophical underpinning for the inductive, top-down engineering approach to materials design.

The flow of information in Olson's diagram constitutes of a set of mappings, each of which involves intermediate models or data visualization, databases, interfaces between models and/or databases, and/or human decisions. This is illustrated in Figure 1.4. We see that certain

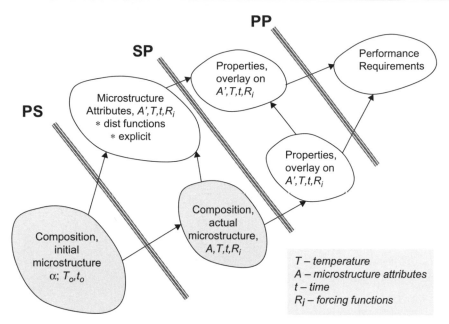

Figure 1.4 Hierarchy of mappings from process route to microstructure to properties to performance.

classes of mappings (models, codes, and heuristics, as well as experiments) are necessary to support materials design:

- *Process-structure (PS) relations:* Used to establish manufacturing constraints, cost factors, and thermodynamic and kinetic feasibilities of process route.

- *Structure-property (SP) relations:* Most often intrinsically hierarchical in nature, these are relations between composition, phases, microstructure morphology, and response functions or properties of relevance to desired performance requirements.

- *Property-performance (PP) relations:* Relations between properties and response functions and specified performance requirements, either through detailed point-by-point computational models or by construction of approximate response surface or surrogate models.

In Figure 1.4, this is further broken down into mappings of vertical (PS, SP, and PP) and lateral type, the latter involving a reduction of degrees of freedom of the material representation and/or associated models. Vertical mappings of SP type are sometimes referred to as *homogenization relations* and normally involve a reduction in the model degrees-of-freedom (condensation) of an equivalent response function or model. Such homogenization may be achieved in numerous ways, ranging from concurrent multiscale models, domain decomposition for discrete to continuous transitions, two-scale homogenization, statistical mechanics of evolutionary systems, "handshaking" methods for informing reduced degree-of-freedom (DOF) models based on results of higher DOF models, variational methods such as the generalized principle of virtual velocities (Needleman and Rice 1980), and a number of other emerging computational multiscale modeling approaches. Decision nodes can be inserted at each vertical or lateral mapping at which designers participate to interpret and negotiate information to be transmitted with the mapping.

The horizontal or lateral mappings at each level are also indicated in Figure 1.4. For example, the actual microstructure can be represented in explicit digital format, or distribution functions of geometric attributes (e.g., grains, phases, particles) can be computed and stored, with a significant reduction in DOF. Following this lateral mapping, microstructures can be reconstructed and then analyzed for properties. Since DOF are condensed in the lateral mapping, information content is reduced and it is necessary to proceed forward with the projection of properties from these reconstructed microstructures. The resulting responses or properties mapped from these reduced DOF microstructure representations differ from those based on simulations (projections) performed on the basis of digital images of actual microstructures. Hence, there is loss of information and propagation of uncertainty associated with such mappings. It is important to note, however, that materials design involves exploration of materials and microstructures that do not exist and therefore reconstructions based on distribution functions are an important starting point for many simulations, anchored by limited numbers of actual microstructures when available.

1.2.2. Multiscale Computational Modeling of Materials

The mappings in Figure 1.2 are distinguished from a hierarchy of material length (and time) scales that are pertinent to materials design. Numerous methods for multiscale modeling have been developed to bridge length scales from atomic to molecular to mesoscopic to continuum and macroscale, with associated time scales ranging from femtoseconds to nanoseconds to seconds, hours, days, and years. There are, of course, practical limitations to concurrent multiscale modeling (linking models at different scales or levels of refinement in an explicit way to facilitate fully coupled bottom-up and top-down modeling) even beyond computational issues. These include, in part, assignment of initial conditions for all evolving microstructure attributes, various model approximations, as well as simplifications in models at each scale to achieve concurrency.

In Figure 1.5, we show how experimental measurements are selectively performed to (i) provide input to models (parameters, initial conditions, configurations), (ii) augment or replace models with measured data to formulate meta-models (e.g., response surfaces and artificial neural networks), and/or (iii) calibrate and validate modeling and simulation methods and tools. Such experiments are only one component of validation of the overall framework. The issue of validation is much more comprehensive, having to do as well with

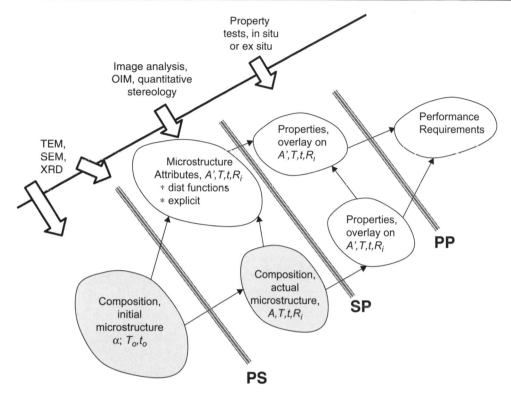

Figure 1.5 Infusion of experiments [Transmission Electron Microscopy (TEM), Scanning Electron Microscopy (SEM), X-Ray Diffraction (XRD), Orientation Imaging Microscopy (OIM)] at different levels of the multilevel materials design hierarchy, indicating how composition, microstructure, and resulting properties can all be informed or calibrated across Process-Structure (PS), Structure-Property (SP) and Property-Performance (PP) mappings.

logic of information flow, consistency of chained models with decision support requirements, model implementation, etc. Validation is discussed in Chapter 4. For some phenomena, such as nonequilibrium material processing, mechanisms may be poorly understood and model uncertainty is so high that design of experiments process route exploration must be carried out experimentally. Historically, that has been the route of materials development. However, when used in combination with process models that can be calibrated by experiment, the ability to search the feasible design space in process-structure mappings can be expanded.

1.2.3. Decision Support in Multilevel Concurrent Design of Materials and Products

This interplay of material processing, characterization, and experiments brings us to a very important perspective about concurrent materials and product design. We do not embrace the notion of *fully automated* simulation-based materials design. Rather, we recognize that only certain classes of materials design problems can be addressed by automated searches of predicted properties of materials that can be synthesized or processed to meet performance requirements. Examples include:

- Combinatorial searches for molecules for pharmaceutical applications (molecular sieves, virus blockers, etc.)

- Design of composites for elastic properties (fiber-reinforced composites, textured polycrystals, etc.)

In such cases, one can identify and explore the space of feasible material structure and then exercise structure-property models. In the first case, atomic-level modeling tools must be used (based on first principles *ab initio* or molecular mechanics), while in the second case, continuum models apply. But the applications that consider materials design as a database search and/or data mining exercise are very limited owing to the many challenges pointed out in Section 1.4.

With regard to materials design objectives, concurrent multiscale modeling schemes or homogenization concepts may be unnecessary in many cases, because the goal is *not* to accurately predict properties but to understand their sensitivity to microstructure and to capture essential dominant mechanisms and their transitions as a function of forcing functions and responses applicable to the given design scenario. *Although seamless "bottom-up" modeling is appealing from a scientific perspective, it is often too idealized and chained with a compilation of approximations that may compromise viability for materials design.* In addressing most applications involving integrated materials and product design, multilevel design demands that models be employed in the range for which they are most appropriate and accurate, using *decision-based protocols* for passing information from one level to the next, as shown in Figure 1.6. The decision support nodes in Figure 1.6 can employ the construct of compromise Decision Support Problems set forth in several reference sources (Mistree, Hughes et al. 1993, Muster and Mistree 1988) and discussed in Chapter 5. Of course, explicit, seamless scale transitions can be utilized for certain mappings as well, but the utility of doing this must be determined by the degree of coupling in the system. Decisions to decouple models executed at different levels in Figure 1.5 and Figure 1.6, as well as at different length and time scales, must consider the *value-of-information* that is compromised in so doing. A metric for value-of-information is assessed by conducting level transitions involving decisions shown in Figure 1.6 and comparing relative performance loss (change of objective function) based on weakly coupled models with representative cases using strongly coupled models (Panchal, Choi et al. 2006, Panchal, Choi et al. 2007), as discussed in Chapter 9. This methodology also pertains to refinement of models framed at a given scale, i.e., variable resolution or fidelity. In general, a hierarchy of models may be employed, with different DOF and at different length and time scales, to inform design decisions.

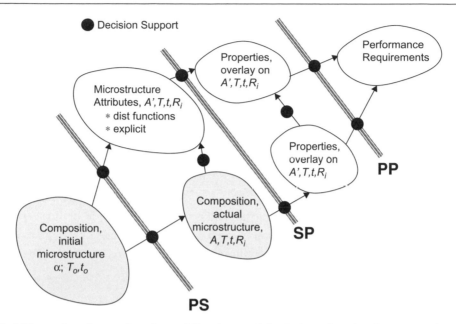

Figure 1.6 Hierarchy of mappings in multilevel materials design, showing how mappings can be accompanied by decisions associated with transfer of information between levels.

In the next section, we distinguish between the notion of *multiscale modeling*, which is inherently a bottom-up exercise, and *multilevel design*. In the latter, the hierarchy of length and time scales, as well as product, assembly, subassembly, and component scales, is considered. The presence of decision nodes at various model and database interfaces in Figure 1.6 highlights a key distinguishing feature of multilevel design—it employs the concept of decision support to link information from models and databases, rather than seeking automated, seamless connectivity. Figure 1.7 illustrates the hierarchy of material length scales below the individual part/component scale, which is a key aspect of concurrent design of the material as a part of the overall product design. This concept of concurrent design of materials and products is broader than just the design of materials to achieve specified properties (although that is an important subset) because the performance requirements may place a complex, conflicting set of demands on the material properties. Moreover, different responses are often coupled to some degree (e.g., ductility and fracture toughness), such that designing to "property sets" that are measured at fixed structure is not always meaningful. To the right of the vertical black line in Figure 1.7, robust systems design methods are well established, and are routinely applied to the design of aircraft, automobiles, refrigerators, etc. (Chen, Allen et al. 1997, Kalsi, Hacker et al. 1999, Rolander, Rambo et al. 2006, Taguchi 1986). On the other hand, methods to address the hierarchy of material length scales on the left side of this figure are not well established. Associated models and model parameters for process-structure and structure-property relations at each scale of material structure (and through the hierarchy of scales) are characterized by

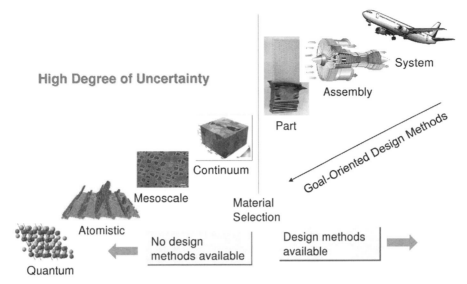

Figure 1.7 Extension of systems-based, top-down materials design from parts, subassemblies, assemblies, and components to hierarchical levels of material structure.

relatively high degrees of uncertainty compared to models and processes on the right side of the diagram. In fact, the uncertainty in material modeling is often difficult to quantify (for example, effects of randomness in material microstructure on variability of properties or responses).

1.3. Context of Systems-based Materials Design

1.3.1. Multiscale Modeling vs. Multilevel Materials Design

It is useful to distinguish the goals and methods of multiscale modeling for the hierarchy of material structure shown on the left in Figure 1.7 from those of materials design. Numerous methods for concurrent multiscale modeling have been developed, as well as a number of other emerging multiscale homogenization approaches. Concurrent multiscale modeling is a specific class of multiscale modeling approaches that involves fully coupled simulation models at multiple scales; this enables both bottom-up prediction of collective responses as a function of microstructure, and top-down assessment of microstructure-scale responses given higher length and time scale behavior. Concurrent modeling of multiple scales is of particular relevance in relating efficient coarse-grain models of response (e.g., finite element analysis of macroscopic structures) to fine-scale behavior (mesoscopic, microscopic, nanoscopic). As such, it is an *analysis tool*. However, it must be borne in mind that distinct objectives can drive the design of each level of hierarchy, and the design process need not rely on chaining or sequencing models at different scales. Concurrent multiscale modeling is not particularly

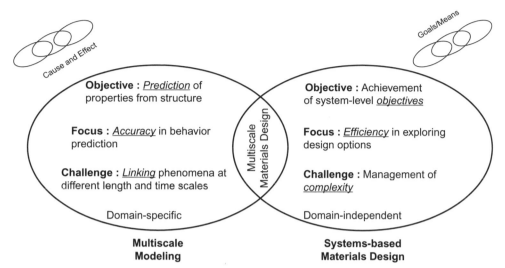

Figure 1.8 Distinctions of multiscale modeling and systems-based multilevel design in the design of materials and products.

attractive for materials design in many cases because the propagation of uncertainty may be too significant to render useful designs; there is also uncertainty associated with approximations involved in the models and methods for scale transitions. This may be viewed as a coupling of models written at different length (and time) scales, and will be discussed in Chapter 9 in terms of an interaction matrix for multiscale/multilevel simulation-based design. In fact, combined bottom-up modeling and top-down experimental strategies that interact at intermediate scales may be of greater utility in materials design (Wang, Kumar et al. 2007).

The key distinctions between multiscale modeling and systems-based materials design are highlighted in Figure 1.8. The primary objective in multiscale modeling is analysis (i.e., prediction of the properties of the material based on the structure), whereas the objective of systems-based materials design is to find material microstructure(s) that satisfy the system-level design objectives. Multiscale modeling represents a deductive approach while systems-based materials design corresponds to an inductive approach, as presented by Olson in Figure 1.8. Hence, the focus of the two activities is different—multiscale modeling efforts emphasize accurate prediction of material properties or responses, while systems-based design efforts focus on efficiency in evaluating design options in the presence of uncertainty. The primary challenge in multiscale modeling is to link phenomena at different length and time scales. In contrast, the primary challenge in systems-based design is the management of uncertainty and complexity of the problem; we address these issues in Chapters 6 through 9. Due to their nature, multiscale modeling approaches are generally domain- and problem-specific, whereas systems-based design approaches are independent of a particular materials design problem.

As previously mentioned, systems-based design of materials employs models in the range for which they are most appropriate and accurate, using decision-based protocols for informing models and decisions based on these analyses. In general, a hierarchy of models at various scales may be employed, with different purposes and at different length and time scales, to inform decisions. The necessity of seamless scale transitions must be determined by the degree of coupling in the system and the utility of information gained by coupling. Hence, concurrent multiscale modeling is not a sufficient (or necessarily even desirable) component of multilevel materials design. At each level of the hierarchy (for each relevant mechanism that relates to response), models and databases (computationally or experimentally generated) are incomplete and sampling is discrete.

Computational materials science, combined with decision-based design, utility theory, and information economics, has promise to extend design exploration beyond human intuition, representing a fruitful area of collaboration between materials engineers, mechanics of materials, applied mathematicians, statisticians, and systems-based design experts. The goal is typically *not* to accurately predict mean values of properties or material responses but rather to understand and quantify sensitivity to microstructure. Predictions must be calibrated to measured responses for mean behaviors and for purposes of validation. In so doing, we should (i) use calibrated models in the range for which they are most appropriate and accurate, (ii) employ the concept of robust design to address uncertainty, and (iii) use decision-based protocols for informing robust design.

1.3.2. *Materials Selection vs. Materials Design*

Returning to the issue of materials selection discussed previously, we are in a position to consider the role of materials selection in concurrent design of materials and products. Interpreting elements of Olson's diagram in Figure 1.2 as comprising levels involved in materials design, it is clear that only the relations between properties and performance (PP mapping in Figure 1.9) constitute materials selection. A set of N properties is typically extracted from reduced-order (lateral) descriptions of the material response, whether obtained by experiment or simulation, and then is related to M performance requirements, comprising a property-performance space of dimension $M + N$. The dimensions of the space for materials selection can then be further reduced by judicious non-dimensionalization of combined properties and performance requirements, which depends on the design and selection problem at hand. For example, the materials selection approach developed by Ashby and colleagues (Ashby 1999, Granta Design Limited 2007) is along these lines. As shown in Figure 1.9, the materials selection problem may be considered the "tip of the iceberg" of the entire materials design process.

We suggest that a simulation-based materials design revolution is underway, in which materials selection is augmented by the design of material microstructure and/or

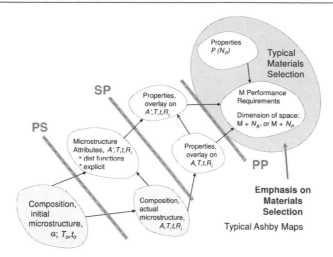

Figure 1.9 Distinguishing the role of materials selection from systems-based materials design.

mesostructure to satisfy specified ranged sets of performance requirements. Often these multiple performance requirements are in conflict in terms of their demands on microstructure, which leads to the need for multi-objective methods to search for satisfactory solutions.

1.3.3. Prospects for Improvement via Systems-based Materials Design

Historically, materials development has not drawn significantly on systematic design principles. Certainly, important methodologies have been developed, largely in terms of improvement of existing materials. Discovery of new materials has been a "hit or miss" proposition. The methodology is typically described by the following empirical route:

- Develop knowledge of initial conditions on composition range and process path to obtain useful material systems.

- Meander through length scales of material structure hierarchy and phenomena and modify by "intelligent perturbation" of process path.

- Move as far as possible toward desirable properties.

What, then, is a realistic assessment of the prospects for improvement via systems design of materials? The following questions offer some perspective. To what degree can designs made on the basis of empirical materials development be replaced by design decisions that are informed by modeling and simulation? It is presently estimated that fewer than 10% of decisions are made on the basis of simulation for many integrated material systems in products. Can the fraction of decisions be increased to 15%, or 30%? Every 10% increase may represent a substantial number of months in the materials development and insertion cycle. It can be argued that a threshold must be crossed to realize the goal of simulation-based integrated design of materials and products. What is that threshold level, and how does it vary among products and material classes? To what extent can multiple phenomena be considered simultaneously rather than sequentially? Very little information is obtained from experiments in the laboratory unless some basic insight is gained regarding sets of experiments that distinguish key physical phenomena in terms of operative time and length scales. Greater understanding leads to more efficient experimental protocols and more rapid material characterization. Modeling and simulation offer much greater acceleration by allowing for mechanisms that govern material response to be explored somewhat independently and in parallel, exercising models at different length and time scales in the hierarchy shown in Figure 1.2. Finally, to what extent can the degree of idealization in the design problem be relaxed to consider microstructures and phenomena in more realistic fashion?

1.4. Multilevel Design—Challenges and Approach

1.4.1. Challenges in Top-Down Materials Design

Typically, engineering design (from a systems perspective) proceeds from the top down through the mapping hierarchy as illustrated in Figures 1.2–1.9. Bottom-up analysis is an important supporting element. Pursuit of top-down materials design has many challenges, among the most prominent being the difficulty of seeking inverse problem (top-down) solutions arising from the nonequilibrium nature of process path-structure relations and structure-property relations. Cases in which inversion is possible are rather limited when process path is considered, such as tailoring of grain orientation distribution (macroscopic texture) to deliver desired macroscopic elastic properties of polycrystalline metals. Other challenges are listed as follows:

- The nonlinear, path-dependent behavior of metals and alloys limits extent of parametric study and parallelization of continuum analyses and engenders dependence upon initial conditions.

- A wide range of local solutions can be realized for specified objective functions in terms of property or performance requirements, leading to nonuniqueness and perhaps large families of possible solutions.

- The role of extreme value distributions (not just mean field averages) of certain microstructure features that control material responses such as fracture and fatigue. In fact, although materials selection is often expressed in terms of property requirements (Ashby 1999), material responses in the presence of evolving (in contrast to stationary) microstructure are often of interest. This means that information is complex and continuous in many cases, rather than discrete property sets.

- Representation of microstructure presents challenges in terms of how much information to store and in what format. Moreover, the goal of materials design is to explore microstructures that do not exist, using computational simulations combined with selected experiments to estimate properties.

- Process capabilities constrain achievable microstructures, and thermodynamics and kinetics (history) considerations limit the range of accessible or feasible microstructures.

- Major sources of uncertainty in processing, microstructure, modeling, etc. must be taken into account, as they can dominate the configuration of the design process and range of acceptable solutions. This consideration typically demands design approaches that are robust against such uncertainty.

- Material models (PS, SP) and design results must be validated, using principles of internal consistency, statistical realizability, and validation of material response at various length scales by direct or indirect measurement (Seepersad, Pederson et al. 2006).

- Tradeoffs between design exploration and computational intensity offer practical limitations on the range of design space considered and the number of iterations involved.

From a practical perspective, process-structure relations are often a weak link in the linear structure shown in Figure 1.2, necessitating a distinction between materials development (assessment of feasible materials and microstructures that can be synthesized or processed) and materials design, which runs the gamut of process-structure-property-performance relations.

1.4.2. Addressing Multilevel Design Challenges

A systems-based design approach is necessitated by many of the challenges associated with concurrent design of materials and products. For example, materials design is inherently multilevel and multifunctional in nature. Most applications require materials that satisfy multiple functions—such as structural load bearing, thermal transport, cost, and long-term stability—and these requirements cannot be defined in isolation from overall system conditions and requirements. These conditions are associated with the operating environment and the component(s) and overall system in which a material is integrated. The material is a subsystem of a larger system that includes parts, assemblies, and physical systems. Materials are themselves hierarchical systems. Desired material properties and performance characteristics often depend on phenomena that operate at different length and time scales, spanning from angstroms to meters and from picoseconds to years. A hierarchy of models applies to a range of length and time scales. Each model is used to inform the formulation of other models on higher length scales that capture the collective behavior of lower length scale subsystems, but it is very difficult to formulate a single model for macroscopic material properties that unifies all of the length scales (McDowell and Story 1998). For example, first-principles models can be used on atomistic and molecular levels to predict structure and properties of ideal designs, but they are too computationally expensive and often too idealized to model real materials with highly heterogeneous structures that strongly influence their macroscopic properties. On the other hand, continuum mechanics models are useful for describing properties at a macroscopic scale relevant to many engineering applications, but they are inappropriate for fine-scale dynamic phenomena that require resolution of discrete defects, atoms, etc.

While it is extremely challenging to develop physics-based models that embody relevant process-structure-property relations on different length and time scales for diverse functionality, the complexity and restricted domains of application of these models limit their explicit integration across length and time scales. Instead, they must be linked in a manner that facilitates exploration of the systems-level design space by a collaborative team of experts. Distributing analysis and synthesis activities also leverages the extensive domain-specific knowledge and expertise of various material and product designers who may be specialized according to length and time scales, classes of materials, and domains of functionality. A fundamental role of each domain-specific expert is to make decisions that involve synthesizing and identifying solution alternatives to achieve desirable tradeoffs between sets of conflicting material property goals. However, material subsystems are interdependent, and the individual decisions associated with them rely on information and solutions generated by other decision makers at other levels of the hierarchy. In the end, preferable *systems-level* solutions are sought, and they are not necessarily obtained by "optimizing" each subsystem individually. Therefore, it is critical to establish multi-objective *decision protocols* for individual designers as well as standards, tools, and mathematical techniques for *interfacing* individual decisions and facilitating information flow among multiple experts.

Since materials are typically hierarchical, heterogeneous systems characterized by a certain degree of randomness, it is not reasonable or sufficient to adopt a deterministic approach to materials design. Parameters of a given model are subject to variation associated with spatial variability of microstructure and variability due to processing. Furthermore, uncertainty is associated with model-based predictions for several reasons. Models inevitably incorporate assumptions and approximations that impact the precision and accuracy of predictions. Uncertainty may be magnified when a model is utilized near the limits of its intended domain of applicability and when information propagates through a series of models. To facilitate exploration of a broad design space, approximate or surrogate models may be utilized, but fidelity may be sacrificed for computational efficiency. Experimental data for conditioning or validating models may be sparse and may be affected by measurement errors. Often, it is expensive or impossible to remove these sources of variability, but their impact on model predictions and final system performance can be profound. Therefore, systems-level design methods need to account for the many sources of uncertainty and facilitate the synthesis of *robust solutions* that are relatively insensitive to them.

It is fully recognized that material structure and property databases for multicomponent material systems serve as foundational elements of materials design, whether based on experiments or simulations. The subject of databases has received much attention in materials design literature to date. It does not receive particular emphasis in this book, however, in view of the fact that databases are considered instruments of informing design decisions, as are models and simulations; accordingly, we implicitly include databases as constituting mappings in Figure 1.9. Databases dominate only certain classes of materials design problems, such as combinatorial design. As is the case with modeling and simulation, databases should also convey uncertainty associated with their elements (assumptions, methods, approximations) to facilitate systems-based robust design. Historically, this aspect has not received much attention.

Finally, it is necessary to establish a computing infrastructure for integrating heterogeneous, distributed software applications and databases in a simulation-based materials design process. An effective *computing infrastructure* needs to automate the details of executing and linking various models, freeing a designer to build upon previous model-based developments and to concentrate on higher-level design issues. The computing infrastructure should be easily extensible and platform independent. It also needs to archive and organize large amounts of data and facilitate real-time data sharing and visualization as well as systematic communication, translation, and search-based retrieval of design information. Tools are needed for online collaboration, communication, and project management, including real-time data sharing. We suggest a framework based on the eXtensible Distributed Product Realization (X-DPR) environment for systems-based materials design. We expand on this framework in Chapter 10.

1.4.3. Organization of the Book

In this book, we focus on various aspects of systems-based concurrent design of materials and products. The flow of information between different chapters is shown in Figure 1.10. In Chapter 2, a review of existing efforts related to materials design is provided. Chapter 3 contains an overview of the overall design framework presented in the book. The essential constituents of the framework (as shown in Figure 1.11), including robust design, complexity management, and a distributed design framework, are introduced in that chapter. Two examples of materials design problems are also discussed in Chapter 3; these examples are used throughout the book to illustrate the design framework. This concludes the first part of the book.

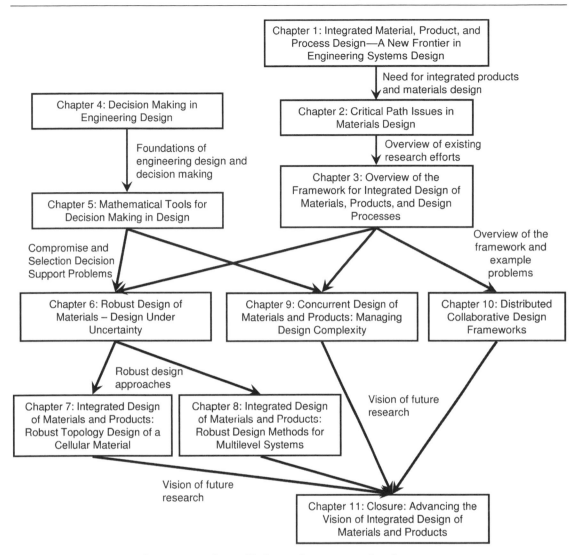

Figure 1.10 Flow of information among the chapters.

The second part contains a detailed discussion of the fundamental constructs on which the framework is based. In Chapter 4, the fundamentals of engineering design processes are discussed, which include a general model of design. The difference between analysis and synthesis is highlighted from a design perspective. An overview of the decision-based perspective to design is provided. The design framework presented in the book is centered on this perspective. Following a discussion of pursuing design for multiple functions, an overview of multi-objective decision making in design is discussed. In Chapter 5, details of two specific multi-objective decision-making constructs are presented. These include the Compromise Decision Support Problem (cDSP) and the Selection Decision Support Problem

Norton, R.L., 2006. Machine Design: An Integrated Approach. Pearson Education, Upper Saddle River, NJ.

Olson, G.B., 1997. Computational design of hierarchically structured materials. Science 277 (5330), 1237–1242.

Olson, G.B., 2000. Designing a new material world. Science 288 (5468), 993–998.

Pahl, G., Beitz, W., 1996. Engineering Design: A Systematic Approach, K. Wallace, trans. Springer, New York.

Panchal, J.H., Choi, H.-J., Allen, J.K., McDowell, D.L., Mistree, F., 2006. Designing design processes for integrated materials and products realization: A multifunctional energetic structural materials example. Design Automation Conference, Philadelphia, PA. Paper Number: DETC2006-99449.

Panchal, J.H., Choi, H.-J., Allen, J.K., McDowell, D.L., Mistree, F., 2007. A systems-based approach for integrated design of materials, products, and design process chains. J. Comput. Aided Mater. Des. 14 (1), 265–293.

Pollock IV, T.M., Allison, J.E., Backman, D.G., Boyce, M.C., Gersh, M., Holm, E.A., LeSar, R., Long, M., Powell, A.C., Schirra, J.J., Whitis, D.D., Woodward, C., 2008. Integrated Computational Materials Engineering: A Transformational Discipline for Improved Competitiveness and National Security. National Materials Advisory Board, NAE, National Academies Press Report Number: ISBN-10: 0-309-11999-5.

Rolander, N., Rambo, J., Joshi, Y., Allen, J.K., Mistree, F., 2006. Robust design of turbulent convective systems using the proper orthogonal decomposition. ASME J. Mech. Des. Spec. Issue on Robust and Risk-Based Des. 128, 844–855.

Seepersad, C.C., 2004. A robust topological preliminary design exploration method with materials design applications. PhD Dissertation, G.W. Woodruff School of Mechanical Engineering, Georgia Institute of Technology, Atlanta, GA.

Seepersad, C.C., Pederson, K., Emblemsvag, J., Bailey, R.R., Allen, J.K., Mistree, F., 2006. The validation square: How does one verify and validate a design method? In: Chen, W., Lewis, K., Schmidt, L. (Eds.), Decision Making in Engineering Design. American Society of Mechanical Engineers, New York.

Shigley, J.E., Mischke, C.R., 1989. Mechanical Engineering Design, fifth ed. McGraw-Hill, New York.

Taguchi, G., 1986. Introduction to Quality Engineering. Asian Productivity Organization, Tokyo.

Wang, A.-J., Kumar, R.S., Shenoy, M.M., McDowell, D.L., 2007. Microstructure-based multiscale constitutive modeling of g-g' nickel-base superalloys. Int. J. Multiscale Comput. Eng. 4 (5–6), 663–692.

CHAPTER 2

Critical Path Issues in Materials Design

Concurrent design of materials and products is a compelling, transformative technology for 21st-century competitiveness. It also serves as an interdisciplinary platform for instruction of new generations of materials scientists and engineers. As pointed out in Section 1.1, the field of *Integrated Computational Materials Engineering* is rapidly moving in this direction in view of the potential advances in product quality and performance, as well as cost reduction. Accordingly, product design and materials development are not mutually exclusive and independent activities but synergistic components of an integrated product, process, and materials design endeavor. This challenge involves a philosophical and cultural shift toward inductive, goal-oriented synthesis of products *and* their constituent materials and processing paths. A systems-based strategy is essential. Several critical path issues are outlined in this chapter, with an emphasis on the limitations of current capabilities and the associated research and development opportunities:

- Adequate models and experimental data on different length and time scales for a diverse set of (multiple) functions that material systems must deliver;

- Techniques for characterizing and managing uncertainty in material models applied to processing paths and structure-property relations, as well as resulting design specifications;

- Tools for linking diverse modeling and simulation tools and methods and related data across length and time scales, functional domains, and material classes; and

- Systems design methods and tools that bridge or integrate the design of materials, manufacturing processes, and products/components.

2.1. The Need for Material Models and Databases

Integrated Computational Materials Engineering (ICME) has been enunciated by a National Academy of Engineering (NAE) National Materials Advisory Board study group (Pollock, Allison et al. 2008) as an approach to concurrent design of materials and products. Emphasis is placed on linking models and databases for a set of material structure-property relations at multiple length and time scales to address problems relevant to specific products and

DOI: 10.1016/B978-1-85617-662-0.00002-8

applications. ICME hearkens back to Olson's hierarchical scheme, shown in Figure 1.2 in Chapter 1 (Olson 1997), with the understanding that top-down strategies are essential to supporting goal/means design of materials to meet application-specific performance requirements. Implications of this vision were clarified much earlier at a 1998 workshop sponsored by the National Science Foundation (NSF) (McDowell and Story 1998) entitled "New Directions in Materials Design Science and Engineering (MDS&E)," involving a collaborative effort of U.S. academic and research communities. The participants concluded that a change of culture is necessary in U.S. universities and industries to cultivate and develop the concepts of simulation-based design of materials necessary to support integrated design of material and products. It also forecasted that the 21st century global economy would usher in a revolution of the materials supply/development industry and realization of true virtual manufacturing capabilities (not just geometric modeling) and realistic material behavior. It was recommended to establish a national road map addressing (1) databases for enabling materials design, (2) developing principles of systems design and the prospects for hierarchical materials systems, and (3) identifying opportunities and deficiencies in science-based modeling, simulation, and characterization "tools" to support concurrent design of materials and products.

Thermodynamics is a fundamental building block for simulation-supported materials design (Asta, Ozolins et al. 2001, Liu and Chen 2007, Olson 1997, van de Walle, Ghosh et al. 2007, Wang et al. 2004), providing information regarding stable and metastable phases, characterization of structures and energies of interfaces, and driving forces (transition states) for rearrangement of structure due to thermally activated processes. As such, it facilitates preliminary design exploration for candidate solutions to concurrent material and product design problems. First-principles calculations are indispensible in this regard, and support exploration of multicomponent systems for which little if any empirical understanding has been established (cf. Cuitino, Stainier et al. 2001, Goddard, Cagin et al. 2001, Haslam, Moldovan et al. 2002, Shenderova, Brenner et al. 1998). Database mining approaches have been combined with these sorts of tools (cf. Liu, Chen et al. 2006, Rajan 2005) to rapidly explore candidate solutions, particularly for design of materials with one dominant level of hierarchy. For multiple levels of hierarchy, multiscale modeling methods (cf. Cuitino, Stainier et al. 2001, Liu, Karpov et al. 2004, McDowell 2008, Ortiz, Cuitino et al. 2001) have been developed with input from digital representations of microstructures (Dawson, Miller et al. 2005). Current efforts in each of these areas are rather extensive.

It is widely acknowledged that improvement in fidelity and accuracy of material models is one element of simulation-based design of materials. However, it is a difficult matter to prioritize which phenomena and classes of models should receive investment for various types of materials design problems. A defining characteristic of materials modeling that has received relatively little attention (perhaps due to a conventional focus on deterministic modeling) is uncertainty. In many cases, it dominates considerations regarding the configuration of a simulation-assisted design framework. Uncertainty is discussed in the next section.

2.2. Characterizing and Managing Uncertainty in Materials Modeling and Design

In materials design applications, uncertainty can be both stochastic and epistemic. Stochastic (i.e., aleatory) uncertainty stems from stochastic variability and inherent randomness of material processing and morphology, as manifested in heterogeneous, randomly distributed microstructure attributes and defects. Epistemic (i.e., model) uncertainty stems from limits to the knowledge captured in models (model idealization and approximation of reality) and databases. This manifests itself in the limited fidelity and accuracy of predictions, lack of information, and modeling errors due to interpolations, approximations, convergence, assumptions, and other factors related to methods of obtaining approximate solutions. Characterizing and managing both types of uncertainty are essential in pursuing materials design applications.

Simulation-based design in the presence of uncertainty is an active research topic for the engineering design community, but significant challenges hinder direct transfer of this body of knowledge to *materials* design under uncertainty. Close collaboration with materials engineers and scientists is essential to identify and to address the nuances of uncertainty related to designing materials. Allen and coauthors, in their review (Allen, Seepersad et al. 2006) of robust design capabilities for multidisciplinary and multiscale applications, identify several challenges relevant to materials design, including techniques for addressing model uncertainty (Du and Chen 2000, Du and Chen 2002), propagation of uncertainty through a series of models, each of which may represent a different scale and/or discipline (Gu, Renaud et al. 2000, Gu, Renaud et al. 2006), collaborative decision making under uncertainty (Chang and Ward 1995, Chang, Ward et al. 1994, Chen and Lewis 1999, Kalsi, Hacker et al. 2001), and multidisciplinary/multiscale optimization under uncertainty (Gu, Renaud et al. 2002, Kokkolaras, Mourelatos et al. 2006, Liu, Chen et al. 2006, Mavris, Bandte et al. 1999). However, as noted by Allen and coauthors and in Chapter 1, there are several aspects that render materials design particularly challenging for successful application of robust design and simulation-based design under uncertainty techniques. These include the propagation of information (and uncertainty) through linked models, some of which may magnify the effects of uncertainty or exhibit significant sensitivity to heterogeneities in input data. Models for material process-structure and structure-property relations are often highly sensitive to assumptions regarding their underlying physics, as well as sparse input data with limited precision due to cost and experimental error. Material models must account for the inherent variability of phenomena manifested at microstructural length scales, including stochastic spatial variations of microstructure morphology. Finally, the difficulty of computationally linking models across the full range of relevant scales, for most applications, means that materials design is not a fully automated endeavor but an interactive enterprise involving materials designers who make decisions with the *assistance* of computers.

Accordingly, *materials design under uncertainty* is still emerging as a research field. McDowell and Olson (McDowell and Olson 2008), for example, highlight the role of uncertainty in materials design and cite the need for robust materials design solutions that are relatively insensitive to variations in material structure at various scales. Noteworthy recent work includes the multiscale design approach of Yin and coauthors (Yin, Lee et al. 2008), who model uncertainty in material microstructure with a random field approach and facilitate uncertainty propagation with a reduced order approach. Zabaras and coauthors are also developing methods for analyzing the effect of topological uncertainties in microstructure on homogenized properties as part of a set of advanced computational techniques for materials-by-design and stochastic optimization of materials deformation processes (Acharjee and Zabaras 2007, Sankaran and Zabaras 2006, Zabaras 2006). Patera and coauthors (Cuong, Veroy et al. 2005, Huynh and Patera 2007) investigate the use of reduced-order models for materials design applications. They focus on the development of reduced-basis output bound methods for quantifying uncertainty of predictions of these reduced-order models that result from the approximations inherent in the models themselves. In this work, materials design applications include application of modeling concepts of linear elastic fracture and Helmholtz elasticity. Zohdi (Zohdi 2003) has investigated the impact of material uncertainty (specifically random dispersion of particulates in a homogeneous matrix) on macroscopic effective properties. He applied a genetic algorithm for designing these types of materials by finding values of design variables (volume fractions, geometries of particulates, etc.) that minimize deviation from desired effective properties. Choi and coauthors (Choi, Grandhi et al. 2006) use polynomial chaos expansion procedures for quantifying material uncertainty—specifically, for quantifying continuum level material properties as uncorrelated random variables—in robust design processes for mechanical systems.

Uncertainty characterization and management is also a significant issue for several large-scale federal initiatives in the emerging materials design field. For example, the Defense Advanced Research Projects Agency (DARPA) Accelerated Insertion of Materials (AIM)[1] program focused on creating tools and methods to accelerate the insertion of new materials (materials development and certification) into production hardware. Part of the AIM program focused on metallic structures, namely nickel-base superalloys for gas turbine engine disks, with the goal of maximizing disk spin speed while minimizing mass (Apelian, Alleyne et al. 2004, Jou, Voorhees et al. 2004). Part of the design process involved predicting and experimentally validating spatial distributions of forged microstructures and associated distributions of properties such as yield strength (i.e., location-specific materials design of components). Another component of the AIM program focused on composite materials for aerospace applications. In the AIM composites program, Hahn and coinvestigators (Cregger, Caiazzo et al. 2004, Hahn 2001) addressed propagation of uncertainty and variability and associated designer confidence in model predictions. They utilized tools such as analysis of variance, sensitivity analysis, structural reliability methods, Monte Carlo analysis, regression, and

---

[1] http://www.darpa.mil/dso/thrusts/matdev/aim/overview.html

robust design to address multiple sources of uncertainty, including uncertainty with respect to input material properties (accuracy, repeatability, and availability of sufficient data), modeling (accuracy, assumptions, interpolation, and extrapolation), and synthesis of experimental and model-based data. Of particular interest were methods for quantifying error bounds on predictions and for validating and updating models.

Designer confidence in the performance of tailored materials is one of the most critical challenges facing the materials design community. As noted in the National Materials Advisory Board's summary of the status and promise of Integrated Computational Materials Engineering (National Materials Advisory Board 2008), engineers need qualified, predictive models for which the accuracy and precision of predictions are known with some level of certainty. While techniques are available for sensitivity analysis and uncertainty quantification, significant work is needed to enhance their efficiency for iterative materials *design* and to infuse these techniques throughout the fields of computational mechanics and materials science. Furthermore, issues of propagation of uncertainty across multiple models (and scales), model-based uncertainty, and uncertainty associated with sparse experimental data and its integration into computational models continue to challenge materials scientists and systems engineers. Significant additional work is needed in these areas to more rapidly qualify new material designs and to shift workload and decision making based on costly and time-consuming physical experimentation to less costly computational modeling and design. In this book, fundamentals of robust design and uncertainty management are discussed in Chapter 6 and incorporated within robust design methodologies for materials applications in Chapters 7, 8, and 9.

2.3. Multiscale Linkage of Material Models in Materials Design

In materials design applications, it is essential to implement not only a design methodology for tailoring multiscale material systems, but also the underlying computing infrastructure and protocols for linking diverse analysis tools, data, and knowledge across scales, functional domains, and material classes. As discussed in Chapter 1, materials design at multiple levels of hierarchy is a much broader activity than multiscale modeling. The need to integrate diverse tools is widely recognized in the science and engineering communities and even more acutely in the materials design field. The NSF's Office of Cyberinfrastructure,[2] for example, was created in 2005 to support the development of data repositories, visualization tools, high-speed computing facilities, and other resources that enhance their access and usability for scientists and engineers, including those involved in multiscale simulation and design applications such as materials design (Oden, Belytschko et al. 2006). Also, the National Institute of Standards and Technology (NIST) has compiled a database of materials

---

[2] http://www.nsf.gov/od/oci/about.jsp

properties.[3] This database supports the development of measurement standards and protocols for hardware and software applications. NIST has also expressed interest in expanding this database to provide coordinated "data on demand" for materials design applications, with an emphasis on standardization, interoperability for industrial and academic users, and responsiveness to research needs for a variety of data, such as nonequilibrium working data (Green 2007). Both industrial and academic researchers have expressed a need for such databases, which may be too costly for a single research group to establish or maintain.

In the absence of publicly available, widely applicable, and comprehensive computing infrastructure or databases for materials design applications, some materials designers have created their own computational frameworks for multilevel materials design applications. For example, in a *Materials by Design*® initiative for advanced steels, Olson (Olson 1997) has utilized the THERMOCALC database and software system to integrate models on a hierarchy of scales from quantum scale codes to continuum level models. Similarly, Liu and coauthors (Liu, Chen et al. 2004, Liu, Chen et al. 2006) have developed a computing framework for multilevel materials design called MatCASE (Materials Computation and Simulation Environment), which integrates software from atomic-scale first-principles calculations for predicting interfacial energies, lattice parameters, and elastic constants to finite element analysis for evaluating mechanical responses of proposed microstructures.

While most materials design computing frameworks are aimed at "hard" computing—the use of physics-based models and databases of experimental data to inform materials design—there is also an increasing interest in "soft" computing applications for materials design. Soft computing includes the use of statistical inference and artificial intelligence tools for identifying statistical or heuristic relationships between parameters in a materials design process. Rajan and coauthors (Liu, Chen et al. 2006) have coupled such an approach, termed *material informatics*, with hard physical modeling and experimentation (namely, the MatCASE developed by Liu and coauthors [Liu, Chen et al. 2004]) for the design of new alloys and catalysts. In similar work, LeClair and collaborators (Chen, Cao et al. 1998, Jackson, Pawlak et al. 1998, Villars, Brandenburg et al. 2000) have used rough sets and neural networks, combined with materials databases and first-principles calculations, to predict structure-property relations and compound formation. Although these soft computing techniques can enhance a materials designer's insight and perspective, they cannot replace hard computing techniques for the realization of new materials and seem to be most useful when paired with them.

The difficulty with most of these hard- and soft-computing tools is that they are customized for specific applications and therefore not easily transferable. Materials designers often find themselves building computing frameworks from scratch for new applications. As noted in

---

[3] http://www.msel.nist.gov/dataontheweb.html

Chapter 1, materials designers need computing frameworks that are platform independent and easily extensible so that models and databases can be easily linked together and combined with commonly available tools for search, visualization, and communication. This type of infrastructure frees the materials designer to concentrate on higher-level design and analysis issues rather than time-consuming details of model integration. In addition, the materials engineer/modeler interplays with the design process by defining models, inputs and outputs, and various sources of uncertainty.

Commercial design integration software is available for sharing and integrating models, databases, and software applications, automating their execution, and streamlining their integration with one another and with design exploration and optimization tools. Two examples are Isight®, a product of Engineous software company (now a subsidiary of Dassault Systemes), and ModelCenter® from Phoenix Integration. Both software tools allow users to perform process integration by linking models and other software applications via a drag-and-drop graphical user interface. They also provide a suite of built-in design exploration tools for optimization, statistical analysis, trade-off studies, design of experiments, and response surface modeling. These tools have proven useful to materials designers. For example, Isight was utilized as the information management and integration infrastructure for the DARPA-sponsored AIM initiative at General Electric (GE) Aviation (Backman, Wei et al. 2006). In this application, Isight served as the infrastructure for a designer knowledge base (DKB) system that incorporated a variety of computational materials simulation and design tools for rapid exploration of new materials, with a focus on nickel-based superalloys for the aerospace industry. The DKB facilitated data storage, integration of physics-based models and empirical data, uncertainty analysis, and design automation tools for generating tradeoff curves. Isight has also been utilized by Questek, LLC, for similar DARPA applications (Green 2007), although QuesTek has also developed its own materials modeling platform, Computational Materials Dynamics (www.questek.com).

One of the disadvantages of these commercially available design integration software tools is the lack of an open-source infrastructure, which means that it is difficult to tailor the infrastructure itself to the needs of materials designers. The motivation and foundation for a more open, extensible framework is presented in Chapter 10.

2.4. A Systems Perspective for Integrated Product, Process, and Materials Design

While increasingly sophisticated materials models, databases, and computing infrastructures are *necessary* components of any materials design effort, they are not *sufficient*. Materials design is fundamentally a *synthesis* process in which a set of requirements or goals for a material (and its parent component, assembly, or system) are translated into suitable material structures and corresponding processing paths. It also requires a systems perspective for bridging multilevel tools, databases, and models that cannot be unified into a single, integrated "model" for relevant material properties. Methods are needed for coordinating the decisions of distributed material and product designers who may possess expertise in understanding and modeling phenomena at specific length or time scales, functional domains, or material classes. As noted by McDowell and Olson in recent overviews (McDowell 2007, McDowell and Olson 2008), it is not sufficient to gather materials design resources—models, databases, and experts—without developing methodologies for strategically coordinating them for the design of material structure, processing paths, and parent products.

Effective materials design methodologies are required to bridge the gap between materials design and cutting edge product/system design applications in aerospace, automotive, biomedical, and other sectors. These methodologies will need to facilitate distributed design exploration by managing interdependencies between distributed decision makers. Nonunique and perhaps large families of potential solutions will need to be explored in a strategic manner that facilitates (1) exploration of the relationships between processing capabilities and the materials design space, (2) rapid identification of variables that most strongly affect material properties and performance parameters of interest, (3) management of uncertainty from random and heterogeneous material systems, sparse experimental data or information, and limiting assumptions underlying materials models, and (4) management of the tradeoff between increasingly comprehensive, detailed design space exploration, experimental costs, and computational expense.

While a growing body of research has embraced the challenge of pursuing materials design, very few have addressed the challenges inherent in developing a comprehensive materials design methodology that is applicable to broad classes of materials and applications. Olson (Olson 1997a, Olson 1997b, Olson 2000) advocates a systems approach to materials design that accounts for process-structure-property-performance relations for the design of advanced steels and other materials. The approach is based on utilizing a heterogeneous set of computational models and experimental tools that are linked via computational thermodynamics tools. It builds on expert intuition regarding initial conditions and potential solutions that enable minimal iterations through the design system, typically on the order of two or three such iterates; moreover, design support tools are not clearly prescribed or utilized. Subbarayan and Raj (Subbarayan and Raj 1999) follow Olson's lead to illustrate how the design of a simple tungsten filament is an example of the design of a nonhierarchical system, comprised of subsystems that share linking variables both within and across levels of the hierarchy. As with Olson's work, formal strategies for exploring the design spaces and coordinating distributed decision making remain open questions. Lu and Deng (Lu and Deng 2004) use variable dependency graphs to formalize the relationships between subsystems in the filament design problem introduced by Subbarayan and Raj. Gall and Horstemeyer (Gall and Horstemeyer 2000) illustrate how the systems philosophy of Subbarayan and Raj and Olson can be customized for design of a cast component using a mechanism-based, multiscale modeling approach (specifically, an internal state variable constitutive model embedded within a nonlinear finite element analysis) that allows a component designer to quantitatively investigate relationships between component-level and microstructure-level parameters and features.

Adams, Kalidindi, and coauthors (Adams, Henrie et al. 2001, Adams, Kalidindi et al. 2005, Kalidindi, Houskamp et al. 2004) have developed an approach for microstructure-sensitive design that enables the incorporation of crystallographic texture as a variable in the design of engineered components with targeted elastic properties. They represent certain aspects of microstructure (namely, the orientation distribution of grains in a polycrystal) in a Fourier space and create a space of potential microstructures called the material hull. Optimal microstructures are identified by intersecting isoproperty surfaces with the hull. The utility of the approach has been demonstrated via application to the design of a compliant beam (Adams, Henrie et al. 2001), an orthotropic plate (Kalidindi, Houskamp et al. 2004), and other components. However, the focus has been on problems for which structure-property relations can be inverted in some manner, which is possible in relating low-order moments of texture to elastic properties. Application of these ideas is unclear for problems involving nonequilibrium evolution of microstructure and associated nonlinearities and path dependence over several length and time scales which preclude such invertibility.

In addition to the work of Adams, Kalidindi, and coauthors, several researchers have focused on different aspects of microstructure-sensitive design. For example, Kulkarni and coauthors (Kulkarni, Krishnamurthy et al. 2004) use genetic algorithms to design the microstructure (volume fraction and radius of particles) of aluminum alloys for targeted properties (strength and ductility). Also, significant effort has been focused on designing composite materials by tailoring either their microstructure attributes (e.g., volume fractions and compositions) or their "architectural layouts" (e.g., lamina and orientations) (Gurdal, Haftka et al. 1999, Jones 1998). Topology optimization techniques have been used to design microstructures with prescribed elastic and thermoelastic properties (Hyun and Torquato 2002, Sigmund 1994, Sigmund 1995, Sigmund 2000, Sigmund and Torquato 1997, Wang and Zhou 2004). Although it is an important first step to incorporate topology and microstructure in the design of engineered components, it is important for materials design to traverse a broader range of length and time scales and to incorporate influential domains such as materials processing and economics.

As emphasized in Section 1.3.2, materials selection has formed the basis for the conventional conception of design *with* materials in most textbooks and engineering practice to date. For example, Ashby (Ashby 2005) focuses on the design engineer's task of selecting materials for specific applications but emphasizes properties of existing materials rather than providing tools for designing new materials that are tailored for a specific application. In many respects, the material selection approach is similar to the combinatorial methodologies that have been proposed for materials design, sometimes applied at even the molecular level (sieves, virus blockers, etc.). Whereas the materials selection approach involves databases of the properties of previously developed and qualified materials, combinatorial methodologies produce libraries of properties for hundreds or thousands of distinct compounds that have been synthesized and screened rapidly by an automated method (e.g., vapor deposition). Symyx technologies has pioneered combinatorial materials discovery methodologies (e.g., Danielson, Devenney et al. 1998a, Danielson, Devenney et al. 1998b, McFarland and Weinberg 1999). Combinatorial approaches have also been applied to materials design applications by Rajan and coauthors in the form of material informatics and data mining (Rajan 2005, Suh, Rajagopalan et al. 2003). Though they have their place in materials design, combinatorial materials design approaches primarily extend the domain of materials selection to strategic searching of databases of parameters that reflect influential structure-property relations or material properties of interest. Systematic materials design methods are needed to leverage the power of computational materials models to support simulation-based design of new materials and exploration of the dependence of macroscopic properties of interest on interconnected phenomena on multiple scales in process-structure and structure-property relations.

Although systems-based approaches seem to be emerging as the preferred paradigm for materials design, they are still in their infancy. Systems-based materials design methodologies (complete with computational and mathematical infrastructures for implementing them) are presented in Chapters 7, 8, and 9.

2.5. The Need for an Integrated Product-Materials Design Methodology

For the past decade, momentum has been building for the emerging field of simulation-assisted materials design. In contrast to the trial-and-error techniques of traditional, experimental materials science and the purely analytical approach of computational materials science, materials design emphasizes *design engineering* for systematically synthesizing materials that are tailored for specific product applications and *integration* of a broad range of materials modeling tools, systems design methodologies and design exploration tools, and empirical databases for exploring material process-structure-property-performance relations across length scales and functional domains. Although some of the foundations for meeting this grand challenge have been laid, a tremendous amount of research remains in the quest to create integrated product-materials design environments that are applicable to broad classes of materials. As summarized in this chapter, research has commenced on increasing the stockpile of materials models and databases, gauging the accuracy and precision of their predictions, linking models and other tools across length scales with computing infrastructures, and developing systems design methodologies for integrating and linking the various tools with decision makers in a hierarchical materials design effort. In the next chapter, we provide an overview of some initial tools that were developed to meet these challenges.

References

Acharjee, S., Zabaras, N., 2007. A non-intrusive stochastic galerkin approach for modeling uncertainty propagation in deformation processes. Comput. Struct. 85 (5–6), 244–254.

Adams, B.L., Henrie, A., Henrie, B., Lyon, M., Kalidindi, S.R., Garmestani, H., 2001. Microstructure-sensitive design of a compliant beam. J. Mech. Phys. Solids 49 (8), 1639–1663.

Adams, B.L., Kalidindi, S.R., Fullwood, D.T., 2005. Microstructure Sensitive Design for Performance Optimization. Brigham Young University Academic Publishing, Provo, UT.

Allen, J.K., Seepersad, C.C., Choi, H.-J., Mistree, F., 2006. Robust design for multiscale and multidisciplinary applications. ASME J. Mech. Des. 128 (4), 832–843.

Apelian, D., Alleyne, A., Handwerker, C.A., Hopkins, D., Isaacs, J.A., Olson, G.B., Vidyanathan, R., Wolf, S.D., 2004. Accelerating Technology Transition: Bridging the Valley of Death for Materials and Processes in Defense Systems. National Materials Advisory Board, NAE, National Academies Press Report Number: ISBN-10: 0-309-09317-1.

Ashby, M.F., 2005. Materials Selection in Mechanical Design. Butterworth-Heinemann, Oxford, UK.

Asta, M., Ozolins, V., Woodward, C., 2001. A first-principles approach to modeling alloy phase equilibria. J. Occup. Med. 53 (9), 16–19.

Backman, D.G., Wei, D.Y., Whitis, D.D., Buczek, M.B., Finnigan, P.M., Gao, D., 2006. ICME at GE: Accelerating the insertion of new materials and processes. J. Occup. Med. 58 (11), 36–41.

Chang, T.S., Ward, A.C., 1995. Conceptual robustness in simultaneous engineering: A formulation in continuous spaces. Res. Eng. Des. 7 (2), 67–85.

Chang, T.S., Ward, A.C., Lee, J., Jacox, E.H., 1994. Conceptual robustness in simultaneous engineering: an extension of taguchi's parameter design. Res. Eng. Des. 6, 211–222.

Chen, C.L.P., Cao, Y., LeClair, S.R., 1998. Materials structure-property prediction using a self-architecting neural network. J. Alloys. Compd. 279, 30–38.

Chen, W., Lewis, K., 1999. A robust design approach for achieving flexibility in multidisciplinary design. AIAA J. 37 (8), 982–989.

Choi, S.-K., Grandhi, R.V., Canfield, R.A., 2006. Robust design of mechanical systems via stochastic expansion. Int. J. Mater. Prod. Technol. 25 (1/2/3), 127–143.

Cregger, S.E., Caiazzo, A., Pugliano, P., Rajagopal, R., Uryasev, S., Accelerated insertion of materials: managing error and uncertainty in structures., 49th International Society for the Advancement of Material and Process Engineering Symposium and Exhibition, 2004, Long Beach, CA, May 17

Cuitino, A.M., Stainier, L., Wang, G., Strachan, A., Cagin, T., Goddard, W.A., Ortiz, M., 2001. A multiscale approach for modeling crystalline solids. J. Comput. Aided Mater. Des. 8 (2–3), 127–149.

Cuong, N.N., Veroy, K., Patera, A.T., 2005. Certified real-time solutions of parametrized partial differential equations. In: Yip, S. (Ed.), Handbook of Materials Modeling. Springer, New York, pp. 1523–1558.

Danielson, E., Devenney, M., Giaquinta, D.M., Golden, J.H., Haushalter, R.C., McFarland, E.W., Poojary, D.M., Reaves, C.M., Weinberg, W.H., Wu, X.D., 1998a. A rare-earth phosphor containing one-dimensional chains identified through combinatorial methods. Science 279 (5352), 837–839.

Danielson, E., Devenney, M., Giaquinta, D.M., Golden, J.H., Haushalter, R.C., McFarland, E.W., Poojary, D.M., Reaves, C.M., Weinberg, W.H., Wu, X.D., 1998b. X-ray powder structure of Sr2CeO4: a new luminescent material discovered by combinatorial chemistry. J. Mol. Struct. 470, 229–235.

Dawson, P., Miller, M., Han, T.-S., Bernier, J., 2005. An accelerated methodology for the evaluation of critical properties in polyphase alloys. Metall. Mater. Trans. A 36 (7), 1627–1641.

Du, X., Chen, W., 2000. Methodology for managing the effect of uncertainty in simulation-based systems design. AIAA J. 38 (8), 1471–1478.

Du, X., Chen, W., 2002. Efficient uncertainty analysis methods for multidisciplinary robust design. AIAA J. 40, 545–552.

Gall, K., Horstemeyer, M.F., 2000. Integration of basic materials research into the design of cast components by a multi-scale methodology. J. Eng. Mater. Technol. 122 (3), 355–362.

Goddard, W.A., Cagin, T., Blanco, M., Vaidehi, N., Dasgupta, S., Floriano, W., Belmares, M., Kua, J., Zamanakos, G., Kashihara, S., Iotov, M., Gao, G., 2001. Strategies for multiscale modeling and simulation of organic materials: Polymers and biopolymers. Comput. Theor. Polym. Sci. 11 (5), 329–343.

Green, M.L., 2007. User-friendly materials databases for the intelligent selection of materials: an industrially responsive materials data center. In Proceedings of the Committee on Integrated Computational Materials Engineering, March 13–14. National Academy of Engineering, Washington, DC. Available at <http://www7.nationalacademies.org/nmab/CICME_home_page.html/>.

Gu, X., Renaud, J.E., Ashe, L.M., Batill, S.M., Budhiraja, A.S., Krajewski, L.J., 2002. Decision-based collaborative optimization. ASME J. Mech. Des. 124 (1), 1–13.

Gu, X., Renaud, J.E., Batill, S.M., Brach, R.M., Budhiraja, A.S., 2000. Worst case propagated uncertainty of multidisciplinary systems in robust design optimization. Struct. Multidiscip. O. 20 (3), 190–213.

Gu, X., Renaud, J.E., Penninger, C.L., 2006. Implicit uncertainty propagation for robust collaborative optimization. ASME J. Mech. Des. 128 (4), 1001–1013.

Gurdal, Z., Haftka, R.T., Hajela, P., 1999. Design and Optimization of Laminated Composite Materials. Wiley, New York.

Hahn, G., 2001. Accelerated insertion of materials-composites. DARPA Workshop, Annapolis, MD, August 27–28

Haslam, A.J., Moldovan, D., Phillpot, S.R., Wolf, D., Gleiter, H., 2002. Combined atomistic and mesoscale simulation of grain growth in nanocrystalline thin films. Comput. Mater. Sci. 23 (1–4), 15–32.

Huynh, D.B.P., Patera, A.T., 2007. Reduced basis approximation and a posteriori error estimation for stress intensity factors. Int. J. Numer. Methods Eng. 72 (10), 1219–1259.

Hyun, S., Torquato, S., 2002. Optimal and manufacturable two-dimensional, kagome-like cellular solids. J. Mater. Res. 17 (1), 137–144.

Jackson, A., Pawlak, Z., LeClair, S.R., 1998. Rough sets applied to the discovery of materials knowledge. J. Alloy. Compd. 279, 14–21.

Jones, R.M., 1998. Mechanics of Composite Materials, second ed.. Brunner-Routledge, New York.

Jou, H.-J., Voorhees, P., Olson, G.B., et al., 2004. Computer simulations for the prediction of microstructure/property variation in aeroturbine disks. In: Green, K.A. (Ed.), Superalloys 2004 (proceedings of the 10th International Symposium on Superalloys). The Minerals, Metals, and Materials Society, Warrendale, PA, pp. 877–886.

Kalidindi, S.R., Houskamp, J.R., Lyons, M., Adams, B.L., 2004. Microstructure sensitive design of an orthotropic plate subjected to tensile load. Int. J. Plast. 20 (8–9), 1561–1575.

Kalsi, M., Hacker, K., Lewis, K., 2001. A comprehensive robust design approach for decision trade-offs in complex systems design. ASME J. Mech. Des. 123 (1), 1–10.

Kokkolaras, M., Mourelatos, Z.P., Papalambros, P.Y., 2006. Design optimization of hierarchically decomposed multilevel systems under uncertainty. ASME J. Mech. Des. 128 (2), 503–508.

Kulkarni, A.J., Krishnamurthy, K., Deshmukh, S.P., Mishra, R.S., 2004. Microstructural optimization of alloys using a genetic algorithm. Mat. Sci. Eng. A 372 (1–2), 213–220.

Liu, H., Chen, W., Kokkolaras, M., Papalambros, P.Y., Kim, H.M., 2006. Probabilistic analytical target cascading: a moment matching formulation for multilevel optimization under uncertainty. ASME J. Mech. Des. 128 (4), 991–1000.

Liu, W.K., Karpov, E.G., Zhang, S., Park, H.S., 2004. An introduction to computational nanomechanics and materials. Comput. Method. Appl. M. 193 (17–20), 1529–1578.

Liu, Z.-K., Chen, L.-Q., 2007. Integration of first-principles calculations, CALPHAD modeling, and phase-field simulations. In: Bozzolo, G., Noebe, R.D., Abel, P.B. (Eds.), Applied Computational Materials Modeling. Springer-Verlag, New York, pp. 171–213.

Liu, Z.-K., Chen, L.-Q., Raghavan, P., Du, Q., Sofo, J.O., Langer, S.A., Wolverton, C., 2004. An integrated framework for multi-scale materials simulation and design. J. Comput. Aided Mat. Des. 11, 183–199.

Liu, Z.-K., Chen, L.-Q., Rajan, K., 2006. Linking length scales via materials informatics. J. Occup. Med. 58 (11), 42–50.

Lu, W.F., Deng, Y.-M., 2004. A system modelling methodology for materials and engineering systems design integration. Mater. Des. 25 (6), 459–469.

Mavris, D.N., Bandte, O., DeLaurentis, D.A., 1999. Robust design simulation: A probabilistic approach to multidisciplinary design. J. Aircr. 36 (1), 298–307.

McDowell, D.L., 2007. Simulation-assisted materials design for the concurrent design of materials and products. J. Occup. Med. 59 (9), 21–25.

McDowell, D.L., 2008. Viscoplasticity of heterogeneous metallic materials. Mater. Sci. Eng. R 62 (3), 67–123.

McDowell, D.L., Olson, G.B., 2008. Concurrent design of hierarchical materials and structures. Sci. Model. Simul. 15 (1–3), 207–240.

McDowell, D.L., Story, T.L., 1998. New directions in materials design science and engineering (MDS&E). Report of a NSF DMR-sponsored workshop, October 19–21.

McFarland, E.W., Weinberg, W.H., 1999. Combinatorial approaches to materials discovery. Trends Biotechnol. 17 (3), 107–115.

National Materials Advisory Board, 2008. Integrated Computational Materials Engineering: A Transformational Discipline for Improved Competitiveness and National Security. National Academies Press, Washington, DC.

Oden, J.T., Belytschko, T., Fish, J., Hughes, T.J.R., Johnson, C., Keyes, D., Laub, A., Petzold, L., Srolovitz, D., Yip, S., 2006. Simulation-based engineering science: revolutionizing engineering science through simulation. In: A report of the National Science Foundation Blue Ribbon Panel on Simulation-Based Engineering Science. National Science Foundation, Washington, DC.

Olson, G.B., 1997a. Computational design of hierarchically structured materials. Science 277, 1237–1242.

Olson, G.B., 1997b. Systems design of hierarchically structured materials: advanced steels. J. Comput. Aided Mat. Des. 4, 143–156.

Olson, G.B., 2000. Designing a new material world. Science 288 (5468), 993–998.

Ortiz, M., Cuitino, A.M., Knap, J., Koslowski, M., 2001. Mixed atomistic-continuum models of material behavior: the art of transcending atomistics and informing continua. MRS Bull. 26, 216–222.

Pollock, T.M., Allison, J.E., Backman, D.G., Boyce, M.C., Gersh, M., Holm, E.A., LeSar, R., Long, M., Powell IV, A.C., Schirra, J.J., Whitis, D.D., Woodward, C., 2008. Integrated Computational Materials Engineering: A Transformational Discipline for Improved Competitiveness and National Security. National Materials Advisory Board, NAE, National Academies Press, Washington, DC Report Number: ISBN-10: 0-309-11999-5.

Rajan, K., 2005. Materials informatics. Mater. Today 8 (10), 38–45.

Sankaran, S., Zabaras, N., 2006. A maximum entropy approach for property prediction of random microstructures. Acta. Mater. 54, 2265–2276.

Shenderova, O.A., Brenner, D.W., Nazarov, A.A., Romano, A.E., Yang, L.H., 1998. Multiscale modeling approach for calculating grain-boundary energies from first principles. Phy. Rev. B 57 (6), R3181–R3184.

Sigmund, O., 1994. Materials with prescribed constitutive parameters: an inverse homogenization problem. Int. J. Solids Struct. 31 (17), 2313–2329.

Sigmund, O., 1995. Tailoring materials with prescribed elastic properties. Mech. Mater 20 (4), 351–368.

Sigmund, O., 2000. A new class of extremal composites. J. Mech. Phys. Solids 48 (2), 397–428.

Sigmund, O., Torquato, S., 1997. Design of materials with extreme thermal expansion using a three-phase topology optimization method. J. Mech. Phys. Solids 45 (6), 1037–1067.

Subbarayan, G., Raj, R., 1999. A methodology for integrating materials science with system engineering. Mat. Des. 20, 1–12.

Suh, C., Rajagopalan, A., Li, X., Rajan, K., 2003. Combinatorial materials design through database science. Mater. Res. Soc. Symp. Proc. 804, 333–342.

van de Walle, A., Ghosh, G., Asta, M., 2007. Ab initio modeling of alloy phase equilibria. In: Bozzolo, G., Noebe, R.D., Abel, P.B. (Eds.), Applied Computational Materials Modeling. Springer-Verlag, New York, pp. 1–34.

Villars, P., Brandenburg, K., Berndt, M., LeClair, S.R., Jackson, A., Pao, Y.-H., Igelnik, B., Oxley, M., Bakshi, B., Chen, P., Iwata, S., 2000. Interplay of large materials databases, semi-empirical methods, neuro-computing, and first-principle calculations for ternary compound former/nonformer prediction. Eng. Appl. Artif. Intel. 13, 497–505.

Wang, M.Y., Zhou, S., 2004. Synthesis of shape and topology of multi-material structures with a phase-field method. J. Comput. Aided Mat. Des. 11 (2–3), 117–138.

Wang, Y., Liu, Z.-K., Chen, L.-Q., 2004. Thermodynamic properties of Al, Ni, NiAl, and Ni3Al from first-principles calculations. Acta. Mater 52 (9), 2665–2671.

Yin, X., Lee, S., Chen, W., Liu, W.K., Horstemeyer, M.F., 2008. A multiscale design approach with random field representation of material uncertainty. ASME IDETC/CIE, Advances in Design Automation Conference, New York. Paper Number: DETC2008-49560.

Zabaras, N., 2006. Advanced computational techniques for Materials-by-Design. Joint Contractors Meeting of the AFOSR Applied Analysis and Computational Mathematics Programs, Long Beach, CA, August 7–11.

Zohdi, T.I., 2003. Constrained inverse formulations in random material design. Com. Meth. Appl. Mech. Eng. 192 (28–30), 3179–3194.

CHAPTER 3

Overview of the Framework for Integrated Design of Materials, Products, and Design Processes

Nomenclature	
n	Number of decision variables
p	Number of equality constraints
q	Number of inequality constraints
m	Constraint functions
x	System variables
$g(x)$	System constraints
d_i^-	Deviation function corresponding to the underachievement of a goal
d_i^+	Deviation function corresponding to the overachievement of a goal
G_i	Goal i
A_i	Achievement of goal i
x_i^{min}	Lower bound on the ith system variable
x_i^{max}	Upper bound on the ith system variable
W_i	Weight assigned to a goal i
f_k	Functions in a lexicographic minimization formulation

In this chapter, we present an overview of the different aspects of the framework. In Section 3.1, we present a discussion of systems-based design of materials as a design process starting from requirements and ending with the design specifications. In Sections 3.2, 3.3, and 3.4, we present overviews of robust design methods, approaches for managing design complexity, and computational framework, respectively (see Figure 3.1). The discussion of robust materials design is followed up in Chapters 6 to 8. The approaches to managing complexity are presented in detail in Chapter 9, and the computational framework is detailed in Chapter 10. In Section 3.5, two materials design examples are presented, which are used throughout the rest of the book.

DOI: 10.1016/B978-1-85617-662-0.00003-X

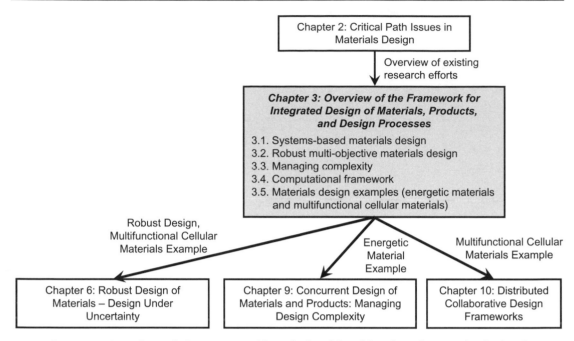

Figure 3.1 Overview of Chapter 3 and its relationship with other chapters in the book.

3.1. Systems-based Materials Design as a Process

The process-structure-property-performance paradigm (Olson 1997), described in Chapter 1 for designing materials, can be readily extended to incorporate the design of the material (composition, process route, morphology, etc.) as part of a larger overall systems design process. Results of modeling performed at multiple length scales and time scales using a variety of methods can be analyzed as necessary to provide decision support for selection of process path and microstructure or mesostructure to deliver a required set of multifunctional, often conflicting properties. For example, applications requiring toughness or energy absorption, such as crash-resistant automotive fenders or blast-resistant panels, promote a combination of high strength and good ductility. However, the interplay of increased toughness with design of the geometry of the fender or panel may be significant, leading to the need to address materials design as part of an overall system of concurrent design of material and product. Moreover, integrated materials solutions that combine multiple functions (sensing, structural, thermal, chemical, and so on) are increasingly attractive for purposes of efficiency and aesthetics. One example is integration of organic optical displays into flexible multilayers. Another example is lightweight composite materials with conductive matrices for electronic applications or electrostatic charge protection. There are a myriad of potential applications in biomaterials and biodevices, including artificial bone, tissue, ligaments, and blood cells, that must interface with biological functions while providing certain other structural, thermal, porous flow, and durability requirements.

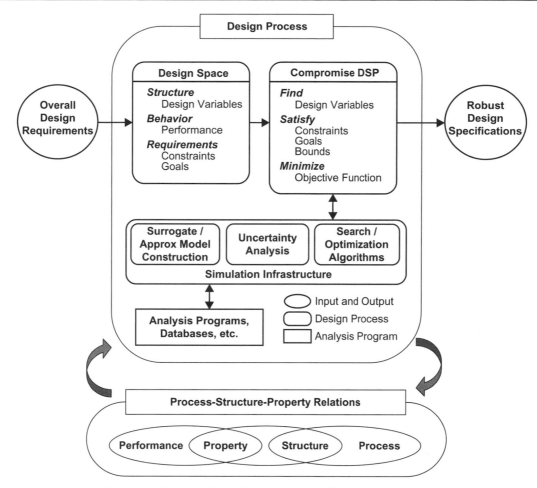

Figure 3.2 Systems-based robust materials design approach.

The inductive goals/means (top-down) engineering approach is relevant to design and contrasts with the usual approach taken in the application of the scientific method, which focuses on deductive cause-and-effect (bottom-up) approaches. A flow chart of the design process, with design requirements as inputs and robust sets of design specifications on design variables as output is shown in Figure 3.2.

We can map Olson's inductive goals/means concept of materials design directly to a systems-based robust design approach, as indicated in Figure 3.2. Effectively, process-structure-property relations inform the designer and map directly into the design process, which facilitates the transformation of overall design requirements into a ranged set of robust specifications for the material system of interest. We emphasize the need for robust design in view of the prevailing uncertainty of the materials design component of the concurrent materials and product design problem, a subject to be treated in detail in this book.

We suggest a hierarchical materials design framework in which mechanistic models are exercised to inform designers regarding decisions that contribute to the goal of robust design. These models add value to input from empirical routes of a more conventional nature and are intended to reduce the level of reliance on physical experiments, thereby reducing cost and time of materials design and development. In fact, "metamodels" are often based on a compendium of experimental results and other types of models or idealizations and are viewed as models in their own right (e.g., interpretations of various measure of material ductility). The approach is characterized by a confluence of engineering science and mechanics, materials science/physics, and systems engineering. Potential benefits of this systems approach include more efficient, concurrent design of materials and components to meet specified performance requirements, the capability to prioritize models and computational methods by the degree of utility in design, prioritizing mechanics and materials science/chemistry/physics phenomena to be modeled, and conducting feasibility studies to establish probable return on investment of new material systems, including specification of design requirements for new materials development.

The foundational components of such a systems-based approach to concurrent product and material design are (1) robust, multi-objective design methods, (2) tools for managing complexity, and (3) distributed computing frameworks. Each of these foundational components is introduced in a separate section of this chapter, followed by more detailed descriptions in subsequent chapters. In Section 3.5, we introduce two example problems that are revisited throughout the text as a means to communicate concepts and methods.

3.2. Robust Multi-objective Materials Design

Most materials design problems of practical interest involve solutions with property/response sets that conflict in terms of their demand upon material structure at various scales. By definition, a multifunctional material is one for which performance dictates multiple property and/or response requirements. For example, a multifunctional material in the mechanical property domain might require target ranges of strength and ductility as conflicting requirements.

Naturally occurring cellular materials offer good examples of multifunctional materials. Examples include bone (see Figure 3.3) and wood. Some of their functions include transport of nutrients, fluid storage and transport, strength, flexibility, and energy absorption/toughness.

As mentioned earlier in this section, we can consider conflicting property requirements in multiple physical domains (such as mechanical, thermal, chemical, magnetic, and electronic). For example, designing materials for gas turbine engine blades involves targets for the following properties (and perhaps more):

- Conductivity (thermal)
- Oxidation resistance (thermo-chemical)

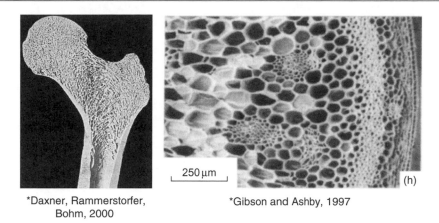

*Daxner, Rammerstorfer,
Bohm, 2000

*Gibson and Ashby, 1997

**Figure 3.3 Multifunctional cellular structures in nature
(Daxner, Rammerstorfer et al. 2000, Gibson and Ashby 1997).**

- Elastic stiffness (mechanical)

- High temperature creep and fatigue resistance (thermomechanical)

The designer might choose to maximize one of these properties/responses, setting constraints on the others, as is typical of conventional design optimization methods. There are several reasons to do otherwise, however. First, the uncertainty of the process-structure-property-performance hierarchy renders optimization of a given response a somewhat academic (idealized) exercise, and the response is often highly sensitive to the fluctuation of design variables. The basic notion of robust design is to instead seek solutions that are relatively insensitive to variations of design variables, as shown in Figure 3.4. Second, design flexibility and the prospect of transferability of the material to other products is substantially enhanced by considering targets or goals for each of the primary properties/responses and letting the designer choose satisfactory solutions rather than optimizing on a single objective. Strictly speaking, optimization for multiple objectives falls into the category of Pareto-optimization, leading to a family of solutions presented to the designer. It is typically a constrained form of the more general goal-based programming methods to be described later for robust design of materials. While at first glance it may appear feasible to reduce multi-objective optimization problems to single-objective optimization problems via combining nondimensional performance metrics, e.g., Ashy maps, one cannot distinguish the sensitivity to variation of individual design variables in such an approach, so treatment of uncertainty is problematic.

Single-objective optimization is appropriate only for cases of limited sensitivity to variation of parameters, well-characterized uncertainty, and a single, dominant functional requirement—this is a demanding set of requirements! Clearly, multi-objective rather than

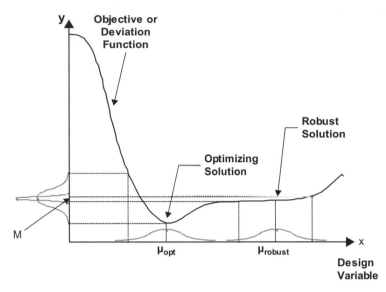

Figure 3.4 Robust design contrasts with optimization by seeking solutions for which the response (objective function) is relatively insensitive to changes of design variable(s).

single-objective design approaches that employ models in multiple physics domains should better serve the purpose of practical systems design of multifunctional materials. The foregoing discussion may be further directed by considering that there are different classes of materials design problems depending on the length scale of the "product" or its primary function:

- *Nanoscale design:* Supported by atomistic simulations and molecular modeling, often seeking discrete solutions using combinatorial search algorithms. Examples include design of molecular transistors, switches, motors, and drug molecules (e.g, virus blockers), and quantum engineering of interfaces such as tailored surfaces or molecular scaffolds and grain or phase boundaries.

- *Microscale design:* Design of *ensembles* or populations of nanostructured building blocks for functional assemblies. Examples include configurations of quantum dots for opto-electronic devices, distributions of coherent nanoparticles for strengthening in alloys, and polymer nanocomposites.

- *Mesoscale design:* Continuum modeling of larger-scale features (grains and phases) that strongly influence material response at the component or system scales. Examples include design of textured alloys and design of integrated circuits with electronic packaging in Very Large Scale Integrated (VLSI) systems.

In transitioning from nanoscale to mesoscale design problems, *complexity* typically increases as measured by interactions among lower-scale entities (many body interactions). Often, model degrees of freedom for material models are reduced in moving upward in length scale, while geometric degrees of freedom increase as required to represent the system. Moreover, the likelihood for requirements of multifunctional behavior increases since functionality is derived from multiple aspects of structure that necessarily get coupled and integrated as the length scale increases. Coupled with the uncertainty associated with a complex hierarchy of process-structure-property-performance relationships, inherent multifunctionality of most product-material systems motivates the need for multi-objective decision support and robust design methods. Constructs are introduced to address these issues in the remainder of this section, with details provided in Chapters 4 through 8.

3.2.1. Multi-objective Decision Making and the Compromise Decision Support Problem

Multi-objective design is motivated by not only the need to consider multifunctional behavior but also the need to consider robustness of candidate solutions. In many cases, robustness establishes preference among candidate solutions, as designers may prefer robust solutions and trade "optimality" or peak multifunctional performance against sensitivity to uncertainty. To facilitate negotiating these tradeoffs, use is made of the compromise Decision Support Problem (cDSP) construct (Mistree, Hughes et al. 1992), in which multiple design objectives are set as goals with associated target values. Solutions are identified by minimizing deviations from these goal targets, according to user preferences and subject to a set of constraints (Chen, Allen et al. 1997, Seepersad, Kumar et al. 2004) (see Figure 3.5). Effectively, we employ this decision-based strategy for utilizing information from simulations, metamodels, or heuristics to facilitate the identification of design specifications that may

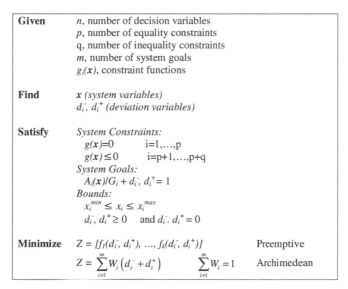

**Figure 3.5 cDSP for multi-objective design
(Mistree, Hughes et al. 1993).**

serve as inputs to another model or decision within a hierarchical structure. This philosophy of decision-based design is described in detail in Chapter 4, followed by an in-depth development of compromise and selection Decision Support Problems (DSPs) for concurrent product and materials design in Chapter 5.

Since the cDSP helps a designer manage multiple conflicting objectives, it is useful for negotiating tradeoffs between nominal performance and performance sensitivity or uncertainty. Resulting designs can be robust against variability associated with process route and initial microstructure, forcing functions, cost factors, design goals, etc. Moreover, the cDSP is a foundational part of new robust design methods that we introduce to deal with uncertainty due to microstructure variability and models (Allen, Seepersad et al. 2006, Chen, Allen et al. 1996) as well as chained sequences of cDSPs in a multilevel (multiscale) context (Choi, Austin et al. 2005). Types of uncertainty are discussed next, followed by an overview of new robust design methods created to manage them.

3.2.2. Types of Uncertainty and Robust Design

Uncertainty in a concurrent product and material design process arises from various sources. We may classify uncertainty as follows (Isukapalli, Roy et al. 1998): (1) natural uncertainty (system variability), (2) model parameter uncertainty, (3) model structural uncertainty, and (4) propagated uncertainty in a process chain. Examples of natural uncertainty include errors induced by processing, operating conditions, etc. Model parameter uncertainty results from incomplete knowledge of model parameters due to insufficient or inaccurate data. Model structural uncertainty is due to insufficient knowledge (approximations and simplifications) about a system. Finally, propagated uncertainty is a result of natural and model uncertainty propagated in a chain of models. Details regarding different types of uncertainty are provided in Chapter 6.

Different types of uncertainty coexist within any system. While some types can dominate a system, others may be essentially negligible. For example, the uncertainty associated with wall thickness of cellular materials is partially natural, irreducible uncertainty (i.e., manufacturing variability) and partially reducible uncertainty (i.e., human measurement error and resulting inaccuracy of data). Variations in wall thickness typically indicate the existence of manufacturing variability while the measurement error is assumed to be negligible. Strictly speaking, however, the uncertainty in wall thickness exhibits both types of uncertainty, and manufacturing variation is the dominant factor in the system.

Another relevant example is measuring variability in a system response based on a limited number of data points. Here, the uncertainty of the system response may be attributed to both natural system variability and data uncertainty due to the limited number of data points. As the amount of data increases, the data-related uncertainty is reduced, and the system variability dominates the measurement interval.

To address these sources of uncertainty in designing materials, there are several possible routes. One is to improve models and characterization of material structure. Another practical approach is to quantify the uncertainty to the extent possible and then design to seek solutions that are less sensitive to variation of microstructure and various sources of uncertainty, according to principles of robust design.

Robust design, proposed by Taguchi (Taguchi 1993), is a method for improving the quality of a product by minimizing the effect of uncertainty on product performance. From Taguchi's perspective, design performance should be on target, with low variability. This philosophy leads to designs that can be quite different from those obtained from optimization. "Optimal" solutions offer performance that may be on target nominally but often deteriorate significantly when conditions or assumptions change. Herbert Simon's philosophy, similar to that of Taguchi, is not to optimize but to "satisfice," or search for solutions that are "good enough" or provide satisfactory performance under a variety of circumstances (Simon 1996).

Three types of robust design are described later in this chapter and illustrated in Figure 3.5. The first two were introduced by Taguchi (Taguchi 1986):

- *Type I Robust Design:* Identify control factor (design variable) values that satisfy a set of performance requirement targets despite variation in noise factors. Methods exist and are extensively used in industry (Allen, Seepersad et al. 2006, Chen, Allen et al. 1997, Chen, Allen et al. 1996).

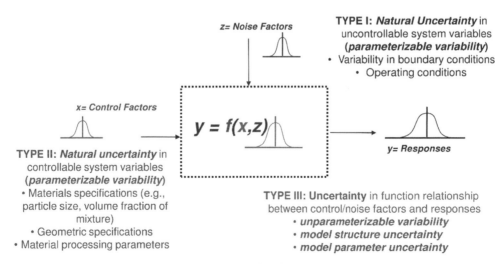

Figure 3.6 Three types of robust design, including insensitivity to uncertainty in noise factors (Type I) and control factors (design variables) (Type II), as well as uncertainty in models (PS and SP) and microstructure (Type III) (Choi 2005).

- *Type II Robust Design:* Identify control factor (design variable) values that satisfy a set of performance requirement targets despite variation in control and noise factors (Chen, Allen et al. 1996).

- *Type III Robust Design:* Identify *adjustable ranges* for control factors (design variables), that satisfy a ranged set of performance requirements and are insensitive to the variability within the model (Choi, Austin et al. 2005).

Although all three types of robust design are useful for materials design, Type III robust design is a particularly important element of designing material systems with intrinsic stochasticity of structure. As in any design process, design of materials is an iterative activity, with initial design exploration aimed at identifying feasible candidate solutions, and subsequent detailed design focusing in greater fidelity and detail on them. In the following section, multilevel design tools are introduced that facilitate efficient design exploration for all three types of robust design.

3.2.3. New Methods for Robust, Multilevel Design of Materials

Because material systems are complex and prone to many of the sources of uncertainty discussed in the foregoing, we seek materials design solutions that are satisficing and robust. The Robust Concept Exploration Method (Chen, Allen et al. 1997) was developed for Types I and II robust design. The Robust Concept Exploration Method (RCEM) is a domain-independent approach for generating robust, multidisciplinary design solutions. Robust solutions to multifunctional design problems are preferable tradeoffs between expected performance and sensitivity of performance due to deviations in design or uncontrollable variables. These solutions may not be global or local optima within the design space, as illustrated in Figure 3.4. RCEM is described in detail in Chapter 6.

Type III robust design, defined by Choi and coauthors (Choi, Austin et al. 2005), considers sensitivity to uncertainty embedded within a model, which can be due to uncertainty in model parameters or due to uncertainty in model structure. Model parameter or structure uncertainty typically differs from the uncertainty associated with noise and control factors, because it could exist in the parameters or structure of constraints, metamodels, engineering equations, and associated simulation or analysis models. The approach for Type III robust design focuses on incorporating Error Margin Indices (Choi 2005) within the RCEM (RCEM-EMI), and it is based on simultaneous incorporation of Types I, II, and III robust design techniques (see Figure 3.7). Type III robust design is typically required to manage (1) inherent variability that is difficult or impossible to parameterize, such as stochastic microstructure, (2) limited data, and (3) limited knowledge in new domains, such as new classes of microstructures.

The application of Types I-III robust design is clarified in Figure 3.7. While the application of traditional Types I-II robust design methods seeks solutions that are insensitive to variations in control or noise parameters, Type III robust design additionally seeks solutions that

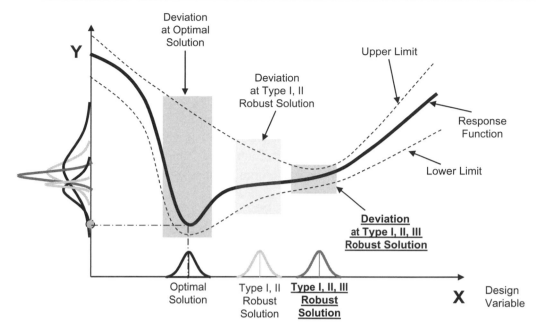

Figure 3.7 Illustration of Types I and II robust design solutions, which not only minimize the nominal value of an objective function but also minimize deviation in an objective function caused by noise factors and changes in design variables, relative to an optimal solution based solely on the extremum of an objective function. Type III robust design also minimizes deviation associated with modeling and microstructural sources of uncertainty (Choi 2005).

have the minimum distance between upper and lower uncertainty bounds on the response function (s) of interest associated with material randomness and model structure/parameter uncertainty. These bounds are determined from the statistics obtained from application of models over a parametric range of feasible microstructures and process conditions relevant to the simulations necessary to support design decisions (Choi 2005). This combination of "flat regions" of the objective function and tight bounds of variability associated with uncertainty of the functional relationship between control/noise and response is a new concept introduced to address realistic process-structure and structure-property relations in materials design. The concept is foundational to the RCEM-EMI, as described in detail in Chapter 8.

To accommodate the hierarchy of length scales and models corresponding to microstructure and pertinent mechanisms controlling material behavior that affect PS and SP relations, another essential ingredient is the top-down design of a system that is insensitive to the propagated uncertainty in the design process—the configuration of the design system. Our approach is termed the Inductive Design Exploration Method (IDEM), for finding solutions that are robust against the propagated uncertainty in a multilevel model chain. The basic idea involves passing down the feasible solution range in an inverse manner, from given final performance range to design space, instead of a bottom-up manner, passing mean and variance (or deviation) of a output to the next model. This approach allows for a designer to find ranged sets of robust solutions among which he or she expresses a preference for a particular solution.

The concept for selecting the most effective robust solution in multilevel robust design is illustrated in Figure 3.8. We consider multiple spaces of different dimension (degree-of-freedom, or DOF), between which models, simulations, or experiments serve as bottom-up mappings that project points in each space (e.g., composition) to the next space (e.g, microstructure), and then on to the final space (e.g., properties/responses). Projected points need not comprise simply connected regions in each space, but we plot them as such for simplicity. In assessing potential top-down solutions, if there exist two candidate solutions, Designs 1 and 2, for which final performance is identical, then which design should we select? Since the projected range of Design 2 in the intermediate space of y lies further from the constraint boundary than Design 1, we contend that Design 2 is better than Design 1; Design 2 is more reliable with regard to potential errors (unquantified uncertainty) in the mapping functions. Current robust design methods cannot address this issue since these methods only focus on the final performance range. The IDEM is discussed in detail in Chapter 8.

The foregoing framework of multi-objective decision support for robust concurrent design of materials can be readily extended to incorporate the design of the material (composition, morphology, etc.) as part of a larger overall systems design process. Results of models at multiple length scales and time scales can be analyzed as necessary to provide decision support for selection of hierarchical morphology and process path to deliver a required set of multifunctional, often conflicting properties.

3.3. Managing Complexity in Multilevel Product and Materials Design

The concurrent design of materials and products provides designers with flexibility to achieve design objectives that were not previously accessible. However, the improved flexibility comes at a cost of increased complexity of the *design process chains* and *the materials simulation models* used for executing the design process. Efforts to reduce the complexity generally result in increased uncertainty. We contend that a systems-based approach is essential for managing both the complexity and the uncertainty in design process chains and simulation models in concurrent material and product design. Our approach is based on systematically *simplifying the design process chains* such that the resulting uncertainty does not significantly affect the overall system performance. Similarly, instead of striving for accurate models for multiscale systems (which are inherently complex), we rely on making design *decisions that are robust to uncertainties* in the models. Accordingly, we primarily pursue hierarchical modeling in the context of design of multiscale systems.

The complexity of design process chains is reduced by selectively ignoring interactions between models and between decisions; for example, we can convert a coupled interaction

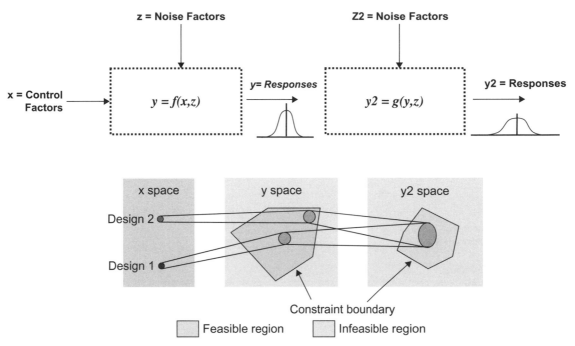

Figure 3.8 Multilevel robust design in which (top) responses *y* from model 1 serve as inputs to model 2, with output *y2*; (bottom) designs 1 and 2 have different model projections (e.g., PS) on the intermediate space, but design 1 is further from the constraint and is therefore preferred (Choi 2005).

(a two-way information flow) into a sequential interaction (a one-way information flow). While complex design processes in which all interactions are considered may lead to better designs, simplified design processes, in which some interactions are ignored, are faster and more resource efficient. Further, not all interactions have equal impact from a decision-making standpoint; some couplings have a significant effect on a designer's decisions, whereas others only have a minor impact. Hence, the proper level of simplification of a complex design process chain is one that reduces the design effort significantly without having a marked effect on the design outcome. In other words, the objective is to identify design processes that are robust to simplification, as opposed to design processes whose simplification results in significant degradation in the outcome. We believe that this is a novel application of the robust design philosophy initially proposed by Taguchi (Taguchi 1986) and later extended by various researchers (Box 1979, Chen 1995, Choi 2005, Choi, Austin et al. 2004, Choi, Austin et al. 2005, Mistree, Seepersad et al. 2002). The fundamental strategy is systematic simplification of *interactions* (between models and decisions) in the design process chains. Explicit consideration of interactions between simulation models and decisions is an essential component in the strategy for designing complex multiscale systems,

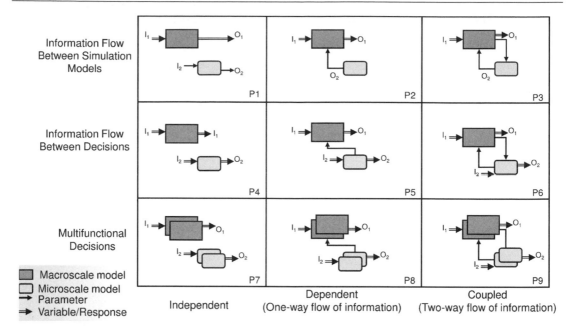

Figure 3.9 Interaction patterns in multilevel design.

particularly for systems with high degree of nonlinearity, uncertainty, and dependence upon initial conditions, as is the case with materials design.

We present a systems-based approach in Chapter 9 that is premised on the assumption that complex systems can be designed efficiently by managing the complexity of design process chains. The approach relies on (1) the use of reusable interaction patterns to model design process chains, and (2) consideration of design process decisions using metrics based on information economics (Lawrence 1999).

We have identified patterns based on interactions between simulation models at multiple scales. These interaction patterns are shown in Figure 3.9, organized in a matrix form. The three columns of the matrix represent three different types of interactions: (1) independent, (2) dependent, and (3) coupled. In the independent scenario, both the microscale and macroscale simulation models can be executed in a parallel fashion. The dependent scenario represents one-way (sequential) flow of information, in which the information generated by the microscale model is fed into the macroscale model. In the coupled scenario, both models need to be executed together, with a two-way flow of information between them.

The classification in the three rows of the matrix is based on design variables and responses associated with different models and multifunctionality. In the first row, the macroscale model has design variables and response variables associated with it, whereas in the second row, both the microscale model and the macroscale model are associated with design and response variables. The third row represents a multifunctional design scenario where at each level, different models are used that predict the system behavior for different functional characteristics (such as thermal, impact, vibration, etc.).

The interaction patterns are provided labels from P1 through P9. The primary advantage of this classification is that each type of interaction pattern is associated with a design process that represents the design exploration loop to be used in the overall system design process. Further, these interaction patterns are domain-independent and hence can be applied to any kind of multilevel design problem. These interaction patterns also have implications on the complexity of design processes. The complexity increases from left-to-right and from top-to-bottom in the matrix of interaction patterns. Interaction Pattern P9 results in the most complex design process, whereas the processes associated with Pattern P1 are the simplest. In moving from the left column to the right column, the complexity increases because of increased coupling. The complexity in the second row is higher than the first row because design exploration needs to be carried out at both scales. The increase in complexity from the second to third row is because of the additional coupling between functions at a given scale in a multifunctional scenario.

Hence, the objective during process simplification is to systematically proceed from Pattern P9 toward Pattern P1. The guiding principle used in determining whether process simplification is appropriate is the answer to the following question: "What is the impact of process simplification on the design decisions?" If the impact on the decision is small and process simplification reduces the design exploration cost drastically, then the designer should go ahead and simplify the design process; otherwise, not. In order to quantify the impact on a decision, metrics based on information economics, specifically value-of-information, have been developed. Value-of-information refers to the benefit of additional information due to preserved coupling per unit cost of computation and extended design time. *Expected value-of-information*, as defined by Howard (Howard 1966) and later applied to the problem of selection of parts from a catalog by Bradley and Agogino (Bradley and Agogino 1994), is given by the difference of the expected utility of the option selected with the benefit of information minus the expected utility without the information. The value-of-information metrics have been extended to determine the potential for improvement in a design solution by Panchal and coauthors (Panchal, Paredis et al. 2007) and applied to materials design problems (Panchal, Choi et al. 2006, Panchal, Choi et al. 2007). The details of the approach are presented in Chapter 9 of this book.

3.4. Computational Framework for Distributed Product and Materials Design

In this section, we present an overview of a computing framework for integrating heterogeneous software resources and facilitating online collaboration, information exchange, and utilization of software resources. The details of this framework are presented in Chapter 10.

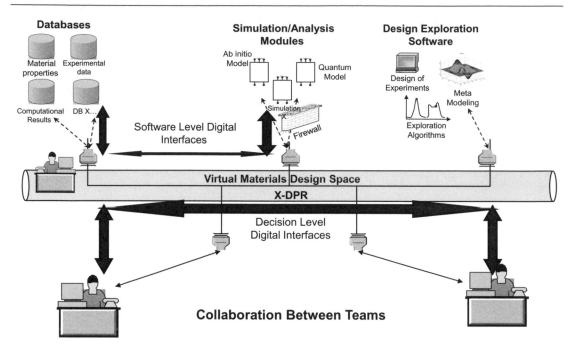

Figure 3.10 A distributed environment for materials design.

Various significant developments are taking place in the domain of distributed computing frameworks. Since it is difficult to provide a detailed discussion of all the technologies, our emphasis will be on a specific instantiation of the framework called eXtensible Distributed Product Realization (X-DPR) (Choi, Panchal et al. 2003, Panchal, Chamberlain et al. 2002). X-DPR addresses the challenge of networking collaborating stakeholders with one another as well as with the myriad analysis codes and domain models that support their decisions.

X-DPR is a computer framework for integrating distributed software resources, referred to as *agents,* over the Internet. The term *agents* refers to software applications that provide some design-related services. Some examples of agents used for designing multifunctional cellular materials are thermal and structural analysis codes, robust design code, design of experiments, and response surface modeling software. Other examples of agents include various materials databases, repositories of experimental data, and computational results from previous design scenarios. X-DPR allows integration of all these heterogeneous software resources over the Internet to form a virtual materials design environment (see Figure 3.10). Through this virtual environment, distributed designers can collaborate to design new materials.

The architecture of the X-DPR framework is shown in Figure 3.11. The framework is designed based on peer-to-peer communication between agents, in which each agent is an independent entity communicating with other agents. The agents require Extensible Markup Language (XML) inputs and provide XML outputs. For example, in the structural

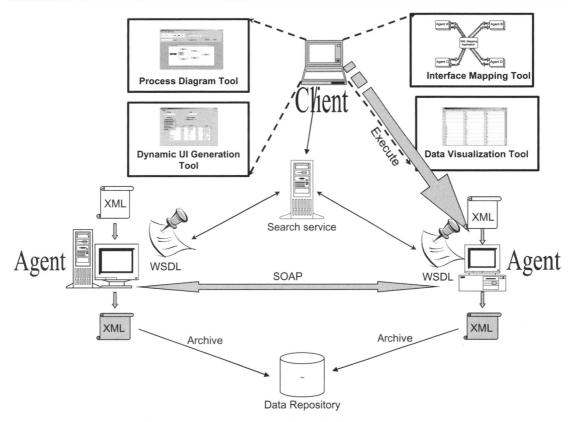

Figure 3.11 Architecture of the X-DPR framework.

design code, the input XML file contains information about the cellular material geometry, boundary conditions, and meshing information. The output XML file contains information about peak stresses and deflection. XML is used for capturing information because it is a standard for representing information on the Internet. Each agent is associated with a Web Service Description Language (WSDL) file that aptly describes the functionality it provides. The WSDL file contains information about inputs and outputs of the agent. The information transfer between these software agents is through an XML-based standard called Simple Object Access Protocol (SOAP). The SOAP standard forms a platform- and language-independent standard for information exchange between software applications. This enables agents to be developed in different languages and on different platforms.

A client application (developed in Java) is used to model and execute a design process both automatically and remotely through the invocation of distributed agents. The four main elements of the client application are (1) a *process diagram tool*, (2) an *interface mapping tool*, (3) a *data visualization tool,* and (4) a *dynamic user interface generation tool*. The *process diagram tool* contains a "whiteboard" on which a process diagram may be created using simple drag-and-drop utilities. For example, the process for designing materials can be modeled graphically in the process diagram tool. The tool equips designers with the ability to search for suitable agents and assign tasks to these agents by interacting with and executing remote agents. The *interface-mapping tool* is developed to provide seamless flow of information between agents. In a generic framework, it is very likely that the output of one agent may not be identical to the required input of another agent. For example, the outputs of the Design of Experiments may not be the same as inputs to structural and thermal analysis codes. In such a scenario, it is difficult to establish structural compatibility of information between the inputs and outputs of agents. The *interface-mapping tool* provides a means of mapping information between the XML-based interfaces of agents. The *data visualization tool,* on the other hand, provides a means for viewing the information contained in XML files. Since the data exchanged in X-DPR is in the XML format, it is inconvenient to view the data directly from XML files containing it. The *data visualization tool* thus provides a user interface for easier interpretation of the data. Finally, the *dynamic user interface generation tool* facilitates remote user interactions with agents. An agent describes user interface elements according to the manner in which stakeholders interact with them. Based on this description, a user interface is generated dynamically at the client side and the required information is obtained. The input obtained from the user is then sent back to the agent for processing. This increases the modularity of the framework by separating the actual engineering information from the user interface used to manipulate it. The framework also contains a search service that is used to locate required agents. The search is performed using a database of WSDL files for all available agents. The details of the framework and its utilization in a cellular materials design example are presented in Chapter 10.

3.5. Materials Design Examples

Elements of a systems-based approach to materials design, as highlighted in the preceding sections, are described in greater detail in subsequent chapters. In this section, we offer two abbreviated examples of mesoscopic materials design that are used throughout the book to demonstrate some of the principles of systems-based robust materials design.

3.5.1. Example 1: Strength and Reaction Initiation Based on Dynamic Plasticity in a Shock-Loaded Energetic Material

Multiphase energetic materials (thermite mixtures) are mixtures of micron scale (or finer) metal and metal oxide (sometimes intermetallic) powders, often with a binder and having intentional porosity. Shock compression of these materials can result in the initiation of exothermic chemical reaction that may or may not propagate in a self-sustained manner at scales well above the mean particle size. Shock compression of spatially resolved particle systems must be modeled to understand the effects of dynamic plastic deformation, particle

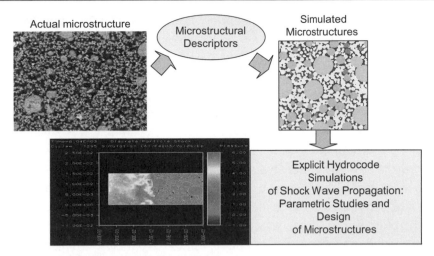

Figure 3.12 Linkage of an actual microstructure to a reconstruction algorithm to support parametric Eulerian hydrocode studies of shock wave propagation through energetic materials and characterization of reaction initiation probability (Austin, McDowell et al. 2006, Benson 1995).

interactions, and pore collapse at the mesoscale on the reaction initiation behavior to design the mixture for specified reaction initiation conditions (shock strength). The simulations involve propagation of shock waves through aluminum-iron oxide thermite systems $(Al + Fe_2O_3)$ composed of micron-size particles suspended in an epoxy binder (Austin, McDowell et al. 2006).

As shown in Figure 3.12, a fully thermomechanically coupled Eulerian code, RAVEN (Benson 1995), is used to model the progression of a shock wave induced by imposed particle velocities on a statistic volume element (SVE) of microstructure. The simulated (reconstructed) microstructures are based on particle size distributions, volume fractions, and levels of porosity measured experimentally for as-processed materials. The measured nearest-neighbor distribution of the Al particles (largest particles in Figure 3.12, approximately 1-2 microns in diameter) is enforced in the simulated mesh using simulated annealing, followed by constrained Poisson point placement of the Fe_2O_3 particles and pores. The epoxy binder then constitutes the remainder. Post-processing enables determination of the shock velocity in the mixture (Hugoniot behavior), the pressure profile throughout the SVE as a function of wavefront position, and the detailed temperature increase throughout the SVE. Appropriate rate- and temperature-dependent constitutive equations are used for the constituents (Klepazcko model (Klepaczko, Sasaki et al. 1993) for 1100 Al, Hasan-Boyce model (Hasan and Boyce 1995) for Epon 828, and an athermal model for elastic-plastic behavior of Fe_2O_3).

The probability of shock-induced reaction initiation is then evaluated from the temperature field at hot spots using the Merzhanov instability criterion for thermal explosion (Merzhanov

1966), which essentially considers whether or not heat generated at hot spots by the reaction can be transferred by conduction to the surrounding material at a rate high enough to quench the reaction. Because the SVE is not large enough relative to the particle size and spacing, it cannot serve as a representative volume element (RVE) for the desired response (the number density of sites for reaction initiation) and it is necessary to build up the statistics for the number density of sites from a set of instantiations or realizations of SVEs for the same microstructure subjected to the same shock-loading condition. It is noted that the RVE size is not based simply on spatial statistics of geometric features, as is commonly implied, but must also be based on analyses of responses to achieve statistical homogeneity or invariance of response with respect to further increases of system size; certain responses (in this case, reaction initiation) depend on extrema of microstructure attributes (i.e., rare events) and therefore require much larger (too large for computational practicality) volumes of material to be analyzed as a single ensemble or RVE. From these simulations, statistics are compiled and the mean response (f_o) and 99% prediction interval for upper (f_1) and lower (f_2) uncertainty bound functions represented using an iterative reweighted generalized linear model, with variance function estimation by maximizing the Pseudolikelihood estimator. In this case, variance is due largely to microstructure stochasticity. For given Al and Fe_2O_3 particle size distributions, for example, the number density of reaction initiation sites as a function of mean void size and volume fraction is shown in Figure 3.13. This facilitates either optimization (based on mean) or Types I-III robust design outlined earlier of a mixture with specified reaction initiation requirements. From the robust design perspective, the desired solutions are characterized by flat regions in mean response that have minimum distance between the upper and lower uncertainty bounds.

To facilitate higher length scale calculations, it is necessary to use the discrete particle simulations to inform extended irreversible thermodynamics continuum internal state variable models with additional flux terms to model delay of macroscopic reaction propagation at length scales of system components.

3.5.2. Example 2: Design of Multifunctional Cellular Materials

Multifunctional materials are integrated systems that serve multiple roles such as structural load bearing, thermal management, energy absorption, or other roles. These multifunctional material systems are compelling for certain applications, including actively cooled supersonic aircraft or spacecraft skins, engine combustor liners, and lightweight structural elements with internal damping. Linear or two-dimensional cellular materials are particularly suitable for multifunctional applications that require not only structural performance but also lightweight thermal or energy absorption capabilities. Certain structural and thermal properties of extruded prismatic cellular honeycomb materials, so-called linear cellular alloys (LCAs) shown in Figure 3.14, are superior to those of metallic foams with equivalent densities.

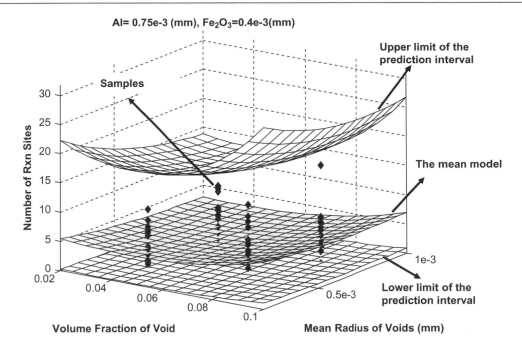

$$f_0(\mathbf{x_0}) = \exp\left\{\hat{\boldsymbol{\beta}}_{converged} \cdot \mathbf{x_0}\right\} - 2$$

$$f_1(\mathbf{x_0}) = \exp\left\{\hat{\boldsymbol{\beta}}_{converged} \cdot \mathbf{x_0} + t_{N-P,1-\alpha/2} \cdot \exp\left(\frac{\hat{\boldsymbol{\theta}}_{converged} \cdot \mathbf{x_0'}}{2}\right)\right\} - 2$$

$$f_2(\mathbf{x_0}) = \exp\left\{\hat{\boldsymbol{\beta}}_{converged} \cdot \mathbf{x_0} - t_{N-P,1-\alpha/2} \cdot \exp\left(\frac{\hat{\boldsymbol{\theta}}_{converged} \cdot \mathbf{x_0'}}{2}\right)\right\} - 2$$

Figure 3.13 Mean response surface model fit to shock simulation results of mixture for the number of reaction sites in SVE as a function of volume fraction and mean radius of voids for a prescribed size distribution of Al and Fe$_2$O$_3$ particles, along with upper and lower uncertainty bounds on response (Choi, Austin et al. 2005). This enables combined Types I-III robust design of the material (Choi 2005).

For example, LCAs exhibit greater in-plane stiffness and strength and out-of-plane specific energy absorption than stochastic metal foams (Evans, Hutchinson et al. 2001, Hayes, Wang et al. 2004). LCAs are advantageous as heat exchangers due to larger surface area density and lower pressure drop, two factors that compensate for lower heat transfer coefficients for laminar forced convection than for turbulent forced convection in stochastic metal foams with comparable relative densities (Lu 1999).

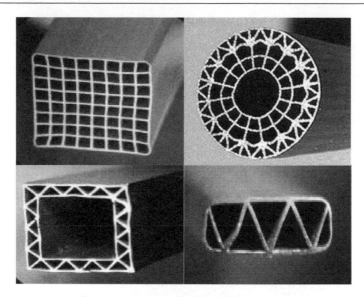

Figure 3.14 Extruded prismatic LCAs.

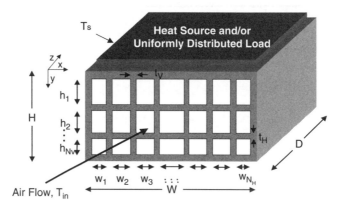

Figure 3.15 Compact, forced convection heat exchanger with graded, rectangular, prismatic, cellular materials.

In addition, the manufacturing process for LCAs facilitates the fabrication of multifunctional cellular materials. Powder slurries are extruded through a die and then exposed to thermal and chemical treatments in a process developed by the Lightweight Structures Group at the Georgia Institute of Technology (Cochran, Lee et al. 2000). Extruded metallic cellular structures can be produced with nearly arbitrary two-dimensional cellular topologies limited only by paste flow and die manufacturability. Wall thicknesses and cell diameters as small as fifty and several hundred micrometers, respectively, have been manufactured.

To better understand the requirements from a design standpoint, consider a potential application such as a structural heat exchanger, as illustrated in Figure 3.15. The structure is *multifunctional*; it is required to have satisfactory performance in more than one domain, including structural, thermal, and impact properties. Thus, a design method for realizing the structures and their constituent materials must support not only *multi-criteria* or multi-objective design with several potentially conflicting objectives, but also *multifunctional* design for which the criteria may be analyzed with multiple domain-specific techniques and software. For example, computational fluid dynamics and finite element codes might be employed to address fluid/thermal and mechanical/structural aspects, respectively. Due to computational demands and the distributed, heterogeneous nature of software and human expertise, it may not be possible to fully integrate all of the contributing multifunctional analyses associated with the systems-level design.

As with most materials design scenarios, the multiple scales are coupled. Performance requirements flow down from the macroscopic product or system level to lower length scales. Conversely, models at higher length scales (e.g., a structure or product in this case) require property predictions that capture the collective behavior of lower length scale structures (e.g., material mesostructure and microstructure). In this case, important macroscopic properties depend not only on dimensions of the material mesostructures but also on their *topologies* because alternative topologies can significantly impact the structural, thermal, and other properties of a cellular structure. For example, desirable heat transfer topologies may not have acceptable structural characteristics and vice versa. A design approach for material mesostructure must facilitate analysis and exploration of topology—not simply dimensional analysis and synthesis.

In addition, the materials must be manufacturable—in this case, with LCA processing techniques. Manufacturability implies constraints on realizable topologies and dimensions as well as expected variation in dimensions, topology (e.g., separated cell walls), and other characteristics like density/porosity or yield strength of the constituent solid material, for example. Certain aspects of the operating environment and manufacturing process may not be tightly controlled. This variation may have a large impact on smaller length scales such as material topology if the magnitude of variation is large relative to characteristic length scales or if the variation is compounded when a large material domain is considered. It is important to design systems with performance that is robust or relatively insensitive to variations in the environment, the manufacturing process, or other factors. For example, it would not be desirable to design and manufacture a structure with compliance or heat transfer rate that is highly sensitive to small changes in magnitude and direction of applied loads or temperature distribution along its boundaries, respectively. Thus, it is important to search for and identify topological designs that embody desirable tradeoffs between multiple, conflicting objectives, including robustness, or variation in performance due to uncontrolled variation in the environment or design parameters themselves. In summary, we are interested in designing mesoscopic cellular materials from a systems perspective in which we consider the topology, the manufacturing process, and the environment of an object as a system and consider the performance of the system in multiple domains.

To address these requirements, a Robust Topological Preliminary Design Exploration Method (RTPDEM) is presented in Chapter 7 of this book. As part of the method, robust topology design methods are introduced as well as multifunctional design methods that facilitate distribution of synthesis activities for highly integrated systems such as cellular material topologies.

References

Allen, J.K., Seepersad, C.C., Choi, H.-J., Mistree, F., 2006. Robust design for multidisciplinary and multiscale applications. ASME J. Mech. Design. 128 (4), 832–843.

Austin, R.A., McDowell, D.L., Benson, D.J., 2006. Numerical simulation of shock wave propagation in spatially resolved microscale particle systems. Model. Simul. Mater. Sci. Eng. 14, 537–561.

Benson, D.J., 1995. A multi-material eulerian formulation for the efficient solution of impact and penetration problems. Comput. Mech. 15, 558–571.

Box, G.E.P., 1979. Robustness in the strategy of scientific model building. In: Launer, R.L., Wilkinson, G.N. (Eds.) Robustness in Statistics. Academic Press, New York, pp. 201–235.

Bradley, S.R., Agogino, A.M., 1994. An intelligent real-time design methodology for component selection: An approach to managing uncertainty. J. Mech. Des. 116, 980–988.

Chen, W., 1995. A Robust Concept Exploration Method for Configuring Complex Systems. Ph.D. dissertation, Mechanical Engineering, Georgia Institute of Technology, Atlanta, GA.

Chen, W., Allen, J., Tsui, K.-L., Mistree, F., 1996. A procedure for robust design: Minimizing variations caused by noise factors and control factors. ASME J. Mech. Des. 118 (4), 478–485.

Chen, W., Allen, J.K., Mistree, F., 1997. A robust concept exploration method for enhancing productivity in concurrent systems design. Concurrent Eng-Res. A. 5 (3), 203–217.

Choi, H.-J., 2005. A Robust Design Method for Model and Propagated Uncertainty. Ph.D. dissertation, The GW Woodruff School of Mechanical Engineering, Georgia Institute of Technology, Atlanta, GA.

Choi, H.-J., Austin, R., Allen, J.K., McDowell, D.L., Mistree, F., 2004. An approach for robust micro-scale materials design under unparameterizable variability. 10th AIAA/ISSMO

Multidisciplinary Analysis and Optimization Conference, Albany, NY. Paper Number: AIAA-2004-4331.

Choi, H.-J., Austin, R., Allen, J.K., McDowell, D.L., Mistree, F., 2005. An approach for robust design of reactive powder metal mixtures based on non-deterministic micro-scale shock simulation. J. Comput-Aided Mater. Des. 12 (1), 57–85.

Choi, H.-J., Panchal, J. H., Rosen, D. W., Allen, J. K., Mistree, F. 2003. Towards a standardized engineering framework for distributed collaborative product realization. ASME 2003 Design Engineering Technical Conferences, Chicago, IL. Paper Number: DETC2003/CIE-48279.

Cochran, J.K., Lee, K.J., McDowell, D.L., Sanders, T.H., 2000. Low-density monolithic metal honeycombs by thermal chemical processing. 4th Conference on Aerospace Materials, Processes, and Environmental Technology, Huntsville, AL.

Daxner, T., Rammerstorfer, F.G., Böhm, H.J., 2000. Adaptation of density distributions for optimising aluminium foam structures. Mater. Sci. Tech. 16 (7–8), 935–939.

Evans, A.G., Hutchinson, J.W., Fleck, N.A., Ashby, M.F., Wadley, H.N.G., 2001. The topological design of multifunctional cellular materials. Prog. Mater. Sci. 46 (3–4), 309–327.

Gibson, L.J., Ashby, M.F., 1997. Cellular Solids: Structure and Properties. Cambridge University Press, Cambridge, UK.

Hasan, O.A., Boyce, M.C., 1995. A constitutive model for the nonlinear viscoelastic viscoplastic behavior of glassy polymers. Polym. Eng. Sci. 35 (4), 331–344.

Hayes, A.M., Wang, A., Dempsey, B.M., McDowell, D.L., 2004. Mechanics of linear cellular alloys. Mech. Mater. 36 (8), 692–713.

Howard, R., 1966. Information value theory. IEEE Trans. Syst. Sci. Cybern. SSC-2 (1), 779–783.

Isukapalli, S.S., Roy, A., Georgopoulos, P.G., 1998. Stochastic response surface methods (SRSM) for uncertainty propagation: Application to environmental and biological systems. Risk Anal. 18 (3), 351–363.

Klepaczko, J.R., Sasaki, T., Kurokawa, T., 1993. On rate sensitivity of polycrystalline aluminum at high strain rates. Trans. Jpn. Soc. Aerosp. Sci. 36 (113), 170–187.

Lawrence, D.B., 1999. The Economic Value of Information. Springer, New York.

Lu, T.J., 1999. Heat transfer efficiency of metal honeycombs. Int. J. Heat Mass Transf. 42 (11), 2031–2040.

Merzhanov, A.G., 1966. On critical conditions for thermal explosion of a hot spot. Combust. Flame 10, 341–348.

Mistree, F., Hughes, O.F., Bras, B.A., 1993. The compromise decision support problem and the adaptive linear programming algorithm. In: Kamat, M.P. (Ed.), Structural optimization: Status and promise. AIAA, Washington, DC, pp. 247–286.

Mistree, F., Seepersad, C.C., Dempsey, B.M., McDowell, D.L., 2002. Robust concept exploration methods in materials design. 9th AIAA/ISSMO Symposium on Multi-disciplinary Analysis and Optimization, Atlanta, GA. Paper Number: AIAA-2002-5568.

Olson, G.B., 1997. Computational design of hierarchically structured materials. Science 277, 1237–1242.

Panchal, J.H., Chamberlain, M., Rosen, D.W., Allen, J., Mistree, F., 2002. A service-based architecture for information and asset utilization in distributed product realization. 9th AIAA/ISSMO Symposium on Multidisciplinary Analysis and Optimization, Atlanta, GA. Paper Number: AIAA-2002-5480.

Panchal, J.H., Choi, H.-J., Allen, J.K., McDowell, D.L., Mistree, F., 2006. Designing design processes for integrated materials and products realization: a multifunctional energetic structural materials example. Design Automation Conference, Philadelphia, PA. Paper Number: DETC2006-99449.

Panchal, J.H., Choi, H.-J., Allen, J.K., McDowell, D.L., Mistree, F., 2007. A systems based approach for integrated design of materials, products, and design process chains. J. Comput-Aided Mater. Des. 14., 265–293.

Panchal, J.H., Paredis, C.J.J., Allen, J.K., Mistree, F., 2007. A value-of-information based approach to simulation model refinement. Eng. Optimiz. 40 (3), 223–251.

Seepersad, C.C., Kumar, R.S., Allen, J.K., Mistree, F., McDowell, D.L., 2004. Multifunctional design of prismatic cellular materials. J. Comput-Aided Mater. Des. 11 (2–3), 163–181.

Simon, H.A., 1996. The Sciences of the Artificial. MIT Press, Cambridge.

Taguchi, G., 1986. Introduction to Quality Engineering. Asian Productivity Organization, Tokyo.

Taguchi, G., 1993. Taguchi on Robust Technology Development: Bringing Quality Engineering Upstream. ASME, New York.

CHAPTER 4

Decision Making in Engineering Design

Nomenclature	
X_j	Design alternatives
A_i	Attributes based on which the alternatives are to be evaluated
$u(A_i(X_j))$	Utility of an alternative X_j for a particular attribute A_i
α, β, γ	Numerical probabilities denoting the preference of an alternative
p_k	Probability associated with different possible outcomes from an alternative

We contend that the concept of materials design is not limited to *selecting* an available material from a database; instead, we actually *tailor* material structure at various scales via associated processing paths to achieve properties and performance levels that are customized for a particular application; see Chapter 1. Accordingly, in this chapter, some fundamental aspects of engineering design are presented. These concepts are utilized throughout the rest of the manuscript to reinforce our contention that materials design is much more than materials selection in design or optimization of subsystems. We begin with a discussion of the goal-oriented nature of design processes in the first section (see Figure 4.1). Engineering design is in the pre-theory stage and as a result, there are several schools of thought. One of these is Decision-Based Design (DBD). Within DBD, there are at least two different perspectives: one articulated by Hazelrigg (Hazelrigg 1996) and the other by Mistree and coauthors

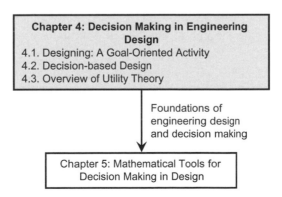

Figure 4.1 Overview of Chapter 4 and its relationship with other chapters in the book.

(Mistree, Smith et al. 1990). In Section 4.2, we present the perspective articulated by Mistree and coauthors. Finally, in Section 4.3, the fundamentals of utility theory as embodied in the decision support constructs embraced by Mistree and coauthors are described as a means of mathematically formulating design decisions in the presence of uncertainty.

4.1. Designing—A Goal-oriented Activity

There appear to be several common notions regarding the meaning of design as it involves development and *selection* of materials. First, there is the concept that design involves selection of the best material for a given application. This perspective focuses attention on material properties since design engineers are accustomed to communicating with materials engineers and developers on the basis of property sets. The Granta LTD Materials Selector Software by Ashby and coauthors (Granta Design Limited 2007) is a prime example. It has been widely employed in undergraduate mechanical, aerospace, civil and materials science, and engineering departments across the globe for instruction in materials selection in design. As discussed in Chapter 1, materials selection is just one aspect of the overall materials design problem.

A second common notion is that materials design comprises the *optimization* of properties or responses obtained from a certain set of material process-structure and/or structure-property simulations in which material microstructure is varied. This notion often compels development of complex, concurrent multiscale models since optimization intrinsically conveys the need for high accuracy of the modeling scheme. It is often pursued by the engineering science community, which employs various micromechanics modeling tools. In reality, only certain limited subproblems are typically amenable to optimization, with the shortcoming that overall system response cannot be optimized in general by independent optimization of subsystems. Moreover, the role of uncertainty must be respected in materials design, since it is so prevalent, which leads more to the notion of Pareto-optimal design involving a family of candidate solutions.

A third conceptualization of materials design is that of materials development in which new materials or variants of existing materials are viewed directly as candidate solutions for new applications. This "Edisonian" mode of materials discovery has been greatly accelerated in recent times with the advent of widely available advanced materials synthesis and processing methods. It is particularly pronounced in the sciences (including materials science, physics, and chemistry) with emphasis on complementary data-mining and combinatorial search methods to produce new candidate materials systems. It appears to be the dominant paradigm in development of nanomaterials and nanostructures. This bottom-up approach may contribute to systems-based materials design in terms of enhancing the material classes available for selection, but integrated multifunctional materials solutions typically require incorporation of nanostructures into larger (mesoscopic or macroscopic) material systems with different scales delivering different functionalities. This necessitates a systems approach. The ultimate reductionist philosophy of the bottom-up approach is that of constructing machines and systems atom-by-atom, a proposition made by the Foresight Institute (Battelle Memorial Institute and Foresight Nanotech Institute 2007), but considered by some as unrealizable for products (Smalley 2001). Thermodynamics and kinetics offer severe constraints on such processes.

In general, serendipitous materials discovery and development approaches promote the notion of design as "art," in that human creativity and intuition differentiate the effectiveness of efforts conducted by various groups and investigators. Systems engineering has often been viewed as a "soft" or qualitative discipline, so it is necessary to devote some attention to the matter of systems design as a quantitative, rather than qualitative, exercise. This is the purpose of this chapter.

While the natural sciences are concerned with how things are, an engineering designer is concerned with how things ought to be in order to attain functional goals (Simon 1996). As pointed out by Braha and Maimon (Braha and Maimon 1997), the distinction between engineering and natural science is that the aims and methodology of engineering differ, i.e., natural sciences are concerned with *analysis* and engineering with *synthesis*; natural science is theory-oriented, while engineering is results-oriented. In the context of materials design, the distinction between engineering design and the physical sciences have been discussed by Olson (Olson 1997, Olson 2000). Engineering approaches have been embodied by many researchers in various definitions of "design." Suh defines design as interplay between what we want to achieve (function) and how we want to achieve it (means) (Suh 1990). Mistree and coauthors view design as the conversion of information that characterizes the needs and requirements for a product into knowledge about the product (Mistree, Smith et al. 1990). That is, starting with desired functional requirements, we should be able to work backwards to explore effective design solutions. Effectively, the goal of design is to transform requirements—generally termed *functions*—into design descriptions (Gero 1990).

4.1.1. Models of Engineering Design

Materials design is a branch of engineering design, with particular nuances related to the hierarchical structure of materials and complex relations between process, structure, and properties. Hence, it is useful to discuss in more general terms the conceptualization of engineering design.

Research in engineering design is categorized into design *philosophies*, *models*, and *methods*. Design theory is a collection of principles that are useful for explaining a design process and provide a foundation for basic understanding required to propose useful methodologies. Design theory explains what design is, whereas design methodology is a collection of procedures, tools, and techniques for designers to use. Design methodology is prescriptive, while design theory is descriptive [9–11]. Design methods have been developed from different viewpoints that emphasize various facets of the overall design process. Some of these views, as summarized by Evbuomwan and coauthors (Evbuomwan, Sivaloganathan et al. 1996), include (1) design as a top-down and bottom-up process, (2) design as an incremental (evolutionary) activity, (3) design as a knowledge-based exploratory activity, (4) design as an investigative (research) process, (5) design as a creative (art) process, (6) design as a rational process, (7) design as a decision making process, (8) design as an iterative process, and (9) design as an interactive process. Although design methods are generally developed with a few of these viewpoints in mind, an ideal design method should support all of these.

Pahl and Beitz (Pahl and Beitz 1996) identify four key phases that are common to any prescriptive model for design. These phases include planning and clarification of task, conceptual design, embodiment design, and detail design. Planning and clarification of task involves identifying the requirements that the outcome of design should fulfill. These requirements are then converted into a statement of the problem to be solved. Conceptual design involves generation of principles used to satisfy the problem statement. Embodiment design involves refinement of the solution for the purpose of eliminating those that are least satisfactory until the final solution remains. During the detail design, all the details of the final design are specified and manufacturing drawings and documentation are produced.

In contrast to the descriptive models of design, prescriptive models exemplify how design should be done and not necessarily how it is done. Most of the prescriptive methods of design are based on the assumption that any design activity consists of three core activities— analysis, synthesis, and evaluation (ASE). *Analysis* is defined as the resolution of anything complex into its elements and the study of these elements and of their relationships. *Synthesis* is the pulling together of parts or elements to produce new effects and to demonstrate that these parts create an order (Pahl and Beitz 1996). A general model of design can be visualized as a feedback loop of synthesis, analysis, and evaluation.

These general ideas for analysis, synthesis, and evaluation are described by Gero (Gero 1990) as a series of transformations of information, starting with the requirements and ending with a description of the design that satisfies the requirements. According to Gero, the key aspects of product information include function (F), structure (S), expected behavior (Be), achieved behavior (Bs), and product description (D) (see Figure 4.2).

- *Function (F)* is the relation between the goal of a human designer and the behavior of the system. According to Pahl and Beitz, a function specifies the relationship between inputs and outputs in terms of energy, material (matter), and signal (information). A function expresses the relationships between inputs and outputs independently of the solution. It facilitates the subsequent search for solutions with a simple,

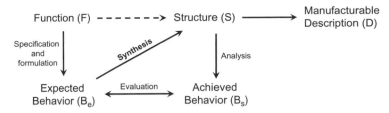

Figure 4.2 Model of design as a process (Gero 1990).

unambiguous function structure. It also allows clear definition of existing or necessary subsystems so that they can be dealt with separately. This reduces the complexity of the overall design problem. Functions are generally described as verb-noun pairs. Some examples of functions include absorb energy, dissipate heat, supply power, transfer material, actuate device, support load, etc.

- The *expected behavior (Be)* represents the physical properties that the artifact should have in order to satisfy the functional requirements (F).

- *Structure (S)* represents the artifact's elements and its relationships. The structure is also called the form of the artifact. It represents the proposed design solution, which includes information about geometry, configuration, and materials.

- The *achieved behavior (Bs)* of the structure is directly derivable using engineering principles.

- The product structure can be converted into a manufacturable *product description* (D).

In terms of these definitions, *analysis* is the transformation of product structure to achieved behavior and *synthesis* is the transformation from expected behavior to structure. *Evaluation* refers to comparison of the expected behavior with achieved behavior. This is an iterative process. The ASE view of design is foundational to many design efforts, such as those of Shimomura (Shimomura, Yoshioka et al. 1998) and Maher (Maher 1990).

In the context of the example of multifunctional cellular materials design discussed in Section 3.5.2, the functions are:

- to support load, and

- to dissipate heat.

The structure is characterized by the following parameters:

- the overall dimensions of the cellular material,

- the location, size, and shape of the cells, and

- the material used in the cellular alloys.

The behavior of the cellular material can be specified in terms of the following parameters:

- overall heat transfer rate, and

- stiffness (or compliance) of the material.

In a general scenario, the process of designing a material starts with a specification of the customer's needs that are translated into functional requirements. The functional requirements are then mapped into the expected behavior using engineering parameters. The *synthesis* transformation refers to the definition (or modification) of the structure based on the expected behavior. The synthesis may be based on knowledge of experts, previous designs, or through computational design approaches. Various simulation models may be used to predict the *achieved behavior* from the structure defined by the synthesis transformation. An example is a finite element model to predict the heat transfer achieved for a given structure and boundary conditions. This finite element model is a way of carrying out the *analysis* transformation in design. The predicted behavior (i.e., the outcome of the analysis) is compared to the expected behavior specified at the start of the design process. The comparison is referred to as *design evaluation*. The specified structure may or may not satisfy all the design requirements. Hence, the designers need to perform an iterative refinement of the structure until the desired performance is achieved. The process is continued until all the design requirements are met.

The model for design presented in Figure 4.2 has been mainly used in the product design domain. However, there are strong parallels between this model and the conceptualization of systems-based materials design by Olson (Olson 1997, Olson 2000). The relationship between these two is presented in Figure 4.3. Applying Gero's model of design to the materials design domain, the function (F) corresponds to the performance of the material, which is derived from the system-level requirements. The Behavior (S) corresponds to the material properties, the Structure (S) corresponds to the material structure, and the Manufacturable Description (D) corresponds to the processing of the material. The transformations in Gero's model of design correspond to the mappings in materials design. Specifically, the transformation "specification and formulation" corresponds to the mapping from performance to properties. The "design synthesis"[1] transformation corresponds to the mapping from the properties to structure. Similarly, the analysis transformation refers to the mapping from structure to properties.

The correspondence between the general model for product design and materials design is important because it allows the materials designers to utilize the methods and approaches developed in product design domain and vice versa. For example, the literature available in *computational design synthesis* can be of great advantage to materials design. In the following section, we present an overview of some of the design synthesis approaches utilized in the product design domain.

---

[1] Note that *design synthesis* is different from *materials synthesis*.

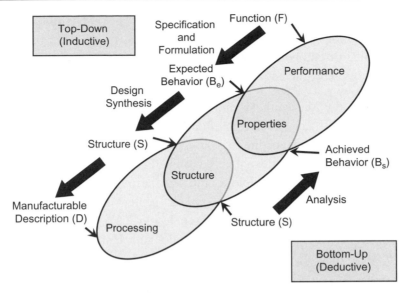

Figure 4.3 The relationship between Olson's diagram (Olson 1997) and Gero's model of design (Gero 1990).

4.1.2. Synthesis Approaches in Engineering Design

Design synthesis refers to that aspect of the design process that involves the creation of alternative solutions and concepts (Antonsson and Cagan 2001, Roozenburg 2002). Computational design synthesis is a special case of design synthesis where synthesis tasks are carried out using computational techniques. Computational design synthesis process consists of four main steps: *representation, generation, evaluation,* and *guidance* (Cagan, Campbell et al. 2005). The representation step involves capturing the form or attributes in the design problem. In the second step, candidate design solutions that are generated using the representation developed in the previous step. The evaluation step involves determining how well a candidate design solution satisfies the design objectives. Based on the gap between the design objectives and the performance of the design solution, the fourth step involves providing guidance to modify the design solution for improved performance.

A parametric optimization method represents a special (and perhaps the simplest) type of synthesis approach where the design space is represented using a set of parameters and the associated bounds on their values. A candidate design solution within the space of possible designs is a single point in the design space. For example, consider the design of an alloy with two constituents. The design parameters can be defined simply as the percentage content of different constituents. The generation step in the design synthesis process involves choosing a feasible composition based on thermodynamics, for example. The evaluation step involves determining the performance of the alloy resulting from the chosen composition. This may be carried out using physical experiments or using simulation models. Finally, depending on the performance targets and the achieved performance, the composition can be modified and a new design solution can be chosen through the guidance step. This guidance can be provided by a simple hill-climbing algorithm or by using sophisticated artificial intelligence (AI)–based algorithms. The guidance can even be provided by an expert designer.

Parametric optimization-based synthesis approaches are most commonly used due to the simplicity in representation and generation of design alternatives. Other approaches differ in the manner in which one or more of the four steps are carried out. For example, in genetic algorithm-based synthesis, (1) the design is represented as chromosomes to which the genetic operators can be applied, and (2) multiple candidate solutions are evaluated simultaneously instead of evaluating one solution at a time. Topology design is another special case of design synthesis in which the representation of the design space is different. The structure is represented as a distribution of material within a domain. The topology design problem is considered in detail in this book in Chapter 7.

These optimization-based methods for synthesis are limited to phases in the design process where the concept has already been fixed and the objective is to determine the best parameter values to maximize the achievement of design objectives. These approaches are not particularly suitable during the initial conceptual phases of design. The multifunctional cellular material discussed earlier in this chapter is a *concept* by which both the functions of supporting load and dissipating heat can be achieved. Other concepts can be used to satisfy these functions, such as a decoupled design where the functions of heat dissipation and load support are satisfied by separate components. The phase in the design process where different concepts are explored and evaluated is referred to as the conceptual design phase. During the conceptual phase, the function-based synthesis approaches are generally used, as opposed to optimization-based approaches.

Function-based synthesis approaches involve formal representations of functions and the relationships between them (Pahl and Wallace 2002, Wood and Greer 2001). These function representations provide designers the capability to perform design synthesis at the function and concept levels. During the design process, the functions are mapped to corresponding concepts. The knowledge about the relationships between functions and concepts can be captured in the form of design catalogs (Roth 2002). A catalog contains abstractions of different components that can be used to satisfy various functions. For example, in machine design, abstractions of standard components such as shafts, cams, levers, screws, gears, and clutches can be used to satisfy functions such as power transmission, velocity reduction, force amplification, etc. A library of such abstract components in the form of design catalogs can be used for design synthesis in future design problems where a new set of functions can be satisfied. Function-based methods have been used significantly for synthesizing dynamic systems, where the function representation is based on bond-graph methods. A function-concept-based design approach for integrated design of materials and products has been proposed by Messer and coauthors (Messer, Panchal et al. 2007). The authors model functions at multiple levels, including both the product level and the materials level. The functions are mapped to multiscale physical phenomena, the phenomena are mapped to solution principles, and solution principles are mapped to solution alternatives characterized by specific properties. These mappings are captured as multilevel design catalogs that can be reused for future design. Function-based methods are commonly pursued in bio-mimetics design.

To summarize, design synthesis refers to the transformation from the expected behavior to structure. More details on engineering design synthesis methods and applications are available in several references, including Antonsson and Cagan (2001) and Chakrabarti (2002). Various types of approaches are available in the engineering design literature. Optimization represents only a subset of methods for design synthesis. According to Cagan and coauthors (Cagan, Campbell et al. 2005), "synthesis as a method contrasts with traditional optimization in that the goal of synthesis is to more broadly capture, emulate, and/or utilize design decisions made by human designers."

4.1.3. Optimum Versus Satisficing Design Solutions

For complex system design problems such as concurrent, integrated materials and product design, the characteristics of design-synthesis problems may in part be summarized by the following descriptive sentences:

- The design problem may itself be loosely defined and open.

- Design involves information that comes from different sources and disciplines.

- Design is multifunctional and is governed by multiple measures of merits and performance.

- All the information required to perform design synthesis may not be available.

- Some of the information used in design may be hard (i.e., available in physically meaningful quantities) and some may be soft (i.e., may be subjective and based on experience).

Complex design problems are not characterized by a single unique solution. In complex real-world design scenarios, design solutions are less than optimal and are called *satisficing* solutions. The term *satisficing* was coined by Simon (Simon 1996) to describe a particular form of less-than-optimal solutions; they are good enough to be acceptable but are neither exact nor optimal. According to Simon, " … engineering design, by showing real world optimization to be impossible, demonstrates engineers are in fact satisficers, persons who accept 'good enough' alternatives, not because they prefer less to more but because they have no choice." In the context of this statement, a designer/decision maker has two choices when faced with a complex real-world problem in design:

- To develop an algorithm, based on a relatively simple model, by means of which an exact optimal solution can be found, and provided the assumptions on which the model is based can be satisfied exactly. However, only rarely will a solution that is optimal for a simple model be optimal in the real world.

- To develop an approximate algorithm or heuristic, based on a relatively complex model that represents the real world more closely than the simple model referred to earlier. The solutions obtained by using the approximate algorithm are *satisficing*.

When the sheer size and complexity of a design problem make an optimal solution unattainable, designers become *satisficers* because the real world demands that solutions be found to all problems. As pragmatists, engineering designers learn quickly to accept solutions that are less than optimal but meet the most important goals and constraints without undue penalties in function, cost, time, and other constraints. This is the essence of designing complex systems, characteristic of integrated design of materials and products. Our approach to integrated materials and product design is based on the *satisficing* view of design. This view is embodied in the Decision Support Problem (DSP) technique (Marston, Allen et al. 2000) discussed in the following section.

4.2. Decision-Based Design

Intuitively, it might seem that advances in modeling and simulation, combined with high-performance computing and trends in information theory and management, would enable a greater degree of automation of the decision process in designing materials and products. However, this prospect of automation is undesirable in our view for several reasons. First, in view of uncertainty, the human designer plays a role in arbitrating the interplay of various factors that serve as input into design decisions. Second, the process of informing decisions is most often imprecise and ill-defined. "To be successful, the engineering design of systems must embrace the notion that many decisions are made during the development process. This is not a controversial position to take. However, adopting the notion that these decisions should be made via a rational explicit process is not consistent with much of the current practice in the engineering of systems" (Buede 2000). Decision-Based Design (DBD) is focused on this formalization of a rational decision making process.

Amongst others, Hazelrigg (Hazelrigg 1998), Muster and Mistree (Muster and Mistree 1988), and Thurston (Thurston 1999), contend that the fundamental premise of DBD is that engineering design is primarily a decision-making process. It is a perspective from which design methods can be developed. Mistree and coauthors suggest that in DBD, decisions serve as markers to identify the progression of a design from initiation to completion (Marston, Allen et al. 2000, Marston and Mistree 1997, Mistree, Smith et al. 1990). Decisions are made by designers and are not just the product of computers and optimization methods, or the application of schemes based on specific analysis tools such as finite element analysis, molecular dynamics, etc. Decisions are means of communication that are characterized by receiving information from many sources and disciplines; they may have both discipline-dependent and discipline-independent features. The role of decisions in design processes is also pointed out by Gero (Gero 1990) in his statement, "[A] prevalent and pervasive view of designing is that it can be modeled using variables and decisions made about what values should be taken by these variables."

The implementation of DBD can take many forms. One of the implementation approaches is the DSP technique. The DSP technique (Muster and Mistree 1988) provides support and a rationale for using human judgment in design synthesis. The assertion is that the process of design, in its most basic sense, is a network of decisions. The DSP technique helps in partitioning the problems in simple terms so that it is possible to find solutions for it while being close to the actual system. According to Kamal (Kamal 1990, Mistree, Smith et al. 1991), all decisions identified in the DSP technique are categorized as *selection*, *compromise,* or a *combination* of these. The authors classify selection and compromise as *primary decisions* and others as *derived decisions.*

Selection decisions involve making a choice between a number of possibilities taking into account a number of measures of merits or attributes. The emphasis in selection is on the acceptance of certain alternatives through the rejection of others. The goal of selection in design is to reduce alternatives to a realistic and manageable number based on different measures of merit, called *attributes,* which represent the functional requirements. An example of a selection decision in the design of composites is resin material selection. Some examples of alternatives in this problem include Resin 9302/CA 9350, Resin 9400/CA9450, and Resin 9405/CA 9470 (Karandikar and Mistree 1992). The attributes of these resins that affect the performance of the composite material are damage resistance, viscosity, resistance to moisture, resistance to chemicals, shrinkage, cost, etc. The attributes may not all be of equal importance with respect to the decision. For some design scenarios, resistance to chemicals may be more important than cost, while for other scenarios, shrinkage and cost may be more important. Some of the attributes may be quantized using hard information (i.e., using numerical values for engineering quantities such as viscosity, shrinkage, etc.) and others may be quantified using soft information (e.g., quantities that are not associated with physical attributes such as aesthetics). The details of the selection decisions (their formulation and solution approaches) are discussed in detail in Chapter 5.

Compromise decisions require that the "right" values (or combination) of design variables be determined, i.e., the system is feasible with respect to constraints and the performance is maximized. The emphasis on compromise is on modification and change by making appropriate tradeoffs. As an example, in addition to the decisions about materials selection, composite materials design problems also involve making compromise decisions associated with the design variables such as the orientation of fibers, size and thickness of fibers, amount of resin, number of layers, etc. All these design variables affect the performance factors of the composite material and the resulting product that is developed using this material. Choosing the appropriate values of these design variables requires a tradeoff between different performance characteristics of the composite material such as strength, stiffness, and fracture toughness. The goal of compromise in design is that of modification of design through iteration based on criteria relevant to the feasibility and performance of the system. One of the primary differences between selection decisions and compromise decisions is that in a selection decision, the alternatives are discrete, whereas in compromise decisions, the design variables are continuous. The details of mathematical formulation of a compromise decision are provided in Chapter 5.

Different combinations of selection and compromise decisions can also occur in design. Decisions can be *independent, dependent, or interdependent.* As the name suggests, independent decisions can be made concurrently without a need for passing information between the decisions. Dependent decisions involve a unidirectional flow of information, and the interdependent decisions involve a bidirectional flow of information. For example, in the case of composite materials design, if the resin and the fiber can be selected independently, that is an example of independent selection decisions. If the selection of materials for resin and fiber is necessary before determining the geometric parameters (compromise decision), then the decisions are sequential. The dependent decisions arise from cases for which upstream decisions only affect downstream decisions. Finally, if the geometric parameters cannot be determined without the knowledge of materials used and at the same time, materials cannot be selected without the knowledge of geometric parameters, it represents an interdependent decision scenario. In such a case, both the decisions need to be made iteratively until the outcomes converge. The flow of information for interdependent decisions is illustrated in Figure 4.4. Interdependent decisions are also called coupled decisions. Since the interdependent decisions involve a bidirectional information flow, the decisions must be made by accounting for the objectives simultaneously. The three types of information flows between decisions are modeled using interaction patterns discussed in Section 3.3.

In Figure 4.4, we illustrate the interactions between only two decisions. Complex material and product design processes may involve many such selection and compromise decisions with associated flows of information, forming a network of decisions (Baskaran 1990). Based on the type of information flow in the network, the relationship between decisions can be modeled in two ways—*hierarchically* and *heterarchically* (Mistree, Smith et al. 1990). In hierarchical decisions, the information flow is clear and the sequence of decisions is well defined. Hierarchical decisions are dominated by independent and dependent decisions.

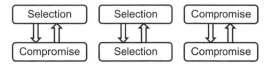

Figure 4.4 Types of interdependent decisions in design.

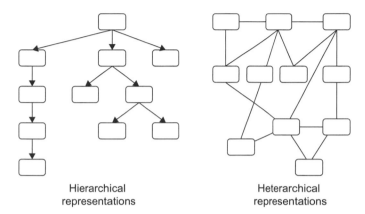

Hierarchical
representations

Heterarchical
representations

Figure 4.5 Hierarchical and heterarchical representations (Mistree, Smith et al. 1990).

In Figure 4.5, a set of decisions with a sequential flow of information is depicted. On the other hand, in the heterarchical relationship between decisions, decisions are highly coupled with each other and it is difficult to define the precedence of decisions. A heterarchical decision scenario is depicted in Figure 4.5, where the sequence in which decisions should be made is not clear. Due to the significant coupling between decisions in the network, large amounts of design iterations are required. In a product development scenario where the system can be easily decomposed into subsystems and requirements for the overall system can be clearly assigned to subsystems at various levels, the decisions are hierarchical. However, in integrated material and product design problems where the requirements for materials and product cannot be decomposed and separated, the decisions are heterarchical.

Interdependent or coupled decisions occur in numerous collaborative and concurrent design scenarios. An example of coupling between decisions from a traditional product development standpoint is the coupling between product design and manufacturing process design. Sambu and coauthors (Sambu, Chen et al. 2004, Sambu 2001) present a strategy for design for manufacturing based on coupled compromise decisions. Such coupling between decisions is particularly relevant in integrated materials and product design problems. The decisions made by materials designers affect the decisions made by product designers, and the decisions made by product designers affect the decisions made by materials designers. This is particularly true in functional materials. For example, in the integrated design of composite materials and structures such as aircraft wings, an important design consideration is to minimize vibration.

Vibration reduction can be achieved by increasing the damping capacity and/or increasing stiffness (Chung 2003). Damping in the entire system can be achieved by (1) designing the material to absorb more vibrational energy, or (2) by modifying the structure to reduce vibration. In other words, the function of vibration reduction is affected by the design of both the material and the structure. A chosen design of the material determines the best design of the structure, and the design of the structure determines the best design of the material. Hence, the decisions associated with the material and the structure are interdependent (coupled). Coupled decision support problems are applied in various problems such as design of composite material structures (Karandikar and Mistree 1993) and components, location-specific design of properties in automotive and engine components, ship design (Baskaran, Bannerot et al. 1989), etc. In this book, we present a coupled materials and product design scenario in Chapter 9.

4.3. Utility Theory

Utility theory is used to facilitate decision making in product realization based on mathematically complete principles that define "rational behavior" for the decision makers and can be used to derive the general characteristics of that behavior (Von Neumann and Morgenstern 1947). A decision involves the evaluation of a set of alternatives and selection of the most preferred alternative. "Utility" represents the decision maker's preference for an outcome, characterized with a set of attributes. In this context, an attribute is equivalent to the response variable that measures the performance of the product. Utility values for attributes generally lie between 0 and 1; a value of 0 denoting an unacceptable outcome and a value of 1 denoting the most preferred outcome. "If an appropriate utility is assigned to each possible outcome and the expected utility of the outcome for each alternative is calculated, then the best course of action is to select the alternative whose outcome has the largest expected utility" (Keeney and Raiffa 1976).

In the terminology of utility theory, a *decision* involves choosing among a set of alternatives $X_1, X_2, \ldots, X_n$. The consequences of selecting a particular alternative are described in terms of a set of attributes $A_1, A_2, \ldots, A_m$ that are common to all the alternatives. The specific values assumed by these attributes for a particular design alternative X_i are indicated as $A_1(X_i)$, $A_2(X_i), \ldots, A_m(X_i)$. The utilities of the alternatives for a particular attribute A_i are indicated as $u(A_i(X_1)), u(A_i(X_2)), \ldots, u(A_i(X_m))$, and the utilities of the alternatives considered with all attributes are indicated as $u(X_1), u(X_2), \ldots, u(X_m)$.

If the values of the attributes for different alternatives are known deterministically, then the alternative with maximum utility can be chosen. However, in general design scenarios, the values of the attributes $A_1(\underline{X}_i), A_2(\underline{X}_i), \ldots, A_m(\underline{X}_i)$, for an alternative $\underline{X}_i$ may not be known with certainty, but probabilities can be assigned to the various possible values of each attribute for each alternative. For example, consider a single-attribute design scenario, where the designer is concerned only with attribute A_1. If the possible values of an attribute are continuous, the consequences of selecting alternative $\underline{X}_i$ may be characterized by a distribution on the attribute A_1 with an associated probability distribution $f_p(A_1(\underline{X}_i))$, where

$$f_p(A_1(\underline{X}_i)) \geq 0 \quad \text{and} \quad \int f_p(A_1(\underline{X}_i))dA_1 = 1 \tag{4.1}$$

If alternative $\underline{X}_i$ leads to a discrete set of j possible outcomes, a probability p_k can be assigned to each possible outcome where

$$p_k \geq 0 \quad \text{and} \quad \sum_k p_k = 1 \tag{4.2}$$

Given this decision model, a decision maker must select the most preferred alternative when the consequences of each alternative are characterized by probability distributions rather than deterministic values for a set of attributes.

Therefore, given an alternative $\underline{X}_i$, if its attribute A_1 has a certain value, we state that the engineering team's utility for the outcome A_1 is $E[u(A_1(\underline{X}_i))]$. If the attribute has uncertainty, the expected utility for the outcome of $A1$ is calculated using its utility function and probability density function $f_p(A_1(\underline{X}_i)$. The expected utility may be calculated as follows:

$$E[u(A_1(\underline{X}_i))] = \int u(A_1(\underline{X}_i))f_p(A_1(\underline{X}_i))dA_1 \tag{4.3}$$

If alternative $\underline{X}_i$ leads to a discrete set of j possible outcomes, the expected utility is

$$E[u(A_1(\underline{X}_i))] = \sum_{k=1}^{b} p_k u_k(A_1(\underline{X}_i)_k) \tag{4.4}$$

The basic properties of utilities that are adopted as assumptions in utility theory are:

$$X_i \succ X_j \quad \text{implies that } u(X_i) > u(X_j) \tag{4.5}$$

$$u(\alpha X_i + (1 - \alpha)X_j) = \alpha u(X_i) + (1 - \alpha)u(X_j) \tag{4.6}$$

where $\succ$ denotes "is preferred to," X_i and X_j are possible alternatives, and α is the numerical probability that X_i is preferred, $(1-\alpha)$ is the probability that X_j is preferred.

The maximization of expected utility can be used as a selection criterion only if the utility functions are developed in this manner. If these two properties hold for a utility function, then the utility function is determined to be a linear transformation. The first property implies that the decision maker has a complete set of preferences. That is, given a set of alternatives $X_1, X_2, \ldots, X_n$, and their attributes $A_1(X_i), A_2(X_i), \ldots, A_m(X_i)$, the engineering team is able to decide a sequence such that $u(X_1) > u(X_2) > u(X_n)$ based on the attributes. The second basic property implies that if activities can be combined with probabilities, then the same must be true with the utilities attached to them. For example, a 50%–50% combination of outcomes X_i and X_j would be the prospect of having X_i occur with a probability of 50% and X_j occur with a probability of 50%. Thus the principle indicates that a designer can state whether he or she prefers the event X_i to the 50%–50% combination of X_j and X_k, $X_i > 0.5X_j + 0.5X_k$, or vice versa. By answering a set of similar questions, called *lotteries,* the utility or difference of utilities can be measured.

The question that arises before using utility functions for decision making is this: *Does a utility function that satisfies the properties discussed previously exist?* Von Neumann and Morgenstern postulated three axioms for utility functions. "Provided that these three axioms are satisfied, there exists a utility function with the above properties and with the desirable property of assigning numerical utilities to all possible outcomes such that the best course of action for the individual is the one with the highest expected utility" (Seepersad 2001). These axioms proposed by Von Neumann and Morgenstern are shown in Table 4.1.

In Table 4.1, α, β, and γ are probabilities. X's are potential outcomes of a decision; $>$ denotes "is preferred to," and $\sim$ denotes indifference. Axiom 1 is a statement of completeness of preferences and a statement of the transitivity of the preferences. Axiom 2:a states that if X_j is preferable to X_i then even a chance of obtaining X_j is preferable to X_i. Axiom 2:b is the dual of Axiom 2:a. Axioms 2:c and 2:d are continuity axioms. No matter how desirable an outcome may be, one can make its influence as weak as needed by giving it a sufficiently

Table 4.1 Von Neumann and Morgenstern axioms of utility
(Von Neumann and Morgenstern 1947).

The System X of Entities, X_1, X_2, X_3, ..., X_n with α and β on the Open Interval (0,1)
Axiom 1
$X_i > X_j$ is a complete ordering of **X**. This means write $X_j < X_i$ when $X_i > X_j$.
(1:a) Then for any two X_i, X_j, one and only one of the three following relations holds: $X_i \sim X_j$, $X_i > X_j$, $X_i < X_j$.
(1:b) If $X_i > X_j$ and $X_j > X_k$, then $X_i > X_k$.
Axiom 2
(2:a) $X_i < X_j$ implies that $X_i < \alpha X_i + (1 - \alpha)X_j$.
(2:b) $X_i > X_j$ implies that $Xi > \alpha X_i + (1 - \alpha)X_j$.
(2:c) $X_i < X_j < X_k$ implies the existence of an α with $\alpha X_i + (1 - \alpha)X_k < X_j$.
(2:d) $X_i > X_j > X_k$ implies the existence of an α with $\alpha X_i + (1 - \alpha)X_k > X_j$.
Axiom 3
(3:a) $\alpha X_i + (1 - \alpha)X_k \sim (1 - \alpha)X_k + \alpha X_i$.
(3:b) $\alpha(\beta X_i + (1 - \beta)X_k + (1 - \alpha)X_k \sim \gamma X_i + (1 - \gamma)X_k$, where $\gamma = \alpha\beta$.

small chance of occurrence. Axiom 3:a states that it is irrelevant in which order the outcomes in a combination are named. Axiom 3:b states that it is irrelevant whether the outcomes are combined in two successive steps with probabilities α, $(1-\alpha)$ and then β, $(1-\beta)$ or in one operation with probabilities γ, $(1-\gamma)$, where $\gamma = \alpha\beta$. These axioms are sufficient to guarantee the existence of a utility function with the desirable property of assigning numerical values to all possible outcomes such that the most preferred course of action is the one with the highest expected utility.

The process of assessment of utility functions for each attribute consists of three steps: (1) identification of the designer's preferences, (2) assessment of the designer's preferences for levels of attributes, and (3) fitting a utility function curve with respect to the levels of attributes. The first step involves determining the general characteristics of the designer's preferences. The preferences can be monotonically increasing (the larger the better), monotonically decreasing (the smaller the better), or non-monotonic, as shown in Figure 4.6. The monotonically increasing utility function can have a convex or concave utility function, depending on the designer's risk-taking nature. A convex utility function implies risk aversion, while a concave utility function implies risk proneness. This is also the case for a monotonically decreasing utility function. In a deterministic context, a convex utility function implies that a decision maker has decreasing marginal utility for an attribute at the direction of preference; and in a probabilistic context, a convex utility function implies that a designer is risk-averse. The utility functions are determined based on the preference equivalence of two options: the *certainty option*, where the attribute values achieved for different alternatives are known for certain, and an *uncertainty option*, where the designers have 50% probability of achieving the lower bound of attribute values and 50% probability of achieving the upper bound. The expected utilities of these two options are the same. In fact, most of the decisions in engineering design are risk-averse.

The discussion in this section up to this point has focused on a single attribute. Design decisions are generally characterized with multiple attributes. The question "How can multi-attribute utility functions be constructed?" is answered by Keeney and Raiffa (Keeney and Raiffa 1976) by developing a method that consists of two stages: (1) assessment of a

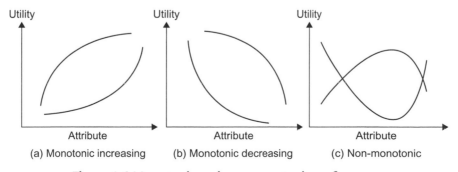

Figure 4.6 Monotonic and non-monotonic preferences.

utility function for each attribute and (2) combination of individual utility functions into a multi-attribute utility function that can be used to evaluate outcomes of alternatives in terms of all the attributes that characterize them. The details of determining multi-attribute utility functions are not discussed in this book; for additional information, refer to Keeney and Raiffa (Keeney and Raiffa 1976). Utility functions are used in the present context for modeling designers' preferences because utility theory is a domain-independent approach used to facilitate design decision making by evaluating preferences under conditions of risk and uncertainty.

4.4. Closing Remarks

In this chapter, we frame the problem domain of concurrent design of materials and products and provide the fundamental concepts of design that are used throughout the rest of the book. In Section 4.1.1, we present a model of engineering design as a framework to consider problems in integrated, concurrent design of materials and products. A general design problem is described using the following keywords: *function, structure, expected behavior,* and *achieved behavior.* The general design process is described using analysis, synthesis, and evaluation. The emerging paradigm of a materials engineer or developer as a designer is key, with a decision-based design framework underpinning his or her involvement in the process of product design.

In Section 4.2, we discuss the decision-based view of design and a specific implementation—the DSP technique. The two important constructs in the DSP technique are compromise Decision Support Problems (cDSPs) and selection Decision Support Problems (sDSPs). These constructs are used to model different types of decisions that occur in integrated product and materials design problems. One of the important aspects of modeling a designer's decision is to quantify his or her preferences towards the achievement of different design objectives. Utility theory, which is discussed in Section 4.3, helps designers in modeling their preferences and allows designers in making tradeoffs between different design objectives in the presence of uncertainty. The cDSPs and sDSPs are used multiple times to formulate the materials design problems in Chapters 6 through 9. Due to the importance of these constructs, we devote the next chapter to the details of mathematical formulation and execution of cDSPs and sDSPs.

References

Antonsson, E.K., Cagan, J. (Eds.), 2001. Formal Engineering Design Synthesis. Cambridge University Press, Cambridge, UK.

Baskaran, E., 1990. A model for the conceptual design of thermal systems: Concurrent decisions in designing the concept., Ph.D. dissertation, Mechanical Engineering Department. University of Houston, Texas.

Baskaran, E., Bannerot, R.B., Mistree, F., 1989. Hierarchical selection decision support problems in conceptual design. Eng. Optimiz. 14, 207–238.

Battelle Memorial Institute and Foresight Nanotech Institute, 2007. Productive nanosystems: A technology roadmap. <http://www.foresight.org/roadmaps/>. Accessed October 8, 2008.

Braha, D., Maimon, O., 1997. The design process: Properties, paradigms and structure. IEEE Trans. Syst. Man Cybern. A Syst. Hum. 27 (2), 146–166.

Buede, D.M., 2000. The Engineering Design of Systems: Models and Methods. Wiley, New York.

Cagan, J., Campbell, M.I., Finger, S., Tomiyama, T., 2005. A framework for computational design synthesis: model and applications. J. Comp. Inf. Sci. Eng. 5 (3), 171–181.

Chakrabarti, A. (Ed.), 2002. Engineering Design Synthesis: Understanding, Approaches, and Tools. SpringerVerlag, London.

Chung, L., 2003. Composite materials for vibration damping. In: Composite Materials: Functional Materials for Modern Mechnologies. Springer, New York, pp. 245–252.

Evbuomwan, N.F.O., Sivaloganathan, S., Jebb, A., 1996. A survey of design philosophies, models, methods, and systems. Proc. Inst. Mech. Eng. 210, 301–319.

Gero, J.S., 1990. Design prototypes: A knowledge representation schema for design. AI Mag. 11 (4), 26–36.

Granta Design Limited, 2007. Granta material intelligence software. <http://www. grantadesign.com/products/mi/>.

Hazelrigg, G., 1996. Systems Engineering: An Approach to Information-Based Design. Prentice Hall, Upper Saddle River, NJ.

Hazelrigg, G.A., 1998. A framework for decision-based engineering design. J. Mech. Design 120 (4), 653–658.

Kamal, S.Z., 1990. The development of heuristic decision support problems for adaptive design. Ph.D. dissertation, Department of Mechanical Engineering, University of Houston, Texas

Karandikar, H., Mistree, F., 1993. Modeling concurrency in the design of composite structures. In: Kamat, M.P. (Ed.), Structural Optimization: Status and Promise. AIAA, Washington, DC, pp. 769–806.

Karandikar, H.M., Mistree, F., 1992. Tailoring composite materials through optimal selection of their constituents. J. Mech. Design 114 (3), 451–458.

Keeney, R.L., Raiffa, H., 1976. Decisions with Multiple Objectives: Preferences and Value Tradeoffs. Wiley, New York.

Maher, M.L., 1990. Process models for design synthesis. AI Mag. 11 (4), 49–58.

Marston, M., Allen, J.K., Mistree, F., 2000. The decision support problem technique: integrating descriptive and normative approaches. Eng. Valuation and Cost Anal. Spec. Issue on Decis. Based Design: Status and Promise 3, 107–129.

Marston, M., Mistree, F., 1997. A decision-based foundation for systems design: A conceptual exposition. CIRP 1997 International Design Seminar Proceedings on Multimedia Technologies for Collaborative Design and Manufacturing. University of Southern California, Los Angeles.

Messer, M., Panchal, J.H., Allen, J.K., McDowell, D.L., Mistree, F., 2007. A function-based approach for integrated design of materials and products concepts. ASME Design Automation Conference, Las Vegas: NV Paper Number: DETC2007/DAC-35743

Mistree, F., Smith, W.F., Bras, B.A., Allen, J.K., Muster, D., 1990. Decision-based design: A contemporary paradigm in ship design. Trans. Soc. Naval Arch. Marine Eng. 98, 565–597.

Mistree, F., Smith, W.F., Kamal, S.Z., Bras, B.A., 1991. Designing decisions: Axioms, models, and marine applications. In: Fourth International Marine Systems Design Conference. Society of Naval Architects of Japan, Kobe, Japan, pp. 1–24.

Muster, D., Mistree, F., 1988. The decision support problem technique in engineering design. Int. J. Appl. Eng. Edu. 4 (1), 22–33.

Olson, G.B., 1997. Computational design of hierarchically structured materials. Science 277, 1237–1242.

Olson, G.B., 2000. Designing a new material world. Science 288 (5468), 993–998.

Pahl, G., Beitz, W., 1996. Engineering Design: A Systematic Approach, second ed. Springer-Verlag, New York.

Pahl, G., Wallace, K., 2002. Using the concept of functions to help synthesize solutions. In: Chakrabarti, A. (Ed.), Engineering Design Synthesis: Understanding, Approaches, and Tools. SpringerVerlag, London, pp. 102–120.

Roozenburg, N.F.M., 2002. Defining synthesis: on the senses and the logic of design synthesis. In: Chakrabarti, A. (Ed.), Engineering Design Synthesis: Understanding, Approaches, and Tools. Springer-Verlag, London.

Roth, K., 2002. Design catalogues and their usage. In: Chakrabarti, A. (Ed.), Engineering Design Synthesis: Understanding, Approaches, and Tools. SpringerVerlag, London, pp. 121–130.

Sambu, S., Chen, Y., Rosen, D., 2004. Geometric tailoring: A design for manufacturing method for rapid prototyping and rapid tooling. ASME J. Mech. Design 126 (4), 571–580.

Sambu, S., 2001. A design for manufacture method for rapid prototyping and rapid tooling, M.S. thesis, GW Woodruff School of Mechanical Engineering, Georgia Institute of Technology, Atlanta, Georgia

Seepersad, C.C., 2001. A utility-based compromise decision support problem with applications in product platform design, M.S. thesis, GW Woodruff School of Mechanical Engineering, Georgia Institute of Technology, Atlanta, GA.

Shimomura, Y., Yoshioka, M., Takeda, H., Umeda, Y., Tomiyama, T., 1998. Representation of design object based on the functional evolution process model. J. Mech. Design 120 (2), 221–229.

Simon, H.A., 1996. The Sciences of the Artificial. MIT Press, Cambridge, MA.

Smalley, R.E., 2001. Of chemistry, love, and nonobots. Sci. Am. 285, 76–77.

Suh, N.P., 1990. Principles of Design. Oxford University Press, Oxford, UK.

Thurston, D.L., 1999. Real and perceived limitations to decision-based design. ASME DETC, Design Theory and Methodology, Las Vegas, NV. Paper Number: DETC99/DTM-8750.

Von Neumann, J., Morgenstern, O., 1947. The Theory of Games and Economic Behavior. Princeton University Press, Princeton, NJ.

Wood, K.L., Greer, J.L., 2001. Function-based synthesis methods in engineering design. In: Antonsson, E.K., Cagan, J. (Eds.) Formal Engineering Design Synthesis. Cambridge University Press, Cambridge, UK, pp. 170–227.

CHAPTER 5

Mathematical Tools for Decision Making in Design

Nomenclature	
P_0	Internal pressure in the pressure vessel
ΔT	Temperature difference between the internal operating temperature and the ambient temperature of the pressure vessel
L	Length of the pressure vessel
L_c	Length of the cylindrical section of the pressure vessel
t_0	Thickness of the middle layer of the cylindrical section of the pressure vessel
t_1	Thickness at the inner and outer layers of the cylindrical section of the pressure vessel
t_2	Thickness at the hemispherical section of the pressure vessel
D	Diameter of the pressure vessel
R	Chamber radius of the pressure vessel
r	Boss opening radius
V	Volume of the pressure vessel
α_1, α_2	Fiber orientation
m	Number of fiber alternatives
n	Number of resin alternatives
I_j	Importance of alternative j
A_{ij}	Rating of alternative i with respect to the attribute j
A_j^{max}	Upper bound of the ratio scale for the jth attribute of an alternative
A_j^{min}	Lower bound of the ratio scale for the jth attribute of an alternative
R_{ij}	Normalized rating of alternative i with respect to attribute j
d^+	Deviation function corresponding to the overachievement of a goal
d^-	Deviation function corresponding to the underachievement of a goal
C	Cost of a composite design alternative
T_{cost}	Target for cost
ρ	Density
T_p	Target for density
E	Longitudinal modulus

DOI: 10.1016/B978-1-85617-662-0.00005-3

T_E	Target for longitudinal modulus
MF_i	Value of the merit function for alternative i
Z	Overall objective function
f_k	Functions in a lexicographic minimization formulation
δ	Change in the rating R_{ij} in attribute j of alternative i
M_i	Materials used in the pressure vessel
σ	Normal stresses
τ	Shear stresses
T_{vol}	Target for the volume of the pressure vessel
w_{sph}	Deflection of the hemispherical section
w_{cyl}	Deflection of the cylindrical section
A_i	Achievement of goal i
W_i	Weight assigned to a goal i
$u(A_i(X_j))$	Utility of an alternative X_j for a particular attribute A_i

As discussed in Chapter 1, designing the material to achieve specified ranged requirements of product performance often involves balancing competing, conflicting objectives with regard to the structure of the material. To offer a designer a tool for mediating these tradeoffs with other models and elements of the design system, the idea of Decision Support Problems (DSPs) can be employed. In this chapter, the detailed mathematical formulations of the decisions are presented. These formulations are utilized in Chapters 5 to 9 of this book (see Figure 5.1). Two types of decision support problems are presented in detail—the selection Decision Support Problem (sDSP; presented in Section 5.2) and the compromise Decision Support Problem (cDSP; see Section 5.3). Before discussing these DSPs, an illustrative example of the integrated design of a pressure vessel and an associated composite material is presented in Section 5.1. The example is used throughout the chapter to illustrate the decision-making constructs.

5.1. An Illustrative Example—Integrated Design of Pressure Vessel and Composite Material

The use of composite materials has provided designers with increased opportunities for tailoring structures and/or materials to meet load requirements. This involves choosing a fiber, a resin, and the proportion and arrangement of these two constituents in the composite material. Since the mechanics and processing of composite materials are well established in the literature [1–6], we use this as a simple example problem domain to illustrate the design decision-making concepts that are used for multilevel materials and product design throughout the rest of the book.

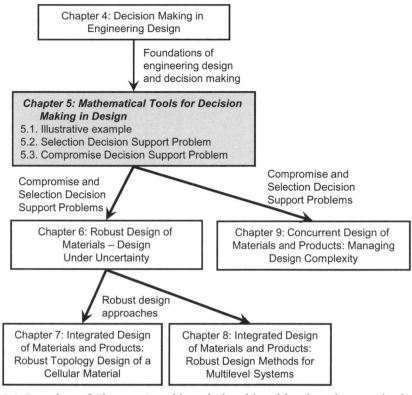

Figure 5.1 Overview of Chapter 5 and its relationship with other chapters in the book.

Consider the design of a composite material cylindrical pressure vessel with hemispherical end closures shown in Figure 5.2, processed via filament winding. The example is discussed in detail by Karandikar and Mistree (Karandikar and Mistree 1992). The pressure vessel is subjected to internal pressure and a constant temperature difference across its wall thickness. The volume of the pressure vessel is required to be 10^7 mm^3. The pressure vessel is to be fabricated from one of three materials: E-glass fiber/epoxy resin, carbon fiber/epoxy resin (Magnamite AS4/3501-6), and Kevlar-49 fiber/epoxy resin. The pressure vessel is to be designed by accounting for various performance criteria, namely (1) maximization of the *pressure vessel performance factor*, (2) minimization of cost, (3) achievement of customer specified volume, (4) carrying high load with high stiffness (resistance to deflection), and (5) avoiding material failure. The *pressure vessel performance factor* (Darms 1963) is defined as (burst pressure) * (volume)/weight. The performance factor is used because the complete chamber may be rated using a single factor, a variety of different pressure vessels, both composite and homogeneous, can be compared, and it approximates the condition of equal factors of safety in all the directions of the pressure vessel. The performance factor is affected by material strength, fiber reinforcement, and stress concentrations in the vessel.

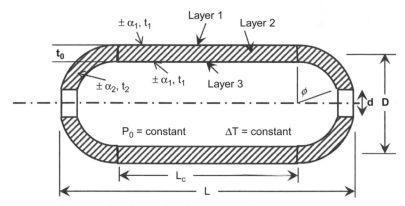

Figure 5.2 Schematic of a composite pressure vessel.

The numerical ranges of performance requirements for the pressure vessel are presented later in this chapter (see Table 5.3). The pressure vessel should not fail under the following loading conditions: internal pressure of 1800 psi and a 100 °C temperature difference between the operating temperature and the ambient temperature.

The loading and the structural configuration of the pressure vessel are as follows:

Loading
• Internal pressure P_0 (1800 psi)
• Constant temperature difference between the internal operating temperature and the ambient temperature ($\Delta T = 100$ °C)
Structural Configuration
• Thickness distribution:
• Hemispherical shell
$$t_2(\phi) = \frac{t_0}{\sin(\phi)} \qquad (5.1)$$
• Cylindrical shell
$$t = t_0 + 2t_1 \qquad (5.2)$$
• Laminate configuration:
• Hemispherical shell: one symmetrical angle-ply laminate, fiber angle α_2 and thickness $t_2(\phi)$
• Cylindrical shell: two symmetrical angle-ply laminates, fiber angles α_1 and α_2 and thicknesses t_0 and $2t_1$

The design of the pressure vessel includes the selection of the material and the determination of the boss opening diameter, chamber diameter, length of the cylindrical section, and the fiber orientation and thickness of each layer.

> | Given: | A set of *feasible alternatives*. |
> | --- | --- |
> | Identify: | The principal *attributes* influencing selection. The *relative importance* of attributes. |
> | Rate: | The alternatives with respect to each attribute. |
> | Rank: | The feasible alternatives in *order of preference* based on attributes and their relative importance. |

Figure 5.3 Word formulation of sDSPs.

5.2. Selection Decision Support Problem

The selection Decision Support Problem (sDSP) facilitates the ranking of design *alternatives* based on multiple *attributes* of varying importance. The design alternatives represent the options available to the designer. The attributes that represent the properties of that influence the decision among design alternatives. For example, if there are m fiber and n resin alternatives available in the composite materials selection problem, then (mxn) fiber-resin combinations can be generated. These (mxn) fiber-resin combinations are the design alternatives. Examples of the attributes include cost, strength, stiffness, and density of each design alternative.

The rank ordering of design alternatives indicates how much one alternative is preferred to another. In the sDSP, either analysis-based or experiment-based hard information and experience-based soft information can be used. The word formulation for an sDSP is given in Figure 5.3.

The procedure for formulating and solving sDSPs is:

Step 1: Describe the alternatives and provide acronyms. Assume that several design alternatives are available. Each alternative is described in words. The advantages and disadvantages of each alternative are set forth and meaningful acronyms are provided. In the composite design example, four choices of fibers (E-glass, Carbon Type I, Carbon Type II, and Kevlar 49) and four choices of resins (Resin 9302/CA 9350 3 phr, Resin 9302/CA 9350 8 phr, Resin 9400/CA 9450, and Resin 9405/CA 9470) are considered. The fibers are labeled from F_1 to F_4 and the resins are labeled R_1 to R_4.

Step 2: Describe each attribute, specify its relative importance, and provide acronyms. Since the alternatives are known, the next step in solving the sDSP is the identification of attributes by which the alternatives are to be judged. An attribute represents a quality of the desired solution and this quality must be quantifiable. The designer should be cautious in ignoring a relevant attribute regardless of its relative importance compared to other attributes. An attribute that is not considered in this step will have no effect on the selection process. Thus the selection process could yield an alternative that will perform well in all aspects except that of the ignored attribute. Therefore, the set of attributes defined must be comprehensive, understandable, unambiguous, and serve the needs of the problem.

The primary function of the fiber in a composite is to carry the load and provide required stiffness to the component. Selection of the fiber is based on the following attributes:

- *Mechanical:* tensile strength, modulus, strain to failure, compression properties, bearing properties (joint strength), shear properties (fiber resin interface bonding), and damage resistance (fatigue, creep, impact, fracture, and toughness).

- *Physical:* density, handling characteristics (safety), machinability, and catenary/twist.

- *Thermal:* coefficient of thermal expansion and thermal stability.

- *Chemical:* resistance to moisture.

- *Economics:* cost.

Selection of a resin for an application is based on the following attributes:

- *Mechanical:* strength, modulus, elongation, and damage resistance (fatigue, creep, impact, fracture, toughness).

- *Physical:* density and processing characteristics such as viscosity, pot life, gel time, fiber wet-out, and cure-cycle sensitivity.

- *Thermal:* glass transition temperature, curing temperature and shrinkage during curing, and thermal stability.

- *Chemical:* resistance to moisture and resistance to chemicals.

- *Economics:* cost.

- *Safety:* health hazard and toxicity.

We need resins with acceptably high strength, modulus, and elongation. These influence the transverse and shear strength of the composite, which in turn influences the burst strength. One can additionally minimize moisture absorption and improve damage resistance properties including fatigue, creep, impact, fracture, and toughness by an appropriate resin choice. In case of resins, processing characteristics are very important. A low viscosity is essential to achieve complete wetting of the reinforcement and for removal of entrapped air and volatile solvents. Suitable viscosities range from 350 cp to 1500 cp (at 77°F). If the resin is too fluid, problems in the control and uniformity of resin content are encountered. The resin will migrate to the outer layers of the winding leaving dry inner layers, a source of premature composite failure. Too viscous a resin, on the other hand, causes fiber fuzzing in the resin bath and feed eye, uneven fiber coating, excessive tension, and air entrapment. A pot life of several hours is a normal requirement to prevent gellation before the winding is completed. Gel time and resin flow under winding tension and during cure are other rheological factors

Table 5.1 Relative importance of attributes in the composite design problem

Selection Attributes	Relative Importance (I_j)
Level 1 Attributes	
Cost	1.0000
Level 2 Attributes	
Longitudinal modulus	0.4667
Longitudinal strength	0.4667
Density	0.0667
Level 3 Attributes	
Fiber: flexibility ratio	0.8570
Fiber: thermal stability	0.1430
Resin: damage resistance	0.0165
Resin: viscosity	0.0500
Resin: pot life	0.2043
Resin: gel time	0.0500
Resin: heat-distortion temperature	0.0500
Resin: shrinkage	0.2043
Resin: resistance to moisture	0.2043
Resin: resistance to chemicals	0.0165
Resin: handling characteristics	0.2043

that need to be considered in the selection process. Low shrinkage is advantageous in reducing internal composite stresses and facilitates the removal of a wound structure from the mandrel. The bonding characteristics and heat distortion temperature are also effectively controlled by the resin. It is noted that models do not exist for all these aspects or phenomena, and hence heuristics or experience must be factored in to augment models that do exist, such as those for elastic stiffness as a function of fiber volume fraction.

The attributes are ranked for their relative importance, I_j. The relative importance of the selection attributes are shown in Table 5.1.

The relative importance is assigned to one of three levels—1 to 3. The achievement of the cost goal is accorded the highest priority; hence it is in Level 1. This is followed by the density, strength, and stiffness goals at the second level. The strength and stiffness goals, however, are given a greater importance by assigning them a higher weight. The other merit function goals for fiber and resin selection are at the third and lowest priority.

Step 3: Specify scales and rate the alternatives with respect to each attribute. There are four types of scales, namely, ratio, interval, ordinal, and composite (Riggs 1977). The choice of a particular type of scale to model an attribute depends on the nature of available information. The ratio scale is used for an attribute for which physically meaningful numbers are available, e.g., cost, density, strength, etc. The ordinal scale is used to model an attribute that can only be qualified in words. An ordinal scale is appropriate for attributes like aesthetic appeal, color, etc. The interval scale is used to transform the quality captured by the ordinal word scale into a numerical interval scale. Interval scales are created for attributes for which only qualitative or soft information is available. Safety, reliability, complexity, simplicity are some examples of attributes measured on an interval scale. The composite scale is used for a generalized attribute that is generated as the result of computations.

All the attributes in the composite design problem can be quantified using ratio scale. The numbers used in a ratio scale are generally analysis- or experiment-based, computable, or measurable and are therefore categorized as hard information. It is important that the ratio scales are established independently of the set of alternatives being considered. It is necessary to specify the upper (A_j^{max} for the jth attribute) and lower (A_j^{min}) bounds for the ratio scale and indicate whether a larger or smaller number indicates preference. For the cost attribute, a smaller number indicates preference, whereas for strength, a larger number indicates preference. Specification of the upper and lower bounds for the ratio scale is imperative. The bounds should indicate the most desirable outcome and the minimum outcome that is acceptable. The bounds should be specified after very careful consideration. For attributes on the ratio scale, the measured or computed number associated with each alternative becomes its rating.

Once the scales are established, the rating, A_{ij}, of alternative i with respect to attribute j begins. For attributes on the ratio scale, the measured or computed number associated with each alternative again becomes its rating. For an attribute on an interval scale, a rating needs to be assigned and justified. The justification of each rating is extremely important, and the set of justifications is called a viewpoint. Ratio scales are seldom converted to interval scales. Ordinal scales must be converted to interval scales to be used in the solution process.

Step 4: Normalize the ratings. The attribute ratings, A_{ij}, are on scales that are not uniform. For example, for some attributes, a larger rating would indicate a preference whereas for others a lower rating would indicate preference. Further, it is unlikely that the upper and lower bounds on the scales are the same. Therefore, it is necessary to convert the attribute ratings to scales that are uniform. This can be achieved using two approaches: (1) by converting the attribute rating, A_{ij}, to a normalized rating, R_{ij}, and (2) by using deviation functions.

Normalized Rating: In the first approach, the normalized scales range from 0 to 1, with a higher number indicating a preference. There are different ways to effect normalization. One way for normalizing an attribute rating for alternative i with respect to attribute j is

$$R_{ij} = \frac{A_{ij} - A_{ij}^{min}}{A_{ij}^{max} - A_{ij}^{min}} \tag{5.3}$$

where A_{ij}^{min} and A_{ij}^{max} represent the lowest and the highest possible values of the alternative rating A_{ij}. The preceding formulation is for the case where the larger value of an attribute rating represents preference. If a smaller value of an attribute rating represents preference, the normalized rating, R_{ij}, is defined as

$$R_{ij} = 1 - \frac{A_{ij} - A_{ij}^{min}}{A_{ij}^{max} - A_{ij}^{min}} \tag{5.4}$$

In cases where the normalized ratings for all the alternatives turn out to be the same, that attribute may be dropped from further consideration.

Normalization Using Deviation Function: In the second approach, a target value corresponding to the desired attribute value is set. Any deviations from the target value are measured as two non-negative deviation variables d^+ and d^-. The deviation variable d^+ represents the overachievement of an attribute with respect to the target value, whereas d^- is the underachievement of the target. For example, in the composite design problem, the target value for cost is set as $T_{cost} = \$1.00$ per kilogram. If the cost of a composite material is $\$1.10$ per kilogram, then the overachievement of the cost attribute is: d_{cost}^+. The relationship between the cost of a composite design alternative, the deviation variables, and the target value is given by

$$C + d_{cost}^- - d_{cost}^+ = T_{cost} \tag{5.5}$$

If the design goal is to achieve a target value for an attribute, both the deviation variables (underachievement, d^-, and overachievement, d^+) are minimized. If the objective is to minimize the attribute value, such as cost, then the overachievement d^+ is minimized and the underachievement d^- is set to zero. Similarly, when the objective is to maximize the attribute such as strength, the designer strives to minimize underachievement and the overachievement variable is set to zero.

The normalization of the attributes associated with the composite design problem is as follows:

Cost:

$$C + d_{cost}^- - d_{cost}^+ = T_{cost} \tag{5.6}$$

where

$$T_{cost} = \$1.0/kg$$

Density:

$$\rho + d_\rho^- - d_\rho^+ = T_\rho \tag{5.7}$$

where

$$T_\rho = 10^{-6} \, kg/mm^3$$

Longitudinal modulus:

$$E_1 + d_E^- - d_E^+ = T_E \tag{5.8}$$

where

$$T_E = 3.0 \times 10^5 \, N/mm^2$$

Longitudinal strength:

$$\sigma_{LU} + d_\sigma^- - d_\sigma^+ = T_\sigma \tag{5.9}$$

where

$$T_\sigma = 10^4 \, N/mm^2$$

Other attributes:

$$A_i + d_i^- - d_i^+ = T \tag{5.10}$$

where

$$T_i = 1$$

The target value for cost T_{cost} is \$1.00/kg. The targets corresponding to density, longitudinal modulus, and longitudinal strength are $T_\rho = 10^{-6} \, kg/mm^3$, $T_E = 3.0 \times 10^5 \, N/mm^2$, and $T_\sigma = 10^4 \, N/mm^2$, respectively.

Step 5: Evaluate the merit function for each alternative. A merit function combines all the individual ratings of the attributes using the proper weights defined in Step 1. There are several methods for modeling the merit function. The most frequently used model, however, is the linear model:

$$MF_i = \sum_{j=1}^{n} I_j R_{ij} \qquad i = 1, \cdots, m \tag{5.11}$$

where

m = number of alternatives
n = number of attributes
I_j = relative importance of the jth attribute
R_{ij} = rating of alternative i for attribute j
MF_i = value of the merit function for alternative i

The linear model for determining the merit function is widely used due to its simplicity. Other types of merit functions that account for nonlinear preferences can also be used. One approach for capturing nonlinear merit function is to use the product of ratings of alternatives.

Another possible approach is to use of utility functions instead of a direct weighted sum of ratings for alternatives.

In the case where deviation functions are used, the overall objective of the selection decision is to minimize the weighted sum of deviation variables:

$$Z = \left\{ d^-_{cost}, (0.0667d^+_\rho + 0.4667d^-_E + 0.4667d^-_\sigma), \left(\sum d^-_k \right) \right\} \tag{5.12}$$

The overall objective function, Z, is a combination of both rank ordering and weighted sum. A weighted sum of the set of attributes within different levels of relative importance attributes (see Table 5.1) are rank-ordered. In the equation given previously, the first level consists of the deviation variable for the cost attribute. The weight assigned to cost is 1. The second level consists of deviation variables associated with density, modulus, and strength. The sum of weights for these three deviation variables associated with density, modulus, and strength (0.0667, 0.4667, and 0.4667 for d^+_ρ, d^-_E and d^-_σ, respectively) is 1.

The value of Z is evaluated for each design alternative. It is equal to a set of weighted deviations corresponding to the different levels of attributes. The design alternatives are compared one level at a time, starting with the first level. An alternative with lower value of a weighted deviation is considered better. If the weighted deviations are equal at a given level, the next level of weighted deviations is compared. For example, consider two design alternatives with $Z_1 = (0.5, 0.3, 0.8)$, and $Z_2 = (0.5, 0.7, 0.5)$. At the first level, both the weighted deviations are equal ($=0.5$). Hence, the second level is compared. At the second level, the first alternative (with a weighted deviation of 0.3) is better than the second alternative (with a weighted deviation of 0.7). Hence, the first alternative is the preferred design option. This kind of a comparison is referred to as *lexicographic minimum*. Formally, a lexicographic minimum is defined as:

Lexicographic Minimum: Given an ordered array **f** of nonnegative elements, f_k's, the solution, given by $\mathbf{f}^{(1)}$, is preferred to $\mathbf{f}^{(2)}$ if $f_k^{(1)} < f_k^{(2)}$ and all higher-order elements (i.e., $f_1, \ldots, f_{k-1}$) are equal. If no other solution is preferred to **f**, then **f** is the lexicographic minimum.

Step 6: Post-solution sensitivity analysis: Post-solution analysis of the sDSP consists of two types of activities, namely, solution validation and sensitivity analysis. Sensitivity analysis includes both assessing the solution's sensitivity to changes in the attribute weights and the solution's sensitivity to changes in attribute ratings. These activities are very important because of the nature and quality (hard or soft) of the information being used.

Validation: Having ranked all the alternatives in order of decreasing merit function values, the problem solver is able to identify the best and some of the better alternatives. Usually, when the number of alternatives is fairly large, the rankings will naturally divide alternatives into several groups of alternatives for which the merit function values are comparable. Alternatives in the same group usually have some characteristics in common. These characteristics should be examined and, if they are desirable, should be included as additional attributes for the selection. This is to assure that no important attribute is left out as a result of which some alternatives are ranked lower than they should have been. Also, a reexamination of the relative weights, attribute ratings, and the numerical calculations is necessary to ensure that no biased judgments of numerical errors occur in any step. Validation of the solution is very important, especially when the highest-ranked alternative is unexpected.

Sensitivity analysis: In applications where the number of alternatives is large, it is very likely that the values of the merit functions of the top two or three alternatives are almost equal. In such cases, it is necessary for a sensitivity analysis be performed. Therefore, the sensitivity analysis consists of determining the effect on the solution of small changes in the relative importance of attributes and to changes in the attribute ratings.

Sensitivity to changes in importance of attributes: During the selection process, the weights for the attributes are derived using judgment that entirely depends on the experience, knowledge, and preference of each individual. For this reason, the sensitivity to the change in the relative weights of attributes needs to be performed. This can be done by reexamining and changing the relative importance of the attributes in or changing the preferences within a comparison and determining the effect of that change on the merit function. The top-ranked alternative that is not affected by small changes in the weights of attributes is the best alternative and should be selected. When the ranking is altered by the changes in the attribute weights, a sensitivity analysis of the attribute ratings may be performed, or the designer may consider including other attributes and then resolve the sDSP.

Sensitivity to the changes in the attribute ratings: As stated before, the ratings may be derived subjectively or directly from the available quantitative information. In the former case, it is possible that errors in ratings occur. Therefore, the sensitivity of the solution to changes in attribute ratings needs to be found. This can be done by studying the change in the merit function value affected by changes in the attribute ratings (e.g., $\pm 5\%$).

Consider a change of $\pm \delta$ in the rating R_{ij} in attribute j of alternative i. The change in the merit function of that alternative will be

$$\delta MF_i = \pm \delta I_j R_{ij} \tag{5.13}$$

The new merit function will be

$$MF_i^{new} = MF_i^{old} + \delta MF_i \tag{5.14}$$

The alternatives are then ranked again, and if the top-ranked alternative remains unchanged, the solution is considered stable. If the top-ranked alternative is changed, the sensitivity of the merit function to other ratings needs to be evaluated further. In some cases, the addition or redefinition of attributes may be necessary.

5.2.1. Utility-based Selection Decision Support Problem

The sDSP presented in the previous section is suitable for making design decisions by accounting for multiple criteria. However, the key limitations of the sDSP construct are that (1) it does not account for uncertainty in the design attributes, and (2) it allows the designers to model only linear preferences for attributes. For example, in the pressure vessel design problem, specific point values of the material properties such as density, strength, resistance to moisture, damage resistance, flexibility ratio, etc., are used. These properties are invariably associated with some uncertainty. A robust decision can be made by accounting for these uncertainties in the sDSP. The preferences for the design attributes are provided in terms of the target values. For example, a target value of cost is provided and the deviation from this target cost value is linearly dependent on the cost. A nonlinear preference, which is normally the case for risk-averse or risk-prone designers, for the cost attribute cannot be modeled using the goal formulation presented in the previous section.

To address these limitations, Fernández and coauthors (Fernández, Seepersad et al. 2005) present a Utility-Based Selection Decision Support Problem (u-sDSP), which is a synthesis of the sDSP and utility theory (discussed in Chapter 4). The u-sDSP offers a means for representing designer preferences and identifying preferred alternatives, especially when uncertainty must be considered with respect to the performance of design alternatives. The structure of the sDSP and the u-sDSP are compared in Table 5.2.

The differences between the sDSP and the u-sDSP are in the quantification of uncertainty associated with the attributes, assessment of the designer's utility functions and selection of most promising design alternative based on expected utility. Instead of simply rating each alternative with respect to each attribute, as is the case with the sDSP, the u-sDSP allows for the characterization of each alternative with respect to performance variability, manifested as ranges or probability distributions of performance. Typically, such measures of variability can be determined from analysis, experimentation, and historical data. For example, instead of specifying a deterministic point value for the tensile strength of the material, a probability distribution or a uniformly distributed range of likely strengths could be specified.

In the u-sDSP, utility functions are used to model the preference for attributes. The attribute values are mapped to a utility function whose values lie between 0 and 1. For example, the preference for tensile strength can be mapped to a monotonically increasing function, higher values of tensile strength corresponding to higher utilities. The preference for cost can be mapped to a monotonically decreasing utility function, higher values of cost corresponding to lower utility for cost. The details of utility function are provided in Chapter 4. Using these utility functions for individual attributes, the expected utility for each design alternative is obtained. The design alternative with maximum expected utility is selected by the designer as the best solution in the presence of uncertain attributes.

Table 5.2 Comparison of sDSP and u-sDSP word formulations
(Fernández, Seepersad et al. 2005)

Selection Decision Support Problem	Utility-Based Selection Decision Support Problem
Given: A set of feasible design alternatives. *Identify:* The principal attributes influencing selection. The relative importance of each attribute. *Rate:* The alternatives with respect to each attribute. *Rank:* The feasible design alternatives in order of preference based on the attributes and their relative importance.	*Given:* A set of feasible design alternatives. *Identify:* The principal attributes influencing selection. The uncertainties associated with each attribute. *Assess:* Designer's utility with respect to each attribute and with respect to combination of attributes. *Evaluate:* Each design alternative using designer's utility functions. *Rank:* Most promising design alternative based on expected utility.

Further details on the u-sDSP are provided by Fernández and coauthors (Fernández, Seepersad et al. 2005). The authors illustrate the u-sDSP method using example of selection of rapid prototyping technology and a material.

5.3. Compromise Decision Support Problem

In the previous section, the emphasis was on selection decisions where the focus is on selecting a design alternative from a set of available alternatives. The second type of decisions that are common in the design process involves a tradeoff (compromise) between multiple design objectives. To formalize the compromise decisions, a compromise Decision Support Problem (cDSP) is used, which is illustrated in this section.

5.3.1. Overview of the Compromise Decision Support Problem

The *compromise Decision Support Problem (cDSP)* involves improvement of a design alternative through modification. cDSPs refer to a class of constrained, multi-objective optimization problems that have widespread engineering applications. It is a mathematical construct that is used to determine the values of design variables that satisfy a set of constraints and achieve a set of conflicting goals as closely as possible. The cDSP is a hybrid formulation in that it incorporates concepts from both traditional mathematical programming and goal programming (GP). The terms *cDSP* and *Mathematical Programming* are synonymous to the extent that they refer to system constraints that must be satisfied for feasibility. They differ in the manner in which the goodness of the solution is modeled and evaluated. It is similar to goal programming in that the multiple objectives are transformed into system goals (involving both system and deviation variables) and the deviation function is solely a function of the goal deviation variables. This is in contrast to traditional mathematical programming where multiple objectives are modeled as a weighted function of the system variables only. The concept of system constraints, however, is retained from the traditional constrained optimization formulation. Special emphasis is placed on the bounds on the system variables unlike in traditional mathematical programming and goal programming. The word formulation of a cDSP is as follows:

Given
• A design alternative • Assumptions used to model the domain of interest • The system parameters • The goals for the design
Find
• The values of the independent *system variables* (they describe the attributes of a product/material) • The values of the *deviation variables* (they indicate the extent to which the goals are achieved)
Satisfy
• The *system constraints* that must be satisfied for the solution to be feasible • The *system goals* that must achieve a specified target value as much as possible
Bounds
• The lower and upper bounds on the system variables
Minimize
• The *deviation function* which is a measure of the deviation of the system performance from that implied by the set of goals and their associated priority levels or relative weights

The details of the different parts of the cDSP are discussed in detail in the context of the integrated pressure vessel and composite material design problem. The detailed formulation of the cDSP for the example problem is shown in Table 5.3.

The four elements of the cDSP listed previously (given, find, satisfy, and minimize) are discussed next.

5.3.2. Details of the Compromise Decision Support Problem

5.3.2.1. System Variables and System Constraints

A *design variable* is one that can be controlled independently by the designer. In the design problem, the length of the pressure vessel, the chamber radius, and boss opening radius are examples of design variables. There are 10 design variables in the problem. A variable can either be continuous (with a lower bound and an upper bound), discrete, or Boolean (1 if TRUE,

Table 5.3 The cDSP for integrated pressure vessel and composite materials design

Given

- Choice of materials for the pressure vessel:
  - E-glass fiber/epoxy resin (G/E)
  - Carbon fiber/epoxy resin (C/E)
  - Kevlar-49 fiber/epoxy resin (K/E)
  - Pressure vessel construction: Layer sequence and fiber orientation ($\pm\alpha$) in each layer
  - Loading: Internal pressure, P_0, and temperature difference, ΔT, across the thickness

Find

- Material to be used:
  - E-glass/epoxy, M_1
  - Carbon/epoxy, M_2
  - Kevlar-49/epoxy, M_3
- Geometry:
  - Length of the pressure vessel, L
  - Chamber radius of the pressure vessel, R
  - Boss opening radius, r
- Hemispherical sections:
  - Thickness at $\phi = 90°$, t_0
  - Fiber orientation, α_2
- Cylindrical section:
  - Thickness of the inner and outer layers, t_1
  - Fiber orientation in the inner and outer layers, α_1

Satisfy

System Constraints

- Selection constraint: only one material can be used

$$M_1 + M_2 + M_3 = 1 \tag{5.15}$$

- Failure for each layer:
  - Bonding break failure

$$\left(\frac{\sigma_{11}}{\sigma_{11BB}}\right)^2 + \left(\frac{\sigma_{22}}{\sigma_{22BB}}\right)^2 + \left(\tau_{12BB}\right)^2 \leq 1 \tag{5.16}$$

- Break failure of the fiber

$$\left(\frac{\sigma_{11}}{\sigma_{11FB}}\right)^2 \leq 1 \tag{5.17}$$

- Geometry:
  - Vessel radius is greater than the boss opening radius

$$R - r \geq 0 \tag{5.18}$$

Table 5.3 (Continued)

- Minimum fiber angle permissible in the hemispherical section

$$\alpha_2 - \sin^{-1}\left(\frac{r}{R}\right) \geq 0 \tag{5.19}$$

- Minimum fiber angle permissible in the cylindrical section

$$\alpha_1 - \tan^{-1}\left(\frac{2R}{L - 2R}\right) \geq 0 \tag{5.20}$$

- Limits on the diameter (or radius) ratios

$$
\begin{aligned}
10r - R &\geq 0 \\
5r - R &\leq 0 \\
L - 4R &\geq 0 \\
L - 6R &\leq 0
\end{aligned}
\tag{5.21}
$$

- Restrictions on deviation variables

$$d_i^- \cdot d_i^+ = 0 \quad i = 1,\ldots,7 \tag{5.22}$$

System Goals
- Merit function value for each material; $i = 1,\ldots,3$; $k = 1,\ldots,3$

$$\sum_{j=1}^{6}(R_{ij}l_j)M_i + d_k^- - d_k^+ = 1 \tag{5.23}$$

- Meet target volume for the pressure vessel ($T_{Vol} = 10^7\,\text{mm}^3$)

$$V + d_4^- - d_4^+ = T_{Vol} \tag{5.24}$$

- Deflections of the hemispherical and cylindrical sections should match at the interface

$$|w_{sph} - w_{cyl}| + d_5^- - d_5^+ = 0 \tag{5.25}$$

- The pressure vessel performance factor should meet the target value T_{perf} ($10^7\,\text{mm}$)

$$\frac{P_0 V}{Weight} + d_6^- - d_6^+ = T_{perf} \tag{5.26}$$

- The material cost of the pressure vessel should meet the target cost T_{cost}

$$Weight \times (Specific\ Cost) + d_7^- - d_7^+ = T_{cost} \tag{5.27}$$

(Continued)

Table 5.3 (Continued)

Bounds on each variable	
$M_1, M_2, M_3 = 0,1$	(5.28)
$0.254 \le t_0 \le 20.32; 0.254 \le t_1 \le 10.16$	(5.29)
$0.0 \le \alpha_1 \le 1.5708; \quad 0.2094 \le \alpha_2 \le 0.4363$	(5.30)
$38.1 \le R \le 304.8; \quad 12.7 \le r \le 76.2; \quad 152.4 \le L \le 914.4$	(5.31)
$0 \le d_i^-, d_i^+ \quad i = 1, \cdots, 7$	(5.32)

Minimize

• Deviation function

$$Z = \{(d_4^- + d_4^+), d_7^+, (0.5d_5^+ + 0.5d_6^+), (d_1^- + d_2^- + d_3^-)\} \tag{5.33}$$

0 if FALSE). Design variables are, by their nature, independent of the other variables and can be changed as required by the designer to alter the design of the system.

Consider a set of n design variables represented by X. Each member of the set X represents an axis of an n-dimensional space. This space is called the *design space*. Each point in the design space refers to one distinct design. The design space defines all possible design options available to a designer.

A *system constraint* is a constraint placed on the design. Examples of these constraints in a concurrent material and product design problem include strength constraints, stiffness constraints, and dimensional constraints. In the pressure vessel design problem, there are various constraints—avoiding bonding break failure, avoiding break failure of the fiber, vessel radius greater than the boss radius, lower bound on minimum fiber angles, etc. All the system constraints must be satisfied for the design to be *feasible*. A design is infeasible if it violates any system constraint. Mathematically, system constraints are functions of design variables only. The subset of the design space that consists of all feasible design points (bounded by the system constraints) is called the *feasible design space*.

The concepts of design space and system constraint are illustrated in Figure 5.4 using the pressure vessel design problem. In the figure, two design variables fiber orientation (α_2) and chamber radius (R) are shown. For illustration purposes, it is assumed that that all the other design variables are fixed. Lines AA' and BB' represent the lower and upper bounds on the chamber radius. Similarly, lines CC' and DD' are bounds on fiber orientation. In addition to the bounds, a constraint corresponding to the minimum fiber angle permissible in the hemispherical section is also shown (curve EE'). Note that the constraints can be linear or nonlinear. The arrows associated with the constraints and bounds depict the region in which they are valid. The region in the 2D space where all constraints and bounds are valid is the feasible design space. In Figure 5.4, the feasible design space is the shaded region PQRST.

In the cDSP shown in Table 5.3, there are three binary variables associated with the material—M_1, M_2, and M_3 corresponding to the three material alternatives. Since a material is either selected or not selected, the variables are either 1 or 0. The other seven design variables (L, R, r, t_0, α_2, t_1, and α_1) correspond to the geometric parameters. In the cDSP, there are various system constraints. Since only one of the materials can be selected, it has to be ensured that only one of the variables has a nonzero value. Hence, there is a corresponding system constraint that the sum of the three material variables (M_1, M_2, and M_3) is 1 (see Equation 5.15). The constraints in Equations 5.16 and 5.17 correspond to the failure criteria for the composite pressure vessel. In this formulation, a variation of the Halpin-Tsai failure criterion is used (Puck and Schneider 1969). The criterion includes the checking of bond break (BB) and fiber break (FB) in each layer of the laminate. In these equations, σ_{11} is the normal stress parallel to the fiber direction, σ_{22} is the normal stress perpendicular to the fiber direction, and τ_{12} is the shear stress. The stresses on the structure due to the loading are obtained by using classical laminate analysis and membrane theory. The hemispherical portion is checked for failure only at $\phi = 90$ degrees.

Constraints are also imposed to model the desired geometrical relationships between certain variables. An obvious one is that the chamber diameter has to be greater than the boss opening

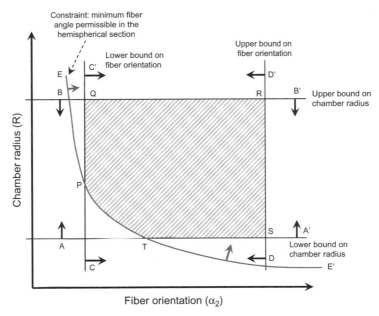

Figure 5.4 A design space represented by two variables, bounds, and one constraint.

diameter and this is modeled in Equation 5.18. The winding angle in the hemispherical section is limited by the ratio of the boss opening and chamber diameters and in the cylindrical section it is limited by the ratio of chamber diameters and the length of the section. This is stated in Equations 5.19 and 5.20. It has been determined experimentally, (see Lark (1977), that better pressure vessel performance in terms of strength and volume to weight ratio can be expected when the boss opening diameter ratio is between 1/10 and 1/5 and the length-to–chamber diameter ratio is between 1 and 3. The constraints in Equation 5.21 are included to maintain these ratios within empirically determined values.

5.3.2.2. Deviation Variables and System Goals

In the cDSP, a design objective is modeled as a *system goal*. It relates the target value for the objective, T_i, to the actual attainment of the goal, $A_i(X)$. The target values are determined based on the design requirements whereas the actual attainment $A_i(X)$ is determined for each point in the design space. For example, in the pressure vessel design problem, the goals include meeting the target value of volume ($T_{vol} = 10^7 \, \text{mm}^3$), minimization of deflections of hemispherical and cylindrical sections (i.e., the target values are zero), meeting the target pressure vessel performance factor ($T_{pref} = 10^7 \, \text{mm}$), where the performance factor is defined as (burst pressure) * (volume)/weight.

Depending on the type of goal, three conditions need to be considered:

- $A_i(X) \leq T_i$; we wish to achieve a value of $A_i(X)$ that is equal to or less than T_i.

- $A_i(X) \geq T_i$; we wish to achieve a value of $A_i(X)$ that is equal to or greater than T_i.

- $A_i(X) = T_i$; we would like the value of $A_i(X)$ to equal G_i.

The difference between the achievement and the target values, *deviation variables* are used. The deviation variables are discussed next. Consider the third condition; namely, we would like the value of $A_i(X)$ to equal G_i. The deviation from the goals is defined as

$$d_i = T_i - A_i(X) \tag{5.34}$$

The deviation variable d_i can be negative or positive. In order to simplify the solution algorithm, the deviation variable d is replaced by two nonnegative variables, d_i^- and d_i^+, such that

$$d = d_i^- - d_i^+ \tag{5.35}$$

where

$$d_i^- \cdot d_i^+ = 0 \quad \text{and} \quad d_i^-, d_i^+ \geq 0$$

The preceding ensures that the deviation variables never take on negative values. The system goal becomes

$$A_i(X) + d_i^- - d_i^+ = T_i \qquad i = 1, 2, \cdots, m \tag{5.36}$$

where

$$d_i^- \cdot d_i^+ = 0 \quad \text{and} \quad d_i^-, d_i^+ \geq 0.$$

The deviation variables d_i^- and d_i^+ are used to allow the designer a certain degree of latitude in making decisions. The deviation variables therefore relate the actual achievement of the goals to the target values. The product of the two deviation variables associated with a goal is always zero. Consider three possible scenarios:

Scenario 1—Underachievement: In this scenario the achievement of the goal is less than the target value, $(A_i \leq T_i)$. For example, if the volume of the pressure vessel is $5 \times 10^6 \text{mm}^3$ (which is less than the target value $T_{vol} = 10^7 \text{mm}^3$), the volume goal is underachieved. In the case of underachievement, the deviation variable $d_i^+ = 0$ whereas $d_i^- > 0$.

Scenario 2—Overachievement: If the achievement of a goal is greater than the target value $(A_i \geq T_i)$, the goal is overachieved. For example, if the material cost of the pressure vessel is $1.50 per kilogram, which is greater than the target value $T_{cost} = \$1.00$ per kilogram, the goal has been overachieved[1]. In this scenario, $d_i^- = 0$ and $d_i^+ > 0$.

Scenario 3—Exact achievement: Finally, in a scenario where the achievement of the goal is the same as the target value $(A_i = T_i)$, then both of the deviation variables are equal to zero; $d_i^- = 0$ and $d_i^+ = 0$.

The value of the *ith* deviation variable is determined by the degree to which the *ith* goal is achieved. It depends upon the value of $A_i(X)$ alone (since T_i is fixed by the designer). The system goal represents an equation for a family of parallel functions. In Figure 5.5, a target value of a goal is shown by line TT'. The points on this line represent different combinations of the design variables (fiber orientation and chamber radius) that result in the target value of the goal. Two lines parallel to the target TT' are shown—one corresponding to a goal that is lower than the target value (line UU'), and another corresponding to a goal that is higher than the target (line OO'). For the line UU', the underachievement d_1^- is nonzero and the overachievement d_1^+ is zero. For OO', it is the other way around.

Since all points on line TT' lie outside the feasible region, achieving the target value of this goal is not possible. It is also not possible to overachieve the target. Hence, a compromise

---

[1] Note that overachievement means only that the numerical value is greater than the target value. It does not necessarily mean that the design is better.

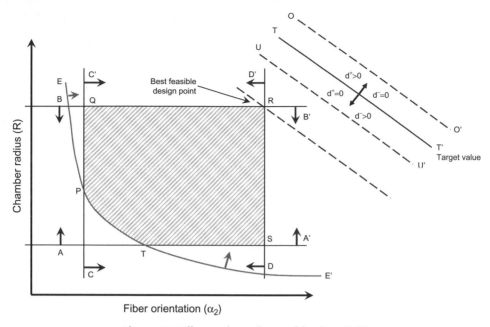

Figure 5.5 Illustration of a goal in the cDSP.

is desired. The feasible design point that minimizes the underachievement (d^-) of the goal is point R. Hence, based on the constraint, bounds and goal shown in the figure, point R is the best compromise and hence represents the solution of the cDSP.

Formulating design objectives as system goals: The following rules are used to formulate the system goals in a way that ensures that all the deviation variables will range within the same values (0 and 1 in this case).

(a) *Maximization Goals:* To maximize the achievement, $A_i(X)$, choose a target value T_i that is greater or equal to the maximum expected value of $A_i(X)$.

Transform the expression into a system goal by adding and subtracting the corresponding deviation variables (which in this case will range between zero and 1), i.e.,

$$\frac{A_i(X)}{T_i} + d_i^- - d_i^+ = 1 \qquad (5.37)$$

In this case, the overachievement variable, d_i^+, will always be zero. Then, minimize the underachievement deviation, d_i^-, to ensure that the performance of the design will be as close as possible to the desired goal.

(b) *Minimization Goals:* To minimize the achievement, $A_i(X)$, choose a target value, T_i, that is less than or equal to the minimum expected value of $A_i(X)$. Transform the expression into a system goal (note the inversion of G and A) and flip the signs of the deviation variables (to account for the inversion). The deviation variables will vary between 0 and 1, i.e.,

$$\frac{T_i}{A_i(X)} - d_i^- + d_i^+ = 1 \tag{5.38}$$

The underachievement deviation, d_i^-, will be zero. Minimizing the overachievement deviation, d_i^+, will ensure that the performance of the design is as close as possible to the desired goal.

(c) *Target Achievement Goals:* If it is desired that $A_i(X) = T_i$, then the sum of the deviation variables $(d_i^- + d_i^+)$ is minimized.

5.3.2.3. The Deviation Function

In the cDSP, the aim is to minimize the difference between that which is desired (target values for the goals) and that which can be achieved. This is done by minimizing the deviation function, $Z(d^-,d^+)$, which is always written in terms of the deviation variables.

A designer sets a target value for each of the goals. It may not be possible to obtain a design that satisfies the target values for all the goals. Therefore, a compromise solution must be accepted by the designer. It is desirable, however, to obtain a design whose performance matches the target values as closely as possible. The difference between the goals and achievement is expressed by a combination of appropriate deviation variables, $Z(d^-,d^+)$. This deviation function provides an indication of the extent to which specific goals are achieved.

All goals may not be equally important to a designer and the formulations are classified as *Archimedean or Preemptive*—based on the manner in which importance is assigned to satisficing the goals. These formulations are discussed here:

The *Archimedean* formulation is essentially a weighted sum formulation. Each goal is assigned weight, which quantifies the importance of this goal with respect to the other goals. The most general form of the deviation function for m goals in the Archimedean formulation is

$$Z(d^-, d^+) = \sum (W_i^- d_i^- + W_i^+ d_i^+) \qquad i = 1,\dots,m \tag{5.39}$$

where the weights, W_1, W_2, …, W_m, reflect the level of importance of each of the goals. In this formulation, the weights, W_i, satisfy the following conditions:

$$\sum_{i=1}^{m} W_i = 1 \quad \text{and} \quad W_i \geq 0 \qquad \text{for all i} \tag{5.40}$$

The *Preemptive* formulation is utilized when assigning weights is not feasible because enough information about the problem is not available. This is particularly true in the earlier exploratory stages of design when the main objective is to refine the design options rather than to choose one best design. As discussed earlier in this section, the deviation function for the preemptive formulation is written as

$$Z = [f_1(d_i^-, d_i^+), \ldots f_k(d_i^-, d_i^+)] \tag{5.41}$$

The overall deviation function consists of a set of k functions $f_1, \ldots, f_k$ that are ranked in terms of their importance (f_1 being most important and f_k the least important). Each of these functions f_i can be a weighted sum of deviation variables d_i^+ and d_i^-. For example, in the pressure vessel example, the deviation function Z consists of the following three functions of the deviation variables:

$$f_1 = d_4^- + d_4^+ \tag{5.42}$$

$$f_2 = d_7^+ \tag{5.43}$$

$$f_3 = 0.5d_5^+ + 0.5d_6^- \tag{5.44}$$

$$f_4 = d_1^- + d_2^- + d_3^- \tag{5.45}$$

Function f_1 consists of deviation variables associated with the volume goal. This implies that the volume goal is the most important. The second function consists of the deviation variables associated with the cost goal, which has the second priority.

The design alternatives are selected based on their deviation functions. First, the value of f_1 is compared. The design alternative with a lower value of f_1 is preferred to the one with a higher value. If the value of f_1 is same for multiple design alternatives, then the value of f_2 is compared. The process is continued until the best alternative is identified. This type of rank ordering is called *lexicographic minimization*. The term *lexicographic ordering* is derived from the order in which words appear in a dictionary. In a dictionary, the ordering is based on the comparison of the first letter, then the second letter, then the third, and so on.

5.3.3. Utility-based Compromise Decision Support Problem

The utility-based cDSP is an adaptation of the standard cDSP discussed in the previous section. The details of this formulation are provided by Seepersad (Seepersad 2001). In the cDSP, the objectives are formulated using multi-attribute utility theory. Each goal is formulated as

$$E[u_i(A_i(x))] + d_i^- - d_i^+ = T_i \tag{5.46}$$

where $E[]$ is the expectation operator u is a single-attribute utility function, and d_i, A_i, and T_i are deviation variables, goal values, and goal target values, respectively, for goal i. The target value T_i is the ideal value of the utility function, which is 1.

The utility-based cDSP accounts for multiple goals by combining single-attribute utility functions into a multi-attribute utility function:

$$U(X) = \sum_{k=1}^{m} k_i u_i(A_i(X)) \tag{5.47}$$

where k_i are scaling constants. The deviation or objective function in the utility-based cDSP is formulated to maximize expected overall utility, which is equivalent to minimizing deviation from the ideal target value for expected overall utility (i.e., 1). According to convention in the cDSP, the latter formulation is chosen for the utility-based cDSP, i.e.,

$$Z = 1 - E[U(X)] \tag{5.48}$$

The utility-based cDSP is utilized by Seepersad and coauthors (Seepersad, Mistree et al. 2005) for the design of product platform of absorption chillers.

5.4. Closing Remarks

Integrated materials and product design problems are often characterized by decisions involving various design alternatives and multiple conflicting objectives. It is important to utilize available information for making best design decisions possible in the presence of uncertainty. In this chapter, two design decision-making constructs are presented, the sDSP and the cDSP. Emphasis is mainly placed on the formulation of these decisions. The sDSP has a significant potential for application in material selection based on multiple objectives in the presence of uncertainty. The solution of the cDSP can be carried out using various optimization techniques commonly available in the non-linear optimization literature.

These decision constructs are used throughout the rest of the book. Specifically, in Chapter 6, the cDSP construct forms the core of a robust design method called the Robust Concept Exploration Method (RCEM). The RCEM method is extended to robust topology and material exploration in Chapter 7. In Chapter 8, the RCEM method is extended to account for uncertainty in simulation models and the uncertainty propagated in a chain of models. Finally, the u-cDSP construct is used in Chapter 9 to formulate both material level and product level decisions in an energetic structural material design example.

References

Darms, S.J., 1963. Optimum design of filament-wound rocket motor cases. Directorate of Materials and Processes, Aeronautical Systems Division, Wright Patterson AFB. Report Number: ASD-TDR-63-396.

Fernández, M.G., Seepersad, C.C., Rosen, D., Allen, J.K., Mistree, F., 2005. Decision support in concurrent engineering: the utility-based selection decision support problem. Concurrent Eng.-Res. A. 13 (1), 13–28.

Karandikar, H.M., Mistree, F., 1992. An approach for concurrent and integrated material selection and dimensional synthesis. J. Mech. Design 114, 633–641.

Lark, R.F., 1977, Recent advances in lightweight, filament-wound composite pressure vessel technology. In: Composites in pressure vessels and piping, ASME Energy Technology Conference, Houston, TX, pp. 17–49.

Puck, A., Schneider, W., 1969. On the failure mechanisms and failure criteria of filament-wound glass-fibres in composites. Plast. Polym. 37, 33–44.

Riggs, J.S., 1977. Engineering Economics. McGraw-Hill, New York.

Seepersad, C.C., 2001. The utility-based compromise decision support problem with applications in product platform design, M.S., thesis, The GW Woodruff School of Mechanical Engineering, Georgia Institute of Technology.

Seepersad, C.C., Mistree, F., Allen, J.K., 2005. Designing evolving families of products using the utility-based compromise decision support problem. Int. J. Mass Customisation 1 (1), 37–64.

CHAPTER 6

Robust Design of Materials— Design Under Uncertainty

Nomenclature	
L	Quality loss
k	Constant in the Quality Loss Function
P_s	System performance
T	Target value for performance
y	Response
x	Control factors
z	Noise factors
μ_y	Expected value of y
f	Function relating the control and noise factors to the response
C_{dk}	Design capability index (DCI)
C_{dl}	DCI based on the lower bound on performance requirement
C_{du}	DCI based on the upper bound on performance requirement

Management of uncertainty is crucial in materials design. The degree of uncertainty is often quite substantial in experiments, processing methods, material structure, and model parameters that support concurrent design of materials and products/systems. Potential sources of uncertainty in a system also include human errors, manufacturing or processing variations, variations of operating conditions, inaccurate or insufficient data, model assumptions and idealizations, microstructure variability, and lack of knowledge. Manufacturing variations are manifested as tolerances in part dimensions, as well as material and structure defects such as pores, inclusions, cracks, or variations of microstructure attributes such as phase distributions or volume fractions, grain sizes, and so forth. Operating conditions, such as the ambient temperature, environment, loading history, magnetic field, etc., may also vary. Realistic nanoscale, microscale, or mesoscale computational simulations should address nondeterministic material behavior. The computational burden of such simulations makes it difficult to obtain sufficient data for estimating accurate variances of responses. In developing these models, materials and structural engineers often employ

DOI: 10.1016/B978-1-85617-662-0.00006-5

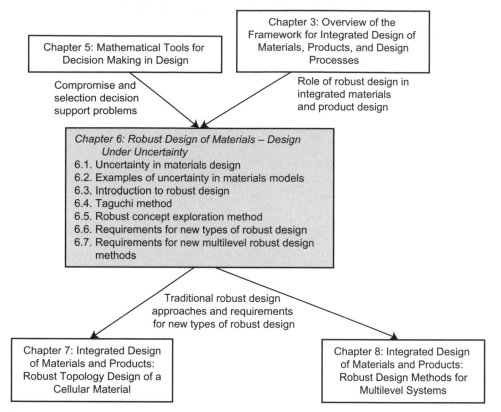

Figure 6.1 Overview of Chapter 6 and its relationship with other chapters in the book.

simplifying assumptions, resulting in additional uncertainty in the design process. These uncertain factors in systems and models must be considered in systems-based concurrent design of materials and products.

In this chapter, different characteristic types of uncertainty associated with material design are defined. Examples of sources of uncertainty are illustrated in the first section. Then, existing approaches for managing uncertainty are discussed. Finally, requirements for new approaches for the management of uncertainty in materials design are addressed. The relationship of this chapter with other chapters in the book is shown in Figure 6.1.

6.1. Uncertainty in Materials Design

Proper management of uncertainty starts from a well-defined classification of uncertainty. Uncertainty management and mitigation strategies depend on the types of uncertainty involved.

6.1.1. Classification of Uncertainty in Material Models

Uncertainty can be classified as either *Aleatory* (irreducible) or *Epistemic* (reducible), depending on the causes of the uncertainty. *Epistemic* uncertainty can be diminished by improvements in measurements and/or model formulation and/or by increasing the accuracy or sample size of data. *Aleatory* uncertainty, on the other hand, is inherent in the physical system and can only be quantified in a statistical sense. Other uncertainty definitions and classifications are available in literature (Ayyub and Chao 1997, Der Kiureghian 1989, Haukass 2003, Nikolaidis 2005, Oberkampf, DeLand et al. 1999). However, these classifications of uncertainty need to be further refined for describing uncertainty in the models in materials science used to predict process-structure and structure-property relations. Relevant types of uncertainty in materials design are categorized as follows, extending the classification of uncertainty types by Isukapalli and coauthors (Isukapalli, Roy et al. 1998):

- *Natural Uncertainty* (NU): uncertainty due to the inherent randomness or unpredictability of a physical system; this is irreducible and can only be quantified in a statistical sense. The variability can be further classified as *parameterizable* and *unparameterizable*. Parameterizable variability can be configured as variance in numeric form, but unparameterizable variability cannot. Randomness of microstructure of materials, for example, gives rise to this kind of uncertainty in resulting properties and responses.

- *Model Parameter Uncertainty* (MPU): this kind of uncertainty reflects incomplete knowledge of model parameters/inputs due to insufficient or inaccurate data; it is reducible by sufficient data or accurate measurements.

- *Model Structure Uncertainty* (MSU): this refers to uncertain model formulation due to approximations and simplifications in a model; it is reducible by improving model formulation. However, it cannot be eliminated. Moreover, it is quite difficult to quantify the relative uncertainty of the structure of several different models.

- *Propagated Uncertainty* (PU): this is uncertainty that is compounded by the combination of all three types of uncertainty in a chain of models that are connected through input-output relations.

These types of uncertainty coexist within any system. While some types can dominate system performance, others may be essentially negligible. Variability in a system response is measured based on acquired samples. Uncertainty of the system response may be attributed to both natural uncertainty and parameter uncertainty due to the limited number of data points. As the amount of data increases, parameter uncertainty is reduced, and the system variability dominates the measurement interval. Multiple types of uncertainty coexist in any system, which makes identification of dominant sources of uncertainty a difficult and important step in a design process.

Based upon this classification of uncertainty, we illustrate each type of uncertainty by a simple example in the next section for better understanding.

6.2. Examples of Uncertainty in Material Models

As an example, consider the uncertainty associated with a microscale discrete particle shock simulation model used for designing energetic materials that have desired reaction initiation characteristics. They are typically three (or more) phase material systems, with two major reactants, porosity, and sometimes a binder phase. An overview of the energetic material design problem appears in Chapter 3. The microscale discrete particle mixture shock simulation is conducted by (1) generating physically realistic microstructures, (2) performing a shock simulation using a hydrocode based on continuum mechanics, and (3) extracting simulation results—the number density of chemical reaction initiation hot spots—from the simulations (Choi, Austin et al. 2005).

6.2.1. Examples of Natural Uncertainty (System Variability)

One of the challenges in designing energetic materials is that the input variables are subject to natural randomness. For example, the volume fraction of each material, a controllable factor (or control factor), could have random variability. Even though a designer determines a volume fraction of a constituent (such as Al) as a design solution, the actual volume fraction of the constituent may vary from the design specification due to variability in material processing. This can produce unexpected performance (reaction initiation) variation in manufactured energetic materials. Additionally, some input variables cannot be controlled or are difficult for a designer to control, and the variability in the uncontrollable factors (noise factors) affect the material's performance. For example, the variability of particle size distributions among the simulation input variables may be difficult or expensive to control. In designing energetic materials, it is necessary to consider the variability in both controllable and uncontrollable factors since the variability could cause significant deviations from the desired performance. These types of uncertainty are categorized under "*parameterizable variability*," which are due to natural randomness and are not reducible.

Another challenge in the design of energetic materials is that the shock simulation includes a noise factor that is hard to parameterize. Experiments or simulations are performed on statistical volume elements (SVEs) or samples with different particle distributions. The natural variability associated with randomness of particle spatial distributions and correlations is very difficult to control. Thus, the system response has pseudorandom variability with a fixed set of shock simulation input parameters, as shown in Figure 6.2. The histogram of responses (the number of reaction initiation sites) of 98 repeated simulations at a given set of input variables is shown in Figure 6.2. This implies that the system has noise factors that have not been modeled.

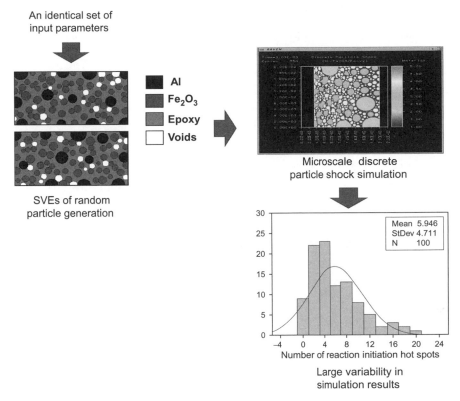

Figure 6.2 Variability in simulation results due to pseudorandom particle generation process.

Each instantiation of an SVE of the material incorporates a "pseudorandom" assignment of particle positions using a constrained Poisson point process (this technique is also known as a random sequential addition process [Torquato 1991]) that prevents particle overlap. This randomness is difficult to parameterize in the system model and causes significant response variance and "*unparameterizable variability.*" Unparameterizable variability in this simulation model must be considered for a designer to ensure that the design will meet specified performance requirements with regard to reaction initiation. This type of uncertainty originating from microstructure variation is a hallmark of the simulation of performance of innumerable material systems. Moreover, it points to the need for an underlying stochastic basis for modeling and simulation in design of real materials and systems.

In the case of the linear cellular alloy (LCA) design problem discussed in Chapter 3, one of the important sources of natural uncertainty is the manufacturing process. While the thermochemical extrusion process offers valuable topological freedom for tailoring metallic or ceramic prismatic cellular materials, it also introduces imperfections in the final parts. Possible imperfections include tolerances or dimensional variation, curved or wrinkled cell walls, cracked or missing cell walls or joints, and variations in porosity and other properties of the cell wall material. For example, the porosity of sintered materials has been found to influence several characteristics including conductivity, strength, and elastic moduli. Analytical relationships have been proposed for each of the properties as a function of porosity (Bocchini 1986). Variations in cell shape have been simulated by comparing the responses of Voronoi honeycombs (generated from a random set of points or nodes separated by a minimum distance) and periodic, hexagonal honeycombs to applied compressive stress. Cell topology, shape, dimensions, and imperfections have a significant impact on mechanical properties. Therefore, it would be advantageous to be able to design the topology, shape, and dimensions of cells and cell walls of prismatic cellular material for realistic fabrication environments in which imperfections are introduced within the cellular structure.

6.2.2. Examples of Model Parameter Uncertainty (MPU)

Model Parameter Uncertainty (MPU) is the incomplete knowledge of parameters or inputs of a mode. For example, in the energetic material design problem, the distribution of particle size is uncertain if there is insufficient sampling. The distribution (log-normal distribution) for Al particle size may be incorrect due to lack of experimental measurements. In this case, we cannot fully rely on the size distribution obtained. In other words, the parameterizable variability measured by sampling is itself uncertain; the mean and variance may not be representative of the actual case or the forms of the particle size distributions may have errors.

Uncertainties in the parameters of a model are different from those of inputs. Here, the parameters are statistical estimators that are introduced for metamodeling. These parameters have no physical meaning; however, the inputs of a system do have physical meaning. Uncertainty in parameters of a model is introduced due to lack of sampling for building a metamodel of a system. For example, we estimate the output distribution based on 100 samples at an identical set of input parameters in Figure 6.2. We need to acquire these samples at all other simulation input sets designed by Design of Experiments (DOE) techniques in order to build the metamodel. The number of input sets with 4 parameters required in the Central Composite Design is 25. Then, the sampling size for obtaining accurate estimators in a metamodel could be very large; the number of samples needed is 2500 in this example. Considering computational intensity of material simulation, obtained estimators in a metamodel may be uncertain due to lack of sampling. Therefore, the combined effect of unparameterizable variability and intensive computation (or expensive experiments) will lead to MPU in a metamodel that characterizes material systems. As discussed in this section, MPU could exist as uncertainty in the distribution of input parameters or uncertain parameters in a metamodel due to lack of sampling or inaccurate measurement.

Moreover, the parameters of models used in the hydrocode simulation itself may be uncertain due to limited experimental data or nonuniqueness of the parameter set owing to an approximate model. For example, models used to predict dynamic elasto-viscoplastic deformation of various phases in energetic material systems have parameters that must be estimated from other kind of data such as high strain rate Hopkinson bar or dynamic gas gun experiments. Owing to the high strain rates, the nearly adiabatic deformation of shocked energetic materials leads to local temperature increases at "hot spots" within the shock front; these local temperatures are used to estimate reaction initiation probability. Temperature dependent material properties are often estimated using data obtained in controlled quasi-static experiments at various temperatures. Dynamic deformation of these material systems is highly complex and of a nonequilibrium character, invoking elements of dislocation multiplication and interaction, phonon drag, and thermal dissipation. There is typically considerable uncertainty associated with material parameters in such cases. This is also often the case in models used to predict process-structure relations for materials involving casting and solidification, primary forming, joining, machining, etc. Another source of uncertainty in computational models is due to lack of convergence on refinement in numerical solutions. This will be addressed in Chapter 9.

6.2.3. Examples of Model Structure Uncertainty (MSU)

Model Structure Uncertainty (MSU) is due to uncertain assumptions and idealizations in the models. In the energetic material shock simulation discussed in the previous section, there are a number of assumptions and simplifications that contribute to MSU.

The actual geometry of the constituent particles in the mixture is quite complex, and it needs to be approximated in the finite element models. The shapes of the iron oxide agglomerates formed by the subparticles are difficult to quantify. The actual geometry of voids in the mixture is difficult to characterize from microscopy images. This complexity of the actual microstructure is simplified so that the cross-sections of all constituents are circular. The plane strain assumption is invoked in order to reduce the true 3D nature of shock loading to a more tractable 2D case, leading to treatment of the particles as cylinders rather than 3D spherical objects. The various phases are assumed to have no friction at contacting interfaces. A compressive shock wave is propagated though the mixture by applying boundary conditions to the SVE, which can vary. The velocity boundary condition is ramped up during the first stage of the simulation according to a quadratic function of time in order to avoid spurious oscillations in the solution associated with instantaneous loading.

With regard to mechanical behavior of constituents, a physically based constitutive model for iron oxide is not available. Therefore, a simple bilinear elastic-plastic model is adopted, consisting of an initial linear elastic response followed by linear isotropic strain-hardening. Strain rate hardening is addressed in the behavior of the Al particles, as well as rate sensitivity. Various models have been proposed for these phenomena, of which a limited set is selected. Phenomena related to the shock-induced reaction initiation among reactants in the mixture (solid-solid and solid-liquid transition states and activation energy barriers) are quite complicated; therefore, a number of assumptions and approximations must be made to idealize the processes involved in thermal explosion. In this model, it is assumed that the reaction rate depends on the temperature only. This assumption neglects (1) the consumption of reactants and (2) any temperature dependence of the pre-exponential factor (Austin 2005). Furthermore, the Merzhanov criterion is used as a model for the linear stability of a thermal explosion (critical nucleus) for reaction initiation.

While the energetic material problem is perhaps on the extreme end of the spectrum of typical materials design problems in terms of model uncertainty, it offers a good practical example of MSU. These assumptions and idealizations are made on the basis of incomplete knowledge of the physical system and often for the sake of computational efficiency. They lead to error in the estimation of the energetic material behaviors based on the simulation results.

6.2.4. Examples of Propagated Uncertainty

Errors that propagate and accumulate along the multiscale simulation chain are important sources of uncertainty, as shown in Figure 6.3. Aforementioned uncertainties of a model at a given scale are propagated and amplified by interfacing with simulation models at other scales.

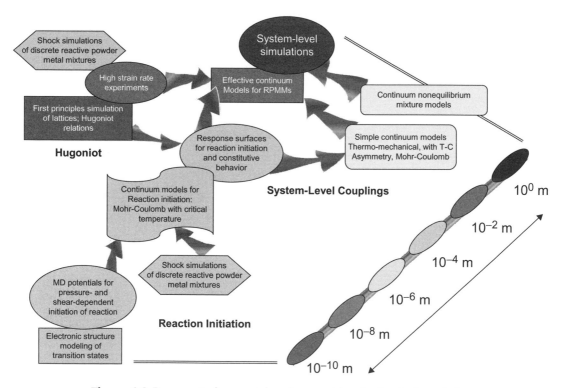

Figure 6.3 Propagated uncertainty in a multiscale simulation chain.

For example, the transition states and associated activation energy barriers for the reaction pathway based on quantum mechanics estimates serve as inputs to a microscale analysis model. Parameters in the potentials obtained in the analysis of the quantum mechanics model are the input parameters to potentials used in molecular dynamics models. The Equation–of–State (EOS) parameters and transition states for reaction initiation obtained in the nanoscale analysis model serve as inputs to the continuum level discrete particle mixture simulations.

In this manner, computational models are interfaced to estimate the overall performance of energetic materials. The uncertainties of the models in the simulation and analysis chain are propagated and accumulated so that the final energetic material system performance estimation may include significant errors. Therefore, mitigation of propagated uncertainty in designing materials and products is critical. The problem can be tackled on two fronts: one can reduce various sources of uncertainty in models and parameters, and one can configure the flow of information in sequential or parallel execution of models in a way that reduces propagated error. However, we cannot eliminate all uncertainty—we can only manage its influence on materials design decisions. As discussed in the next section, this means that we seek designs that are robust against uncertainty rather than pursuing the notion of optimality.

6.3. An Introduction to Robust Design

In engineering design, the concept of robustness is used to mitigate loss of functionality or performance due to reliance on information that is uncertain or difficult to model or compute. In robust design, the goal is to achieve a system level response that is insensitive to variations without removing the underlying sources of those variations (Byrne and Taguchi 1987, Nair 1992, Tsui 1992). This anticipated variability often represents information from other parts of the product realization process, beyond the boundaries of the system to be modeled and designed. For example, a system may be designed for performance (response) that is relatively insensitive to design variations caused by the manufacturing process, materials processing, or even materials supply. It may also be designed in a way that key properties are insensitive to compositional variation or a range of microstructure morphologies due to variation of the thermomechanical process route.

6.3.1. Optimization Versus Robustness

The approach outlined in this book for designing materials is grounded in achieving system robustness in the presence of uncertainty and leads to designs that may differ substantially from those anchored in traditional optimization that inherently do not consider uncertainty. The objective is to achieve "satisficing" solutions (see Section 4.1.3) that provide good performance despite the presence of uncertainty, as opposed to solutions that are optimum in a narrow range of conditions but perform poorly when the conditions change slightly.

Traditional optimal solutions offer performance that may be nominally on target but often deteriorate significantly when conditions or assumptions change, or in the presence of material property variability or degradation in service. In so doing, the designer is drawn toward viewing the material process-structure-property relations as having stochastic character, and away from the conventional deterministic view of materials simulation (one input, one output) or materials selection based on deterministic properties rather than ranged sets of properties.

In robust design, the sensitivity of performance objectives is typically minimized with respect to variations in boundary conditions, dimensions, material properties, and other factors. Simultaneously, the deviation from mean or expected performance is minimized or matched with a target. These two goals are often treated as two separate objectives that arc traded against one another using multi-objective decision-making techniques. Typically, these problems result in Pareto families of solutions for which it is impossible to improve one objective without worsening another; e.g., it is impossible to improve the nominal value of a performance parameter without worsening its sensitivity. A collection of Pareto solutions is called a Pareto set or a Pareto frontier (Pareto 1909).[1] The set may include "optimal" solutions with extreme sensitivity to variation, highly robust solutions with minimal sensitivity but degraded performance, and tradeoff solutions that lie somewhere in between.

The robust design paradigm is illustrated in Figure 6.4. "Optimal" designs tend to be superior for a very limited set of conditions.[2] Robust designs, on the other hand, may offer somewhat reduced nominal performance, but the performance is relatively insensitive to changes in conditions (design variables). The robust design methodologies introduced in this chapter are aimed at identifying these robust tradeoff solutions for product and material design problems.

6.3.2. Types of Robust Design

Conceptually, it is useful to identify different types of robust design based on the source of the variation. In the robust design literature, there are three categories of information that interact with the system model for a process or product: control factors, noise factors, and responses. These are shown in Figure 6.5. Control factors, also known as *design variables,* are parameters which a designer adjusts to move towards a desired product. Noise factors are exogenous parameters that affect the performance of a product or process but are not under a designer's control. Responses are performance measures for the process or product.

---

[1] The concept of a Pareto solution and a Pareto set is borrowed from economics and named for the economist Vilfredo Pareto, who defined an allocation of resources as Pareto-efficient if it is impossible to identify another allocation that makes some people better off without making others worse off.

[2] Robust solutions are rarely *optimal* from a deterministic perspective because some measure of nominal performance has been sacrificed in favor of robustness (i.e., performance insensitivity) and feasibility over a range of conditions. Also, optimal solutions are constrained by one or more active constraints. These active constraints are likely to be violated when design variable values or other input factors and conditions change.

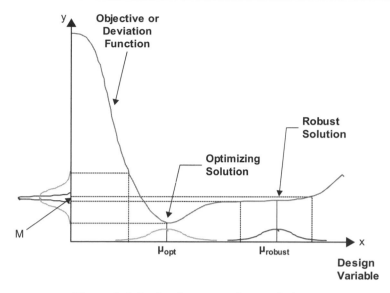

Figure 6.4 Optimal versus robust solutions.

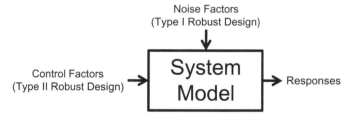

Figure 6.5 A P-diagram showing the input and response in a design product or process. Robust design is classified based on the source of variability.

Historically, two types of robust design have been considered (Chen, Allen et al. 1996):

- *Type I Robust Design* is used to identify control-factor (design-variable) values that satisfy a set of performance requirement targets despite variations in noise factors.

- *Type II Robust Design* is used to identify values of control-factor (design-variable) values that satisfy a set of performance requirement targets despite variation in control factors themselves. For example, in the early stages of design, it is clear that design variable values will change as the design evolves; therefore, it is preferable to identify starting values for which change will have the least possible effect on system performance and thus require minimal iteration as the design process proceeds.

Type I robust design is most commonly encountered in practice. Type II robust design has been documented in the literature (Myers and Montgomery 1995), but its use is not as prevalent in practice as Type I. With these types of robust design, a designer can accommodate uncertainty in the control factors that are generally used as inputs to the simulation models.

In the next section, the foundations of Type I robust design, anchored in the work of G. Taguchi, are presented, followed by criticisms of the Taguchi approach and alternative methods for robust design. We discuss the applications of robustness in the early stages of design, primarily by implementing Type II robustness.

6.4. Taguchi Method—Type I Robust Design

The collection of design principles and methods known as robust design is founded on the philosophy of a Japanese industrial consultant, Genichi Taguchi, who proposed that product design is a more cost-conscious and effective way to realize robust, high-quality products than by tightly controlling manufacturing processes. Taguchi noticed that there are two ways a product may prove to be unsatisfactory—the product may not meet target performance specifications or the variability in the product's performance may be unacceptably large. Taguchi noticed that traditionally design was carried out in two steps—systems design and tolerance design (Byrne and Taguchi 1987). Systems design requires innovation to develop new concepts and engineering knowledge to develop the concepts into design embodiments. The outcomes of the system design step are materials, geometry, and tentative values of product parameters. Tolerance design involves tightening tolerances on product or process parameters whose variations impart large influences on product performance. From Taguchi's perspective, tolerance design is expensive and should be used only when robustness cannot be "designed in" by selecting parameter levels that are least sensitive to variations. In order to reduce the variation in product performance while minimizing the overall product design and manufacturing cost, Taguchi proposed an intermediate *parameter design* step that precedes tolerance design but follows the system design. In the parameter design step, the product design parameters are divided into control factors and noise factors. The values of control factors are under designers' control, whereas the noise factors are not under designers' control. For example, the dimensions of the product are control factors, but the ambient conditions in which a product is used (temperature, humidity, etc.) are noise factors. In the parameter design step, the values of control factors are determined such that any variations in the product performance resulting from the variations in noise factors are minimized.

Further, instead of measuring quality by means of tolerance ranges, Taguchi proposed a Quality Loss Function in which the quality loss, L, is proportional to the square of the deviation of performance, P_s, from a target value, T. If Y represents the actual process response and t is the target value, then

$$L = k(P_s - T)^2 \tag{6.1}$$

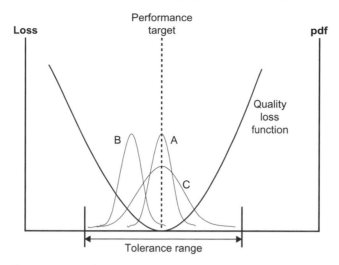

Figure 6.6 The Quality Loss Function and performance ranges for three products, A, B, and C, that are designed and manufactured to lie within the tolerance range. The performance of the products varies through different ranges and the values of mean performance may or may not coincide with the desired performance target.

As shown in Figure 6.6, any deviation from target performance results in a quality loss as does an increase of variability in product performance. The Quality Loss Function represents Taguchi's philosophy of striving to deliver on-target products and processes rather than those that barely satisfy a corporate limit or tolerance level.

Taguchi's robust design approach for parameter design involves clearly separating *control factors* and *noise factors*. Designed experiments, based on orthogonal arrays, are conducted in control and noise factors to evaluate the effect of control factors on nominal response values and sensitivity of responses to variations in noise factors. The intent is to minimize performance deviations from target values while simultaneously bringing mean performance on target, as shown in Figure 6.6. In Taguchi's approach, the overall quality of alternative designs is compared via signal-to-noise ratios that combine measures of mean response and standard deviation of the response. Product or process designs, characterized by specific levels of control factors, are selected that maximize the signal to noise ratio. Note that for all types of robust design, robust solutions may not be "optimal"; conversely, optimal decisions are rarely robust. A corollary of this observation is that designs for optimality are very costly to pursue if variability and uncertainty is addressed by improving models and databases. In this case, designs must insist on tight control over material, product, cost, and customer usage profile.

In robust design, it is important to take advantage of interactions and nonlinear relationships between control and noise factors to dampen the effect of noise factors and thus reduce variation in the response(s). Taguchi provided the initial insight into robust design; however, many improvements have been proposed to extend and refine his methods.

6.4.1. Limitations to the Taguchi Approach to Robust Design

Although Taguchi's robust design principles are advocated widely in industrial and academic settings, his statistical techniques, including orthogonal arrays and the signal-to-noise ratio, have been criticized extensively, and improving the statistical methodology used in robust design has been an active area of research (Myers and Montgomery 1995, Nair 1992, Tsui 1992). In the panel discussion reported by Nair (Nair 1992), practitioners and researchers discuss Taguchi's robust design methodology, the underlying engineering principles and philosophy, and alternative statistical techniques for implementing it. For example, many alternative experimental designs have been proposed, and a significant area for debate and scholarly research has been the comparative advantages of Taguchi's cross arrays (in which control factors and noise factors are varied according to separate plans) versus combined arrays (in which control and noise factors are varied jointly according to a single plan). A key advantage of the combined array approach is that it provides flexibility for the designer to rule out certain effects a priori and thereby accomplish computational savings. Welch and colleagues (Welch, Yu et al. 1990) were the first to propose the combined array that was later expanded upon by Shoemaker and coauthors (Shoemaker, Tsui et al. 1991), Borror and Montgomery (Borror and Montgomery 2000), and others. Today, with many alternative combined array designs available, a systematic approach to selection became necessary and has been provided by Wu and Zhu (Wu and Zhu 2003). Despite the convincing theoretical case for combined arrays, empirical studies suggest that cross arrays provide superior outcomes under a wide range of conditions (Frey and Li 2004, Kunert, Auer et al. 2005). Shoemaker and coworkers (Shoemaker, Tsui et al. 1991) note that combined arrays are successful to the extent that the model fits well.

It has been demonstrated that using the signal-to-noise ratio as the objective in robust design can hide information regarding noise interactions (Box 1988, Shoemaker, Tsui et al. 1991). Nair (Nair 1992) also reports that a panel of statisticians suggest independently modeling the mean response and variability directly or via statistical data transformations (Box 1988, Tsui 1992, Vining and Myers 1990), rather than modeling the signal-to-noise ratio—a practice that discards useful information about the response (particularly by confounding mean response with variance information). In the engineering literature, many authors separate out the objectives of mean-on-target (signal) and variability (noise); Mourelatos and Liang (Mourelatos and Liang 2005) show a Pareto frontier for the tradeoffs among these objectives. Chen and coauthors (Chen, Allen et al. 1996) also construct an approximate function— labeled *quality utility*—for the Pareto efficient frontier to facilitate exploration of alternative robust design solutions for bi-objective problems involving bringing the mean-on-target and minimizing variation.

Murphy and coauthors (Murphy, Tsui et al. 2005) review mathematically rigorous methods for considering multiple responses. Chen and coauthors (Chen, Allen et al. 1996) formulate a robust design problem as a multi-objective decision using the compromise Decision Support Problem (cDSP) (Bras and Mistree 1993). Both control and noise factors are considered as potential sources of variation, and constraints are modeled in a worst-case formulation to ensure feasibility robustness. Separate goals of bringing the mean-on-target and minimizing variation (for each design objective) are included in a goal-programming formulation of the objective function. This provides flexibility for achieving compromises among multiple performance objectives, as well as individual or collective compromises among mean values and variations for all objectives. In more recent work, Chen and coauthors have extended the approach to include alternative formulations of the objective function, such as compromise programming (Chen, Wiecek et al. 1999) and physical programming (Chen, Atul et al. 2000).

Researchers have also developed alternative analysis procedures, including "response modeling" (Welch, Yu et al. 1990), dual response approaches (Vining and Myers 1990), and rejection of predecided criteria in favor of graphical analysis and discovery (Box 1988). Leon and coauthors proposed the performance measure independent of adjustment (PerMIA) for use in robust design optimization (Leon, Shoemaker et al. 1987). An important alternative to conventional analysis of robustness measures is known as "operating window methods"; the operating window is the set of conditions under which the system operates without failure (Clausing and Frey 2005, Joseph and Wu 2002). Robust design methods have been developed based on operating windows to increase the window by applying severe restrictions on the design early in technology development.

6.4.2. Solving Robust Design Problems

Taguchi's robust design principles have also been extended by applying them to simulation-based design. With increasing pressure to reduce development costs and reduce development times, robust design is increasingly conducted by evaluating the relationship between input factors and responses using computer simulations rather than prototypes and physical experiments. Given that computer experiments lack pure error, different experimental design and analysis strategies are recommended (Simpson, Peplinski et al. 2001). In response to this need, the field of design and analysis of computer experiments has grown rapidly in recent decades providing a variety of useful techniques (Santner, Williams et al. 2003).

A number of researchers advocate nonlinear programming approaches for robust design. Ramakrishnan and Rao (Ramakrishnan and Rao 1996) formulate a robust design problem based on Taguchi's Quality Loss Function, using statistical concepts and nonlinear programming. They consider variations in both control and noise factors. Cagan and Williams (Cagan and Williams 1993) establish first-order necessary conditions for robust optimality based on measures of the flatness and curvature of the objective relative to local variations in design variables. Michelena and Agogino (Michelena and Agogino 1994) introduce an approach based on monotonicity analysis for solving robust design problems in which nominal performance values are preferred. Sundaresan and coauthors (Sundaresan, Ishii et al. 1995) introduce a sensitivity index for formulating a nonlinear objective function for robust design.

Since constraints are typically an important aspect of a nonlinear programming problem, several authors have investigated the formulation of constraints for robust design applications. Parkinson and coauthors (Parkinson, Sorensen et al. 1993) coined the term *feasibility robustness* to refer to designs that continue to satisfy constraints and remain within a feasible design space despite variations in control or noise factors. They proposed worst-case, Taylor series–based and linear statistical analysis approaches for calculating the magnitude of variation that is transmitted from control and noise factors to constraints. Yu and Ishii (Yu and Ishii 1994) propose a manufacturing variation pattern approach for adjusting constraints to account for correlated, manufacturing-induced variations. Otto and Antonsson (Otto and Antonsson 1993) adopt a constrained optimization approach for robust design using a modified version of Taguchi's signal-to-noise ratio as the objective function.

Once a robust design problem has been formulated, it must be solved. The computational burden can be significant, particularly for design problems in which a broad design space must be explored. Solution of a robust design problem is distinguished by the need to evaluate not only a nominal value for each response but also the variation of each response due to control or noise factor variation. As shown in Figure 6.5, consider a response, y, which is a function of control factors, x, and noise factors, z, i.e.,

$$y = f(x,z) \tag{6.2}$$

where the function f can be a detailed simulation model, a surrogate model, or a physical system or prototype. The challenge is to estimate the expected value, μ_y, and variance, σ_y^2 of the response. There are many techniques for transmitting or propagating variation from input factors to responses, and each technique has strengths and limitations. Monte Carlo analysis is a simulation-based approach that requires a very large number of experiments (Liu 2001). It is typically very accurate for approximating the distribution of a response, provided that probability distributions are available for the input factors. On the other hand, it is very computationally expensive, especially if there are large numbers of variables or if expensive simulations are needed to evaluate each experimental data point. Du and Chen (Du and Chen 2000) review several approaches for maintaining feasibility robustness and introduce a most probable point (MPP)–based approach that offers accuracy similar to Monte Carlo–based approaches with fewer computations. If only a moderate number of experimental points are affordable, a variety of space-filling experimental designs are available (Koehler and Owen 1996, McKay, Beckman et al. 1979), and experimental designs such as fractional factorials or orthogonal arrays can be used (Myers and Montgomery 1995, Wu 2000). These experimental designs provide adequate estimates of the range of the response rather than its distribution, and they require fewer experimental points. All of these experimental techniques can be used in two ways: (1) to provide estimates of the variation or distribution in responses at a particular design point or (2) to construct surrogate models of the response that can then be used in place of a computationally expensive simulation model for evaluating mean responses and variations (Chen, Allen et al. 1996, Mavris, Bandte et al. 1999, Welch, Yu et al. 1990). All these methods suffer from the problem of size identified by Koch and coauthors (Koch, Allen et al. 1997) in which the number of experiments becomes prohibitively large (given the computational expense of most engineering simulations) as the number of input factors or design variables increases.

An alternative method for propagating variation is by Taylor series expansions (Phadke 1989). A first-order Taylor series expansion, for example, can be used to relate variation in response, Δy, to variation in a noise factor, Δz, or a control factor, Δx, i.e.,

$$\Delta y = \sum_{i=1}^{k} \left| \frac{\partial f}{\partial x_i} \Delta x_i \right| + \sum_{i=1}^{m} \left| \frac{\partial f}{\partial z_i} \Delta z_i \right|, \tag{6.3}$$

where the variation may represent a tolerance range or may be a multiple of the standard deviation. Higher-order Taylor series expansions can be formulated to provide a better approximation of the variation in response, but higher-order expansions also require higher-order partial derivatives of the response function with respect to control and noise factors. Taylor series expansions are relatively accurate for small degrees of variation in control or noise factors but lose their accuracy for larger variations or highly nonlinear functions. A Taylor series expansion requires evaluation of the partial derivative or sensitivity of the response function with respect to changes in control or noise factors. If analytical expressions are available for the sensitivities, this can be a computationally attractive and relatively accurate approach, even for large numbers of control and noise factors (Bisgaard and Ankenman 1995). Alternative approaches include estimating sensitivities using finite difference techniques, automatic differentiation (a feature built into some computer programming languages), and other techniques such as perturbation analysis and likelihood ratio methods (Andradottir 1998). However, these techniques can diminish computational efficiency and accuracy. Sensitivity-based approaches have been proposed for modeling both constraints (Michelena and Agogino 1994, Phadke 1989, Simpson, Peplinski et al. 2001) and objectives (Belegundu and Zhang 1992, Su and Renaud 1997) in robust design.

6.4.3. Solving Materials Design Problems under Uncertainty

While robust design techniques have been applied primarily to product- and system-level design problems, materials design applications present a set of unique challenges for robust design and simulation-based design under uncertainty. In some cases these challenges can be met by extending techniques from the product design domain, but other materials design applications call for novel or specialized approaches. One aspect of robust materials design is the need to capture stochastic spatial variation in microstructure morphology. For example,

Yin and coauthors (Yin, Lee et al. 2008) model uncertainty with a random field approach; Choi and coauthors (Choi, Grandhi et al. 2006) utilize a polynomial chaos expansion procedure; Zabaras and coauthors (Sankaran and Zabaras 2006) apply a maximum entropy technique for modeling topological uncertainties in polycrystalline metallic microstructures. Zohdi (Zohdi 2003) utilizes a straightforward ensemble averaging technique to evaluate macroscopic effective mechanical properties for the design of random particulate media with a genetic algorithm. For other applications, such as some mesostructure topology design efforts, uncertainty in material properties or topology is represented with standard continuous or discrete probability density distributions (e.g., Sandgren and Cameron (2002), Seepersad, Allen et al. (2006)), which are also common approaches for product- and systems-level design.

Another materials design challenge is the propagation of uncertainty through models of material process-structure and structure-property relations that are often highly sensitive to underlying assumptions and to the limitations of sparse input data (often obtained from costly experiments). Some researchers have applied standard Monte Carlo analysis techniques (e.g., for designing advanced composites in the DARPA AIM program (Cregger, Caiazzo et al. 2004 (May 17), Hahn 2001 (August 27–28))) to solve this problem, but the computational expense challenges faced by product-level designers can be even more profound on the materials level. Some materials designers (e.g., Cuong, Veroy et al. (2005), Huynh and Patera (2007), Yin, Lee et al. (2008)) have focused on the use of reduced-order models for reducing the expense of estimating variability in output parameters. These reduced-order models differ from the metamodels often used by product designers because they are derived directly from underlying physics-based models via identification of significant input factors (and elimination of insignificant factors), rather than statistically fitted to data points obtained from the underlying model. Modeling, propagating, and reducing the impact of uncertainty in materials design applications remain open and challenging areas of research.

6.5. Robust Concept Exploration Method (RCEM)—Type II Robust Design

Much of the work on robust design has focused on the detailed design stages. It is usually assumed that a preliminary design—with concrete layout and preliminary design specifications—has already been determined, but exploration of a broad design space and significant adaptations or variations in a system are typically not undertaken or facilitated. However, some authors have focused on infusing robust design techniques in the earlier, more conceptual stages of design when decisions are made that profoundly impact product performance and quality. Primarily, this has been achieved by Type II robust design, which enhances the robustness of design decisions with respect to subsequent variations in designs themselves. Although we do not intend to catalog all of the applications of robustness in the early stages of design, we mention a few interesting design applications here.

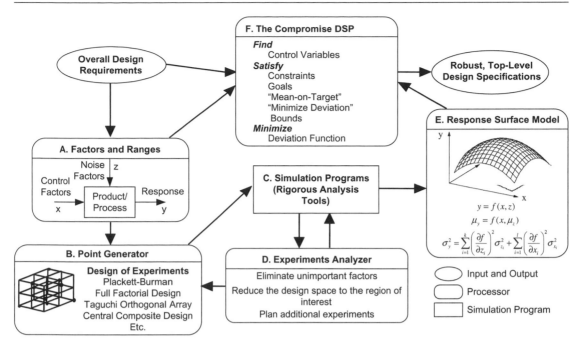

Figure 6.7 Computing infrastructure for the RCEM (modified from Chen 1996).

Chen and coauthors (Chen, Allen et al. 1996, Chen, Allen et al. 1996) formulate the Robust Concept Exploration Method (RCEM)—a domain-independent, systematic robust design approach for the early stages of design—by integrating statistical experimentation and approximate models, robust design techniques, multidisciplinary analyses, and multi-objective decisions. The computing infrastructure of the RCEM is shown in Figure 6.7. The RCEM has been employed successfully for a simple structural problem and the design of a solar-powered irrigation system (Chen 1995), a high-speed civil transport (Chen, Allen et al. 1996), and a general aviation aircraft (Simpson 1996, Simpson, Chen et al. 1999). In addition, RCEM has been extended to facilitate the design of complex systems (Koch, Mavris et al. 2000), hierarchical systems (Koch 1997), and product platforms (Simpson, Chen et al. 1999, Simpson, Maier et al. 2001).

In addition to facilitating the generation of robust, flexible, ranged sets of design specifications, the RCEM framework facilitates exploration of a broad design space. This involves creating an experimental design; i.e., determining the combinations of values of independent design variables at which experiments are carried out. Using the data generated in the experiments, a simple model (metamodel) is fit between the independent design variables (inputs) and the responses (outputs). The simple model is then used in decision making. Response surface methodology has been used as one metamodeling technique within the RCEM framework (Liu 2001, Zohdi 2003). Response surface methodology is a collection of statistical techniques for empirically mapping relationships between independent design variables and their dependent performance functions. Response Surface Methodology is particularly useful if there are large computer run times associated with the simulation of complex systems. This has the advantage that it uncouples the design procedure from the development of the response surface and the response surface may be used repeatedly.

Parkinson (Parkinson, Sorensen et al. 1993) notes that robust design has the ability to "control or absorb variability"; therefore, robust design is useful for problems in which a controlled degree of variability is valuable, such as in the design of product families. Simpson (Simpson 1996) extended the use of robust design principles into a formal methodology to optimize scalable product platforms for families of products, demonstrating its use to design families of electric motors (Simpson, Maier et al. 2001), general aviation aircraft (Simpson, Chen et al. 1999, Simpson, Maier et al. 2001, Simpson, Seepersad et al. 2001), and absorption chillers (Hernandez, Simpson et al. 2001). Robust design is also valuable for the design of flexible systems—"systems designed to maintain a high level of performance … when operating conditions or requirements change in a predictable or unpredictable way" (Olwenik, Brauen et al. 2004); for example, aircraft that can reconfigure their wings in flight are flexible systems, as are robots that adapt to changing environments or tasks.

In the methods and applications discussed so far, a fundamental assumption is that the topology or layout of the design is known before robust design methods are applied. However, Seepersad and colleagues (Seepersad 2004, Seepersad, Allen et al. 2005) have developed a technique for robust topology design in which the topology of the product is adjustable along with parameters that describe that topology. This is quite relevant to effects of material processing on structure-property relations in materials design. For example, Seepersad provides examples for design of simplified scenarios for heat sinks and combustor liners from linear cellular or honeycomb materials that contain a series of channels to facilitate heat flow while maintaining structural integrity. In the robust topology design of these channels, the design variables include the shape of the channels (triangular, rectangular, etc.), the layout, and the specific dimensions. The robust topology design method is useful not only for minimizing the impact of fabrication-related imperfections but also for minimizing the impact of subsequent topological or dimensional adjustments that, in later stages of design, may be required for meeting additional, multifunctional requirements.

In the early stages of a design process, we have explicit information regarding the target value of a system performance. If the target of a performance is uncertain, then the cDSP (and goal programming) that deals with deviation from the target value could produce a different solution than the designer intended. Instead, ranged requirements could be given for a system performance—the performance of a system should be better than a lower requirement limit, smaller than an upper requirement limit, or between lower and upper limits. The approach can be used in the concurrent design of materials and products. For example, in the pressure vessel and composite design material problem discussed in Chapter 4, instead of selecting

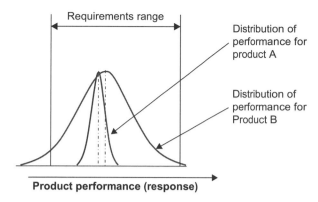

Figure 6.8 Comparing two designs with respect to a range of requirements.

a single set of values for design variables associated with the composite material (orientation of fibers, number of layers, thickness of layers, etc.), a range of these material properties can be determined and passed on to the pressure vessel designer. This provides more freedom to the product design to satisfy the design objectives. Hence, there is a higher likelihood of achieving the system level goals as closely as possible.

There are some cases in the early stages of design when requirements themselves are uncertain and most appropriately expressed as a range rather than a target value, as shown in Figure 6.8. In these cases, it is not appropriate to bring the mean-on-target and minimize variation. Instead, it may be necessary to measure the extent to which a range or distribution of design performance satisfies a ranged set of design requirements. Design capability indices (DCIs) are a set of metrics designed especially for assessing the capability of a ranged set of design specifications for satisfying a ranged set of design requirements. These design capability indices are incorporated as goals in the cDSP within the RCEM framework. The details are described by Chen and coauthors (Chen, Simpson et al. 1999). In further work, Chen and Yuan (Chen and Yuan 1999) introduced a design preference index that allows a designer to specify varying degrees of desirability for ranged sets of performance, rather than specifying precise target values or limits for a range of requirements beyond which designs are considered worthless.

DCIs are used as metrics for system performance and robustness; these are used as goal formulations in the cDSP formulation instead of directly using the means and variances of system performances (Chen, Simpson et al. 1996, Chen, Simpson et al. 1999). The DCIs are mathematical constructs for efficiently determining whether a ranged design specification is capable of satisfying a ranged set of design requirements. The procedure to evaluate the index is illustrated in Figure 6.9; C_{du}, C_{dl}, and C_{dk} in the figure are calculated as

$$C_{dl} = \frac{\mu - LRL}{3\sigma}; C_{du} = \frac{\mu - LRL}{3\sigma}; C_{dk} = \min\{C_{dl}, C_{du}\} \qquad (6.4)$$

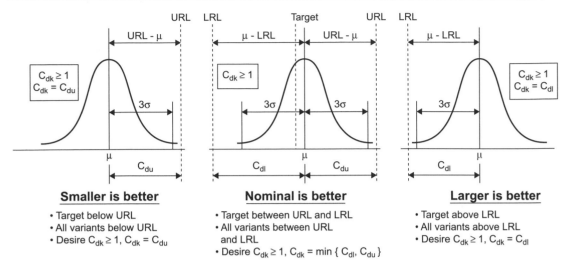

Figure 6.9 DCIs (Chen et al. 1999).

When the DCI is negative, the mean of system performance is outside of the system requirement range. If the index is greater than unity, then the design will meet the requirement satisfactorily. Therefore, a designer's objective is to force the index to unity so that the larger portion of performance deviation falls into the range of design requirements. Forcing the index to unity is achieved by reducing performance deviation and/or locating the mean of performance deviation farther from requirement limits.

In order to use the DCIs, a designer specifies a range of property requirements that need to be satisfied. For example, for a material design problem, a range of strengths, hardness, density, and coefficient of thermal expansion can be provided by the designer in terms of the lower and upper requirements limits (shown as LRL and URL, respectively, in Figure 6.9). Given the range of requirements, the means (μ) and standard deviations (σ) of design variables (e.g., composition, grain sizes, etc.) are determined such that the DCI is close to 1. By forcing the index to unity, the designer can ensure that the range of material design variables enables the achievement of specified ranges of the material property requirements.

The DCIs calculated in this manner are employed in the cDSP for finding a ranged set of design specifications. The Robust Concept Exploration Method with Design Capability Indices (RCEM-DCI) provides the following advantages. With DCIs, a designer can efficiently check whether a family of designs can satisfy design requirements while eliminating the tedious task of evaluating large numbers of discrete or continuous design specifications. In addition, a designer can consider multiple aspects of quality improvement by adjusting the location of the mean of the performance distribution as well as the variation. Finally, the DCI is easy for a designer to compute and understand.

We have reviewed the Taguchi method and its extensions for Type I robust design and RCEM and DCIs for Type II robust design. The advantages of the methods are also discussed. However, these methods cannot fully support robust design of multiscale materials. In the following sections, the limitations of the methods for Type I and II robust design are discussed. Identifying the limitations, we present new types of robust design to augment these methods.

6.6. Requirements for New Types of Robust Design

In Type I and II robust design, parameterizable variability in natural uncertainty has been considered to achieve robust system performance. For example, the variability in impact velocity of the microscale shock simulation (a noise factor) or the variability of the mean size of aluminum particles (a control factor) are considered to have robust reaction initiation of multifunctional energetic structural materials. However, other types of uncertainty, including unparameterizable variability in NU, MPU, and MSU, have not been addressed in Types I and II robust design. For example, the unparameterizable variability due to simulated random microstructures in energetic materials is irreducible variability; therefore, a robust design approach is required. Incomplete estimation of the variability (model parameter uncertainty) due to expensive computation of the shock simulation also requires a robust design approach since it is not a problem of calibrating a model to true data. These examples are typical types of uncertainty in multiscale materials design and should be managed.

Other types of uncertainty, excluding parameterizable variability, are usually embedded within function relationships (models). Uncertainty in a model is typically different from the uncertainty associated with noise and control factors because it may be due to the parameters or structure of constraints, metamodels, engineering equations, and associated simulation or analysis models. In materials design, models are often nonlinear and reflect the time history of environment and boundary conditions; such models are intrinsically nonunique and are often sensitive to initial conditions. Therefore, a new type of robust design—*Type III robust design*—should be established for managing these additional types of uncertainty. Type III robust design considers sensitivity to the uncertainty embedded within a model, as shown in Figure 6.10.

6.7. Requirements for New Multilevel Robust Design Methods

6.7.1. All-in-One versus Multilevel

Another challenge in materials design is the propagated uncertainty resulting from hierarchical information dependency in a multiscale model chain. A simple example of an analysis chain is illustrated in Figure 6.11. Input variables (*x1, x2*) may be parameters related to a material's process route or structure that may be tailored by materials designers. The

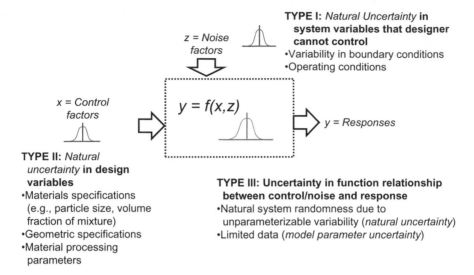

Figure 6.10 Three types of robust design, including insensitivity to uncertainty in noise factors (Type I) and control factors (design variables) (Type II), as well as uncertainty in models and microstructure (Type III).

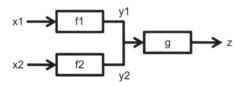

Figure 6.11 An example of information flow in a model chain.

functions (*f1, f2*) may be simulation models for predicting material properties (*y1, y2*) such as modulus of elasticity, ultimate strength, yield strength, the Hugoniot relation, etc. The derived material properties and/or responses are interfaced to a product-level model (*g*), such as a finite element analysis model, and then a system-level response of interest (*z*), such as structural integrity, thermal behavior, etc., is obtained. Uncertainty may be accumulated and amplified through this sequential chain, making the variance of the final response (*z*) unacceptably large. This is an important issue because small variations or errors in input parameters may cause high levels of variability in the system response.

Design exploration methods can be categorized as an "all-in-one approach" (Du and Chen 2002) or as a "multilevel approach," depending on the approach for estimating propagated uncertainty. In the all-in-one approach, the system boundary is large since multiple subsystems are considered as one system. As shown in Figure 6.12, the amount of uncertainty in the final performance (z) due to the uncertainty in random inputs (x1 and x2) is quantified in this approach. If all subsystems are configured as all-in-one, then Types I, II, and III robust design may be present. However, the multiple models in the model chain must be

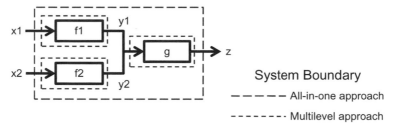

Figure 6.12 System boundary of all-in-one and multilevel approaches.

computationally or mathematically interfaced to effectively construct a single system-level model in order to employ the all-in-one approach. Many commonly employed uncertainty analysis methods (e.g., Monte Carlo simulation, Latin HyperCube sampling, first-order and second-moment analysis, the stochastic response surface method (Isukapalli, Roy et al. 1998), etc.) are applicable *only* if an integrated single model is available for the multiple model chain. Unfortunately, this is only practical for limited classes of materials design problems.

For the all-in-one approach, all computer models within the chain must be integrated and made available simultaneously to a robust design exploration procedure. However, if the decomposed subsystems are controlled by groups from multiple disciplines and are of fundamentally different character (e.g., discrete and continuous, dynamic and thermodynamic) or reside in a distributed environment on heterogeneous computing platforms, then integrating the subsystems for the all-in-one approach is costly even if a computational framework is available for system integration. Moreover, if decisions must be made as part of the I/O structure in connecting models, then quite naturally, the all-in-one approach breaks down. In other words, the need for dynamic reconfigurability of the design process in complex systems design is not addressed satisfactorily by the all-in-one approach. These problems are shared by other design methods as well (Chen, Allen et al. 1996, Parkinson, Sorensen et al. 1993, Sundaresan, Ishii et al. 1995). These drawbacks of the all-in-one approach suggest that a multilevel approach that consists of modular uncertainty analyses along a chain of models is preferable.

6.7.2. Methods for Multilevel Robust Design

Multilevel approaches can be decomposed into methods that consider uncertainty and those that do not. Multilevel approaches which do not consider uncertainty (i.e., multilevel optimization methods) were originally proposed by Schmit and Mehrinfar (Schmit and Mehrinfar 1982) and Sobieszczanski-Sobieski and co-authors (Sobieszczanski-Sobieski, James et al. 1987) in the 1980s. These methods have evolved into Concurrent SubSpace Optimization (CSSO; Ford and Bloebaum, 1993, Wujek et al. 1996) and Bi-Level Integrated Synthesis (BLISS; Sobieszczanski-Sobieski and Kodiyalam, 2001), etc. In these methods, a higher-level optimization calls multiple lower-level optimizations and uses the decision making of the lower-level optimizations for higher-level decision making. A review of these methods is not included in detail in this chapter and can be found in the references. Instead, our review focuses on the multilevel approaches that account for uncertainty (i.e., multilevel robust design methods).

Gu and coauthors (Gu 2000) propose worst-case propagated uncertainty analysis and robust optimization. With their approach, a first-order sensitivity analysis is performed on each subsystem. Final system response deviation is estimated by propagating the results of individual subsystem uncertainty analysis. Du and Chen (Du and Chen 2002) suggest analysis methods that accommodate a generic probabilistic approach instead of using worst-case sensitivity analysis in order to estimate the amount of uncertainty efficiently and accurately. In these approaches, mean and variance (or deviation), instead of the single value of an interface variable in the all-in-one approach, are transferred from one subsystem to another to calculate propagated uncertainty in the model chain. Therefore, the amount of information passed to calculate propagated uncertainty is less than that in the all-in-one approach.

These two approaches still require a large amount of information flow across the boundaries of the subsystems. This is due to the fact that the processes of uncertainty analysis and design exploration are tightly coupled. Whenever uncertainty analysis at a given input point is requested from a design exploration algorithm, the subsystem analyses must be performed sequentially, passing the information related to mean and variance of responses. The mean and variance of the final performance must then be returned to the design exploration algorithm in real time. This sequential uncertainty propagation in the methods becomes more difficult to deal with in a distributed environment. In the Concurrent SubSystem Uncertainty Analysis (CSSUA) method presented by Du and Chen, the associated subsystem analysis computations are parallelized to identify the mean of the linking variables. However, since the robust optimization process is sequential and the uncertainty analysis process is a subprocess of the optimization process, the uncertainty analysis, which consumes most of the computing power, cannot be fully parallelized.

Another multilevel robust design method is probabilistic Analytical Target Cascading (Liu, Chen et al. 2006), which is rooted in a multilevel optimization method, Analytical Target Cascading (ATC; Hyung Min and Tao, 2003). In ATC, a higher-level problem solution defines the target values for lower-level design problems. Collecting the lower-level decision making results, a designer makes higher-level decisions by minimizing gaps between the targets from higher-level problems and the performance achieved from lower-level problems. In probabilistic ATC, gaps between targets and performance in terms of means and variances are considered to identify robust multilevel design solutions. In probabilistic ATC, design targets at a level of a problem are derived from its super-level, which is well matched to the goal/means (inductive) materials design approach. This method eliminates bottom-up uncertainty propagation though a multilevel chain since the problem is decomposed and design exploration is executed only at each level. However, this requires the exchange of a large amount of decision-making information about the various levels of the problem. Each evaluation at a super-level optimization spawns multiple sublevel design optimization problems. Moreover, a super-level iteration must wait for the results of sublevel optimization; therefore, multilevel optimizations are highly interdependent and cannot be parallelized.

As a summary, in this chapter, the types of uncertainty associated with multilevel materials design are discussed. The types of uncertainty are classified into natural uncertainty (parameterizable or unparameterizable), model parameter uncertainty, model structure uncertainty, and propagated uncertainty. Existing robust design methods are valid only for the parameterizable natural uncertainty but not for other types of uncertainty. This motivates a new type of method for (Type III) robust design. Further, previously developed multilevel robust design methods are not suitable for considering propagated uncertainty in multiscale materials design, giving rise to another requirement of multilevel robust design methods. Our answers to these challenges are discussed in Chapter 8.

References

Andradottir, S., 1998. "Simulation Optimization". In: John, E. (Ed.), Handbook of Simulation. Wiley and Sons, Inc., NY.

Austin, R., 2005, "Numerical Simulation of the Shock Compression of Microscale Reactive Particle Systems", M.S. Thesis, Woodruff School of Mechanical Engineering, Georgia Institute of Technology, Atlanta.

Ayyub, B.M., Chao, R.-J., 1997. "Uncertainty Modeling in Civil Engineering with Structural and Reliability Applications". In: Ayyub, B.M. (Ed.), Uncertainty Modeling and Analysis in Civil Engineering. New York, CRC Press, pp. 3–32.

Belegundu, A.D., Zhang, S., 1992. "Robustness of Design through Minimum Sensitivity,". Journal of Mechanical Design 114, 213–217.

Bisgaard, S., Ankenman, B., 1995. "Analytic Parameter Design,". Quality Engineering 8, 75–91.

Bocchini, G.F., 1986. "The Influence of Porosity on the Characteristics of Sintered Materials,". The International Journal of Powder Metallurgy 22 (3), 185–202.

Borror, C.M., Montgomery, D.C., 2000. "Mixed Resolution Designs as Alternatives to Taguchi Inner/Outer Array Designs for Robust Design Problems,". Quality and Reliability International 16, 117–127.

Box, G., 1988. "Signal-to-Noise Ratios, Performance Criteria, and Transformations". Technometrics 30, 1–18.

Bras, B.A., Mistree, F., 1993. "Robust Design Using Compromise Decision Support Problems,". Engineering Optimization 2 (3), 213–239.

Byrne, D.M., Taguchi, S., 1987. "The Taguchi Approach to Parameter Design,". Quality Progress 20 (12), 19–26.

Cagan, J., Williams, B.C., 1993. First-Order Necessary Conditions for Robust Optimality. In: ASME Advances in Design Automation, Vol. 65-1. ASME DE, Albuquerque, NM, pp. 539–549.

Chen, W., 1995, "A Robust Concept Exploration Method for Configuring Complex Systems", PhD Dissertation, G. W. Woodruff School of Mechanical Engineering, Georgia Institute of Technology, Atlanta, GA.

Chen, W., Allen, J.K., Mavris, D., Mistree, F., 1996. "A Concept Exploration Method for Determining Robust Top-Level Specifications,". Engineering Optimization 26 (2), 137–158.

Chen, W., Allen, J.K., Tsui, K.-L., Mistree, F., 1996. "A Procedure for Robust Design: Minimizing Variations Caused by Noise Factors and Control Factors,". ASME Journal of Mechanical Design 118 (4), 478–485.

Chen, W., Atul, S., Messac, A., Sundararaj, G., 2000. "Exploration of the Effectiveness of Physical Programming in Robust Design,". ASME Journal of Mechanical Design 122 (2), 155–163.

Chen, W., Simpson, T. W., Allen, J. K., and Mistree, F., 1996, "Using Design Capability Indices to Satisfy a Ranged Set of Design Requirements", in *Advances in Design Automation*, Irvine, CA, pp. ASME Paper no. 96-DETC/DAC-1090.

Chen, W., Simpson, T.W., Allen, J.K., Mistree, F., 1999. "Satisfying Ranged Sets of Design Requirements Using Design Capability Indices as Metrics,". Engineering Optimization 31, 615–639.

Chen, W., Wiecek, M., Zhang, J., 1999. "Quality Utility: A Compromise Programming Approach to Robust Design,". Journal of Mechanical Design 121, 179–187.

Chen, W., Yuan, C., 1999. "A Probabilistic-Based Design Model for Achieving Flexibility in Design,". ASME Journal of Mechanical Design 121 (1), 77–83.

Choi, H.-J., Austin, R., Allen, J.K., McDowell, D.L., Mistree, F., Benson, D.J., 2005. "An Approach for Robust Design of Reactive Powder Metal Mixtures Based on Non-deterministic Micro-Scale Shock Simulation,". Journal of Computer-Aided Materials Design 12 (1), 57–85.

Choi, S.-K., Grandhi, R.V., Canfield, R.A., 2006. Robust Design of Mechanical Systems via Stochastic Expansion. International Journal of Materials and Product Technology 25 (1/2/3), 127–143.

Clausing, D., Frey, D.D., 2005. Improving Systems Reliability by Failure-Mode-Avoidance Including Four Concept Design Strategies. Systems Engineering (INCOSE) 8 (3), 245–261.

Cregger, S. E., Caiazzo, A., Pugliano, P., Rajagopal, R., and Uryasev, S., 2004 (May 17), "Accelerated Insertion of Materials-Managing Error and Uncertainty in Structures", in *49th International Society for the Advancement of Material and Process Engineering Symposium and Exhibition*, Long Beach, CA.

Cuong, N.N., Veroy, K., Patera, A.T., 2005. Certified Real-Time Solutions of Parametrized Partial Differential Equations. In: Yip, S. (Ed.), Handbook of Materials Modeling. Springer, pp. 1523–1558.

Der Kiureghian, A., 1989. Measures of Structural Safety under Imperfect State of Knowledge. J. Structural Eng 115 (5), 1119–1139.

Du, X., Chen, W., 2000. "Towards a Better Understanding of Modeling Feasibility Robustness in Engineering Design,". Journal of Mechancial Design 122, 385–394.

Du, X., Chen, W., 2002. "Efficient Uncertainty Analysis Methods for Multidisciplinary Robust Design,". AIAA Journal 40 (3), 545–552.

Ford, J.M., Bloebaum, C.L., 1993. Decomposition Method for Concurrent Design of Mixed Discrete/Continuous Systems. ASME, New York, NY, USA, Albuquerque, NM, USA pp. 367-376.

Frey, D. D. and Li, X., 2004, "Validating Robust Parameter Design Methods", in *ASME Design Technical Conferences*, Salt Lake City, UT, pp. Paper Number DETC2004-57518.

Gu, X., Renaud, J.E., Batill, S.M., Brach, R.M., Budhiraja, A.S., 2000. "Worst Case Propagated Uncertainty of Multidisciplinary Systems in Robust Design Optimization,". Structural and Multidisciplinary Optimization 20 (3), 190–213.

Hahn, G., 2001 (August 27-28), "Accelerated Insertion of Materials-Composites", in *DARPA Workshop*, Annapolis, MD.

Haukass, T., 2003. Types of Uncertainties, Elementary Data Analysis, Set Theory, Reliability and Structural Safety, Lecture Notes. University of British Columbia.

Hernandez, G., Simpson, T.W., Allen, J.K., Bascaran, E., Avila, L.F., Salinas, R.F., 2001. "Robust Design Modeling of Products and Families,". Journal of Mechanical Design 123 (2), 183–190.

Huynh, D.B.P., Patera, A.T., 2007. "Reduced Basis Approximation and A Posteriori Error Estimation for Stress Intensity Factors,". International Journal of Numerical Methods in Engineering 72 (10), 1219–1259.

Hyung Min, K., Michelena, N.F., Papalambros, P.Y., Tao, J., 2003. Target cascading in optimal system design, Transactions of the ASME. Journal of Mechanical Design 125 (3), 474–480.

Hyung Min, K., Rideout, D.G., Papalambros, P.Y., Stein, J.L., 2003. Analytical target cascading in automotive vehicle design, Transactions of the ASME. Journal of Mechanical Design 125 (3), 481–489.

Isukapalli, S.S., Roy, A., Georgopoulos, P.G., 1998. "Stochastic Response Surface Methods (SRSMs) for Uncertainty Propagation: Application to Environmental and Biological Systems,". Risk Analysis 18 (3), 351–363.

Joseph, V.R., Wu, C.F.J., 2002. "Operating Windows Experiments: A Novel Approach to Quality Improvement,". Journal of Quality Technology 34, 345–354.

Koch, P. N., 1997, "Hierarchical Modeling and Robust Synthesis for the Preliminary Design of Large Scale Complex Systems", PhD Dissertation, The G.W. Woodruff School of Mechanical Engineering, Georgia Institute of Technology, Atlanta, GA.

Koch, P.N., Allen, J.K., Mistree, F., Mavris, D., 1997. The Problem of Size in Robust DesignPaper Number: DETC97/DAC-3983. In: ASME Advances in Design Automation. ASME, Sacramento, CA. Paper Number: DETC97/DAC-3983

Koch, P.N., Mavris, D., Mistree, F., 2000. "Multi-level, Partitioned Response Surfaces for Modeling Complex Systems,". AIAA Journal 38 (5), 875–881.

Koehler, J.R., Owen, A.B., 1996. Computer Experiments. In: Ghosh, S., Rao, C.R. (Eds.) Handbook of Statistics. Elsevier Science, NY, pp. 261–308.

Kunert, J., Auer, C., Erdbrugge, M., Gobel, R., 2005. An Experiment to Compare the Combined Array and the Product Array for Robust Parameter Design. Journal of Quality Technology pp. (in press).

Leon, R.V., Shoemaker, A.C., Kacker, R.N., 1987. "Performance measures independent of adjustment: an explanation and extension of Taguchi's signal-to-noise ratios (with discussions),". Technometrics 29 (3), 253–285.

Liu, H., Chen, W., Kokkolaras, M., Papalambros, P.Y., Kim, H.M., 2006. Probabilistic analytical target cascading: A moment matching formulation for multilevel optimization under uncertainty, Journal of Mechanical Design. Transactions of the ASME 128 (4), 991–1000.

Liu, J.S., 2001. Monte Carlo Strategies in Scientific Computing. Springer, NY.

Mavris, D.N., Bandte, O., DeLaurentis, D.A., 1999. "Robust Design Simulation: A Probabilistic Approach to Multidisciplinary Design,". Journal of Aircraft 36 (1), 298–307.

McKay, M.D., Beckman, R.J., Conover, W.J., 1979. "A Comparison of Three Methods for Selecting Values of Input Variables in the Analysis of Output from a Computer Code,". Quality Engineering 11 (3), 417–425.

Michelena, N.F., Agogino, A.M., 1994. "Formal Solution of N-Type Robust Parameter Design Problems with Stochastic Noise Factors,". Journal of Mechanical Design 116, 501–507.

Mourelatos, Z. P. and Liang, J., 2005, "A Methodology for Trading-Off Performance and Robustness under Uncertainty", in *ASME Design Technical Conferences*, Long Beach, CA, pp. Paper Number DETC2005/DAC-85019.

Murphy, T.E., Tsui, K.-L., Allen, J.K., 2005. "A Review of Robust Design Methods for Multiple Responses,". Research in Engineering Design 15, 201–215.

Myers, R.H., Montgomery, D.C., 1995. Response Surface Methodology: Process and Product Optimization Using Designed Experiments. Wiley, New York.

Nair, V.N., 1992. "Taguchi's Parameter Design: A Panel Discussion,". Technometrics 34, 127–161.

Nikolaidis, E., 2005. Types of Uncertainty in Design Decision Making. In: Nikolaidis, E., nGhiocel, D.M., Singhal, S. (Eds.) Enigneering Design Reliability Handbook. CRC Press, New York, pp. 8.1–8.20.

Oberkampf, W. L., DeLand, S. M., Rutherford, B. M., Diegert, K. V., and Alvin, K. F., 1999, "A New Methodology for the Estimation of Total Uncertainty in Computational Simulation", in *AIAA*, pp. 99-1612. Paper Number: Paper Number 99-1612.

Olwenik, A., Brauen, T., Ferguson, S., Lewis, K., 2004. "A Framework for Flexible Systems and Its Implementation in Multiattribute Decision Making,". Journal of Mechancial Design 126, 412–419.

Otto, K.N., Antonsson, E.K., 1993. "Extensions to the Taguchi Method of Product Design,". ASME Journal of Mechanical Design 115 (1), 5–13.

Pareto, V., 1909, *Manuel D'Economie Politique*.

Parkinson, A., Sorensen, C., Pourhassan, N., 1993. "A General Approach for Robust Optimal Design,". ASME Journal of Mechanical Design 115 (1), 74–80.

Parkinson, A., Sorensen, C., Pourhassan, N., 1993. "General approach for robust optimal design,". Transactions of the ASME 115 (1), 74–80.

Phadke, M.S., 1989. Quality Engineering Using Robust Design. Prentice Hall, Englewood Cliffs, NJ.

Ramakrishnan, B., Rao, S.S., 1996. "A General Loss Function Based Optimization Procedure for Robust Design,". Engineering Optimization 25, 255–276.

Sandgren, E., Cameron, T.M., 2002. "Robust Design Optimization of Structures through Consideration of Variation,". Computers and Structures 80 (20-21), 1605–1613.

Sankaran, S., Zabaras, N., 2006. "A Maximum Entropy Approach for Property Prediction of Random Microstructures,". Acta Materialia 54, 2265–2276.

Santner, T.J., Williams, B.J., Notz, W.I., 2003. The Design and Analysis of Computer Experiments. Springer, New York.

Schmit, L.A., Mehrinfar, M., 1982. "Multilevel Optimum Design of Structures with Fiber-composite Stiffened-panel Components,". AIAA Journal 20 (1), 138–147.

Seepersad, C. C., 2004, "A Robust Topological Preliminary Design Exploration Method with Materials Applications", PhD Dissertation, The G.W. Woodruff School of Mechanical Engineering, Georgia Institute of Technology, Atlanta, GA.

Seepersad, C.C., Allen, J.K., McDowell, D.L., Mistree, F., 2006. Robust Design of Cellular Materials with Topological and Dimensional Imperfections. ASME Journal of Mechanical Design 128 (6), 1285–1297.

Shoemaker, A.C., Tsui, K.-L., Wu, C.F.J., 1991. "Economical Experimentation Methods for Robust Design,". Technometrics 33, 415–427.

Simpson, T. W., 1996, "A Concept Exploration Method for Product Family Design", PhD Dissertation, The G.W. Woodruff School of Mechanical Engineering, Atlanta, GA.

Simpson, T.W., Chen, W., Allen, J.K., Mistree, F., 1999. Use of the Robust Concept Exploration Method to Facilitate the Design of a Family of Products. In: Roy, U., Usher, J.M., Parsaei, H.R. (Eds.) Simultaneous Engineering: Methodologies and Applications. Gordon and Breach Science Publishers, Amsterdam, The Netherlands, pp. 247–278.

Simpson, T.W., Maier, J.R.A., Mistree, F., 2001. Product Platform Design: Method and Applications. Research in Engineering Design: Method and Application 13, 2–22.

Simpson, T.W., Peplinski, J.D., Koch, P.N., Allen, J.K., 2001. "Metamodels for Computer-based Engineering Design: Survey and Recommendations,". Engineering with Computers 17, 129–150.

Simpson, T.W., Seepersad, C.C., Mistree, F., 2001. "Balancing Commonality and Performance within the Concurrent Design of Multiple Products in a Product Family,". Concurrent Engineering: Research and Applications (CERA) 9 (3), 177–190.

Sobieszczanski-Sobieski, J., James, B.B., Riley, M.F., 1987. "Structural Sizing by Generalized, Multilevel Optimization,". AIAA Journal 25 (1), 139–145.

Sobieszczanski-Sobieski, J., Kodiyalam, S., 2001. "BLISS/S: A new method for two-level structural optimization,". Structural and Multidisciplinary Optimization 21 (1), 1–13.

Su, J., Renaud, J.E., 1997. "Automatic Differentiation in Robust Optimization,". AIAA Journal 36 (6), 1072–1079.

Sundaresan, S., Ishii, K., Houser, D.R., 1995. "Robust optimization procedure with variations on design variables and constraints,". Engineering Optimization 24 (2), 101–109.

Torquato, S., 1991. "Random heterogeneous media: microstructure and improved bounds on effective properties,". Applied Mechanics Review 44, 37–76.

Tsui, K.-L., 1992. "An Overview of Taguchi Method and Newly Developed Statistical Methods for Robust Design,". IIE Transactions 24 (5), 44–57.

Vining, G.G., Myers, R.H., 1990. Combining Taguchi and Response Surface Philosophies: A Dual Response Approach. J Quality Technology 22, 38–45.

Welch, J.T., Yu, T.K., Kand, S.M., Sachs, J., 1990. "Computer Experiments for Quality Control by Parameter Design,". Journal of Quality Technology 22, 15–22.

Wu, C.F.J., Zhu, Y., 2003. "Optimal Selection of Single Arrays for Parameter Design Experiments,". Statistica Sinica 13, 1179–1199.

Wu, C.F.J.H., 2000. MExperiments: Planning, Analysis and Parameter Design Optimization. John Wiley and Sons, New York.

Wujek, B.A., Renaud, J.E., Batill, S.M., Brockman, J.B., 1996. "Concurrent subspace optimization using design variable sharing in a distributed computing environment,". Concurrent Engineering: Research and Applications 4 (4), 77–361.

Yin, X., Lee, S., Chen, W., Liu, W. K., and Horstemeyer, M. F., 2008, "A Multiscale Design Appraoch with Random Field Representation of Material Uncertainty", in *ASME IDETC/CIE, Advances in Design Automation Conference*, New York, NY, Paper Number: DETC2008-49560.

Yu, J., Ishii, K., 1994. Robust Design by Matching the Design with Manufacturing Variation Patterns. In: ASME Design Automation Conference, Vol. 69-2. ASME, DE, Minneapolis, MN, pp. 7–14.

Zohdi, T.I., 2003. "Constrained Inverse Formulations in Random Material Design,". Computer Methods in Applied Mechanics and Engineering 192 (28-30), 3179–3194.

Integrated Design of Materials and Products—Robust Topology Design of a Cellular Material

Nomenclature	
C	Compliance function
C_{ij}	Elastic constant
D	Total depth of an LCA
d_i^+, d_i^-	Deviation variables for goal i
DSP	Decision Support Problem
E_s	Young's modulus for base material
$\tilde{E}_x/E_s$, E_{11}/E_s	Effective elastic compressive stiffness in the first principal direction
$\tilde{E}_y/E_s$, E_{22}/E_s	Effective elastic compressive stiffness in the second principal direction
G_{12}/E_s	Elastic shear stiffness in the in-plane transverse direction
H	Total height of an LCA
h_i	Height of cell row i
k	Thermal conductivity
LCA	Linear cellular alloy
$\dot{m}$	Mass flow rate
MMA	Method of Moving Asymptotes
N_H	Number of columns of cells
N_V	Number of rows of cells
$\dot{Q}$	Total rate of steady state heat transfer
Re	Reynolds number
RTPDEM	Robust Topological Preliminary Design Exploration Method
S	Stress
S_y	Yield strength
t_H	Thickness of horizontal cell walls
T_{in}	Inlet temperature of cooling fluid
T_s	Temperature of heat source
t_V	Thickness of vertical cell walls

DOI: 10.1016/B978-1-85617-662-0.00007-7

v_f	Volume fraction of solid cell wall material
W	Total width of an LCA
w_i	Width of cell column i
W_i	Weight for goal i
X_i	In-plane thickness of cell wall i
$X_{i,L}, X_{i,U}$	Lower and upper limits for X_i
Z	Objective function
α	Coefficient of thermal expansion
Δi	Range of values of metric i due to dimensional imperfections
ΔP	Pressure drop
ΔX_i	Range of potential values for in-plane thickness of cell wall i
Φ_A	Set of acceptable topologies, as specified by the structural designer
ϕ^T	Set of elements in the final topology
μ_i	Mean value of metric i
σ_i	Standard deviation of metric i due to topological imperfections

Tradeoffs are ubiquitous in concurrent design of materials and products. A designer of touring bicycles, for example, may prefer titanium alloys for their strength and light weight, but steel is stiffer, much less expensive, and more easily repaired. Consider a scenario where a designer of a new supersonic passenger aircraft wants an airframe material that is lightweight with high fatigue strength and toughness, low levels of environmental degradation, and various additional properties, all at temperatures that may exceed 500K. As in many other applications, the optimal material does not exist to meet all of these criteria; instead, the designer seeks materials that balance their satisfaction as closely as possible. Materials design, then, is not an elusive hunt for perfect materials, but a strategic development of, or search for, satisfactory materials that perform well enough with respect to several different criteria to support a specific application. Conceptually, it involves the design of the material (composition, process route, etc.) to achieve these tradeoffs. The modification of microstructure or mesoscopic morphology/arrangement by modification of process route constitutes concurrent materials and product design.

In this chapter, we focus on the design of multifunctional structures comprised of cellular materials. These cellular structures offer illustrative examples of the need to explore tradeoffs in designing materials concurrently with products. They are effectively tailored structures with both mesocopic (cells) and microscopic (grains, phases, defects in cell walls) structure. Two-dimensional or prismatic cellular materials are suitable for multifunctional applications that require not only high specific stiffness and strength but also other performance capabilities such as thermal dissipation or energy absorption capabilities. Some combinations of properties, such as effective elastic moduli and effective thermal conductivity, may have similar trends in their dependence on underlying cellular structure (cell topology and relative density). Other properties, such as convective heat transfer rates and in-plane effective elastic buckling stiffness, place conflicting demands on the cellular structure. When designing prismatic cellular materials for multifunctional applications, we observe tradeoffs between conflicting properties such as elastic buckling strength and total rates of steady state heat transfer. These tradeoffs are embodied as significant differences in the spatial distribution and connectivity of material in the corresponding cellular structures. Furthermore, we often observe tradeoffs between nominal property values and their robustness or insensitivity to likely imperfections in the cellular structure. These imperfections are often introduced by manufacturing processes that also place limitations on realizable cell wall thicknesses, aspect ratios, densities, and other features.

In this chapter, we provide an overview of the tradeoffs in the design of prismatic cellular (honeycomb) materials and the multi-objective decision protocols and parametric and topology design methods we utilize to explore these tradeoffs. The chapter begins in Section 7.1 with an example of multifunctional design of cellular structures, for which we apply our mathematical construct for multi-objective decision support, the compromise Decision Support Problem (cDSP). In Section 7.2, the construct is augmented with robust design and topology design techniques to create a Robust Topological Preliminary Design Exploration Method (RTPDEM), which is useful for designing robust cellular structures and for exploring tradeoffs between matching target material properties and minimizing sensitivity to likely defects in the cellular topology. Finally, in Section 7.3, the RTPDEM is applied to a combustor liner example in which multifunctional cellular structures are designed in a collaborative, multistage process. The overview of this chapter is shown in Figure 7.1.

7.1. Multifunctional Design of Prismatic Cellular Structures

Substantial tradeoffs between multifunctional performance requirements, such as strength and steady state heat transfer rates, are well documented for prismatic 2D cellular materials of the type shown in Figure 7.2. Gu and coauthors (Gu, Lu et al. 2001), and Evans and coauthors (Evans, Hutchinson et al. 2001), provide thermomechanical indices that combine measures of total heat transfer rate, pressure drop, and normalized in-plane effective elastic shear stiffness and facilitate comparison of performance of alternative cell topologies via constrained optimization techniques. They note that hexagonal cells exhibit higher ratios of overall heat transfer coefficient to pressure drop than square or triangular cells, making them suitable for thermal applications. In contrast, the relatively high shear stiffness of triangular cells makes them more suitable for many thermomechanical applications, depending on the level of heat flux and the in-plane dimensions of the material. Even within a single functional domain, tradeoffs are evident. For example, square cell topologies exhibit maximum effective elastic stiffness in the principal in-plane directions, for example, but provide only a fraction of the

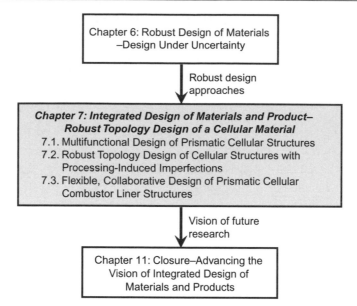

Figure 7.1 Overview of Chapter 7 and its relationship with other chapters in the book.

Figure 7.2 Extruded prismatic 2D cellular materials (LCAs).

transverse effective elastic shear stiffness of triangular cells with equivalent relative densities when subjected to uniform in-plane loading (Hayes, Wang et al. 2004).

In special cases, multiple criteria or performance indices are aligned such that they are maximized or minimized together by a common material structure. For example, Hyun and Torquato (Hyun and Torquato 2000, Hyun and Torquato 2002) found that triangular cell structures have nearly optimal effective shear and bulk elastic moduli and effective thermal conductivity for all relative densities, with Kagome-like cell structures possessing similar properties at low and intermediate relative densities and slightly superior effective shear modulus at intermediate relative densities.

Elasticity, in analogy to steady state heat conduction, is governed by a second-order partial differential equation (Laplacian), and in some cases, such as axial deformation and steady state conductive heat transfer in a 1D bar, the equations assume analogous forms. Therefore,

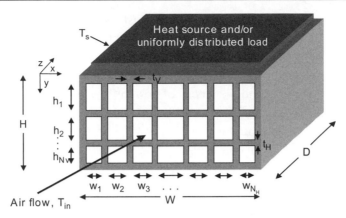

Figure 7.3 An application of prismatic cellular materials as structural heat exchangers (Seepersad, Dempsey et al. 2004).

it is not surprising that elastic stiffness and thermal conductivity can in some cases be optimized by similar cell morphologies (depending on boundary conditions). In contrast, other phenomena place very different demands on cell morphology, especially if those phenomena are scale-dependent and/or nonlocal, such that features at a particular location in a continuous system depend on the state of the continuum at other points at a finite distance. An example is forced convective heat transfer within cellular materials, in which a cooling fluid is forced through the cells, which act as convective passageways to convect heat away from a high heat flux region. Under laminar flow conditions, the convective heat transfer coefficient for a cellular passageway depends on the size and shape of that cell and temperature distributions in the cell walls and cooling fluid that in turn depend on the relative positions and morphologies of neighboring convective passageways.

When designing prismatic cellular materials for multifunctional applications, we observe tradeoffs between performance characteristics such as effective elastic stiffness and total rates of steady state heat transfer. These tradeoffs motivate us to design customized cell morphologies that meet conflicting target performance goals as closely as possible for a specific application, in lieu of the conventional practice of selecting from a library of standardized, periodically repeating cell topologies (e.g., square, hexagonal, triangular) that are optimized based on a single, combined performance metric. Specific geometric constraints of a parent product also may drive the need to customize the structure.

For example, consider the design of prismatic cellular materials (LCAs) for the application illustrated in Figure 7.3. The application requires heat removal along with structural rigidity for cooling high-power electronics. As shown in this figure, the material is expected to withstand distributed structural loads while transferring heat away from a high heat flux source to air that is forced through the cells. Consequently, a designer must seek a compromise between two objectives: (1) maximizing the total rate of steady state heat transfer, and (2) maximizing the overall structural elastic stiffness. If these are regarded as independent targets, the design problem does not fit the mold of a conventional, single-objective, mathematical programming problem. Instead, it must be formulated as a multi-objective problem. Specifically, the design problem is modeled as a cDSP, as shown in Figure 7.4. The challenge is to find the values of design variables that satisfy a set of constraints and bounds and meet a set of conflicting goals as closely as possible. In this case, the design variables include the number of columns of cells, N_H, the number of rows of cells, N_V, the thickness of the horizontal cell walls, t_H, and the thickness of the vertical cell walls, t_V, as illustrated in Figure 7.3, for a heat exchanger with a total height, H, width, W, and depth, D, of 25 cm, 25 cm, and 75 cm, respectively. The constraints include a fan curve that

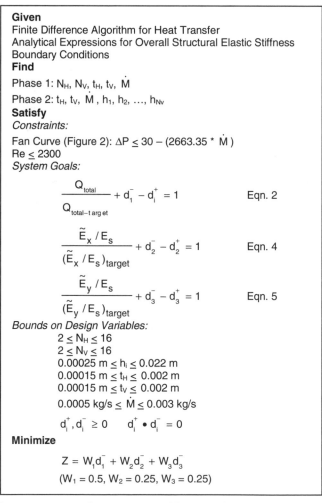

Given
Finite Difference Algorithm for Heat Transfer
Analytical Expressions for Overall Structural Elastic Stiffness
Boundary Conditions
Find
Phase 1: N_H, N_V, t_H, t_V, $\dot{M}$

Phase 2: t_H, t_V, $\dot{M}$, h_1, h_2, ..., h_{N_V}
Satisfy
Constraints:

Fan Curve (Figure 2): $\Delta P \le 30 - (2663.35 * \dot{M})$
Re $\le$ 2300
System Goals:

$$\frac{Q_{total}}{Q_{total-target}} + d_1^- - d_i^+ = 1 \qquad \text{Eqn. 2}$$

$$\frac{\tilde{E}_x / E_s}{(\tilde{E}_x / E_s)_{target}} + d_2^- - d_2^+ = 1 \qquad \text{Eqn. 4}$$

$$\frac{\tilde{E}_y / E_s}{(\tilde{E}_y / E_s)_{target}} + d_3^- - d_3^+ = 1 \qquad \text{Eqn. 5}$$

Bounds on Design Variables:
$2 \le N_H \le 16$
$2 \le N_V \le 16$
$0.00025 \text{ m} \le h_i \le 0.022 \text{ m}$
$0.00015 \text{ m} \le t_H \le 0.002 \text{ m}$
$0.00015 \text{ m} \le t_V \le 0.002 \text{ m}$
$0.0005 \text{ kg/s} \le \dot{M} \le 0.003 \text{ kg/s}$

$d_i^+, d_i^- \ge 0 \qquad d_i^+ \bullet d_i^- = 0$

Minimize

$$Z = W_1 d_1^- + W_2 d_2^- + W_3 d_3^-$$
$$(W_1 = 0.5, W_2 = 0.25, W_3 = 0.25)$$

Figure 7.4 cDSP for a structural heat exchanger (Seepersad, Dempsey et al. 2004).

relates available pressure drop, Δ_P, to the mass flow rate of air, $\dot{m}$, through the prismatic cells, a restriction to ensure laminar flow, and a set of bounds for each design variable. The objective is to minimize the weighted sum, Z, of deviations from targets for effective elastic compressive stiffness, $\tilde{E}_x/E_s$ and $\tilde{E}_y/E_s$ (where E_s is the modulus for the base material), and the total rate of steady state heat transfer, $\dot{Q}$.

As shown in Figure 7.5, the cDSP is solved with a simulation infrastructure that includes an optimization algorithm for searching the design space and a thermal finite difference model and simple analytical estimates for evaluating total rates of steady state heat transfer and effective elastic compressive stiffness, respectively. The cell walls are comprised of copper, and boundary conditions include a heat source temperature of 373K on the top; insulated right, left, and bottom sides; and a working fluid of air entering at room temperature (293K). The flow rate of the working fluid is constrained by the pressure curve of the fan and flow resistance of the channels. Further details of the models, boundary conditions, and underlying assumptions are described in Seepersad, Dempsey et al. (2004).

Sample results for a Pareto family of solutions are illustrated in Figure 7.6 for a trial in which cell wall thicknesses and cell aspect ratios are varied but the number of rows and columns of cells, N_H and N_V, are fixed at 12 and 4, respectively. These values are based on prior explorations of maximum total rate of heat transfer as a function of N_H and N_V for this application. Rectangular cell topology is chosen because it exhibits a higher range of Nusselt numbers than other topologies for steady state laminar flow (Incropera and DeWitt 1996). Results are shown for three different sets of targets for total rate of steady state heat transfer

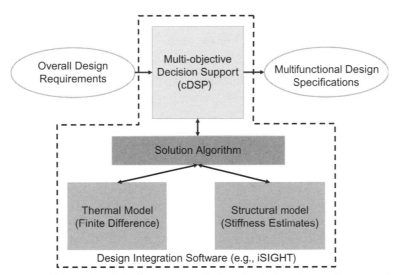

Figure 7.5 Multifunctional design approach for obtaining the design specifications illustrated in Figure 7.4 (Seepersad, Kumar et al. 2005).

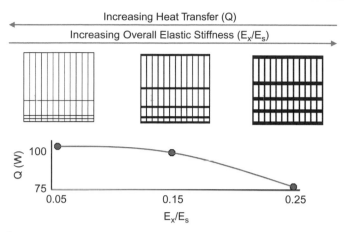

Figure 7.6 A family or pareto set of structural heat exchanger solutions, exhibiting a range of multi-objective tradeoffs.

rate and effective elastic stiffness. For the first design, the total rate of steady state heat transfer is maximized without regard for stiffness values, resulting in a design for a single dominant function. For the second and third designs, targets for effective elastic stiffness are set at 0.15 and 0.25, respectively, of the elastic stiffness of the fully dense solid, resulting in multifunctional designs. As shown in all of the resulting designs, the cells themselves are typically graded when total rates of heat transfer are considered. Specifically, the cells near the heat source (at the top of the structure) tend to elongate to facilitate higher heat transfer rates (Nusselt numbers). The details of this functional grading depend on the location and temperature of the heat source and accompanying thermal gradients throughout the structure, the characteristic fan curve, bulk material properties, and other factors. Visually, the tradeoffs between heat transfer and stiffness capabilities are evident in the cell morphology changes. Thicker cell walls contribute to increased stiffness. However, they also lead to decreased total rates of steady state heat transfer via factors such as reduced surface area for convection and reduced mass flow rate of cooling fluid associated with smaller cell passageways and higher overall pressure drop (resulting in reduced mass flow rate of air according to the fixed fan curve in Figure 7.4). Also, thicker cell walls make the multifunctional designs easier to fabricate.

These sample results illustrate the importance of generating a family of designs that represent a range of tradeoffs between conflicting multifunctional objectives is supported by the sample results included in Figure 7.6. The families embody the scope of the multifunctional design space and offer a range of alternatives for further analysis and design. The tradeoffs are clearly embodied in the morphology of the designs, as well as their performance capabilities. These tradeoffs are a hallmark of materials design. In the next section, we observe similar tradeoffs between expected properties and the sensitivity of those properties to manufacturing-related imperfections.

7.2. Robust Topology Design of Cellular Structures with Processing-Induced Imperfections

The properties and performance of prismatic cellular or honeycomb materials are often sensitive to defects or imperfections associated with the fabrication process. Those imperfections may include tolerances or dimensional variations, curved or wrinkled cell walls, cracked or missing cell walls or joints, and variations in porosity and other properties of the cell wall material. Their occurrence is statistically expected in the normal course of processing and can significantly degrade the multifunctional properties and performance of a cellular material. For example, thermal conductivity, strength, and elastic moduli of the base (cell wall) material have been found to drop exponentially as a function of porosity in the base material (Bocchini 1986). Porosity in the cell walls is a common feature of extruded metal oxide powder slurries, which are reduced in hydrogen to form the base metal. Process-induced variations in cell shape have been found to reduce elastic buckling and plastic yield strength by 25% or more relative to perfectly periodic hexagonal honeycombs, due to higher bending moments and increased stresses in relatively longer cell walls (Gibson and Ashby 1997, Silva, Hayes et al. 1995). The in-plane effective elastic modulus and effective initial compressive yield strength of triangular, hexagonal, and square honeycomb structures are quite sensitive to missing cell walls, decreasing by as much as 50% for only 5% relative density of missing cell walls (Wang and McDowell 2004); the level of sensitivity to these defects has been shown to depend on cell shape (Wang and McDowell 2004).

In most cases, it is not possible to eliminate these defects. An alternative strategy is to design *with* them by seeking robust solutions for properties that are relatively insensitive to stochastic variations in material structure and process paths. Robust solutions are typically obtained by simultaneously minimizing, maximizing, or target-matching the nominal values of performance parameters while minimizing the variation in those parameters due to stochastic variations in the design itself or its environment. Multi-objective methods are useful for obtaining robust solutions by facilitating the search for compromise solutions or tradeoffs between the nominal values and variations in different objectives. The resulting robust solutions are rarely *optimal* from a deterministic perspective because some measure of nominal performance has been sacrificed in favor of robustness (i.e., performance insensitivity) and feasibility over a range of conditions.[1]

When designing prismatic cellular materials with potential imperfections, we observe tradeoffs between non-robust designs (obtained without regard to imperfections) and robust designs. Tradeoffs are also observed between cellular topologies that are designed for robustness against different types of imperfections (Seepersad, Allen et al. 2005). For example, consider

---

[1]   Typically, optimal solutions are constrained by one or more active constraints. These active constraints are likely to be violated when design variable values or other input factors and conditions change.

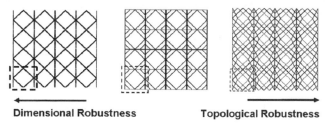

Figure 7.7 Robust prismatic cellular materials designs. periodic unit cells are highlighted with dashed boxes (Seepersad, Allen et al. 2005).

the prismatic cellular structures illustrated in Figure 7.7. During the fabrication process, these materials are subjected to dimensional and topological imperfections, namely in terms of dimensional tolerances and damaged cell walls. All of the materials exhibit similar values of in-plane effective elastic shear and compressive stiffness, but the sensitivity of these properties to imperfections differs significantly from material to material. Robustness to dimensional imperfections is higher for the leftmost design because it contains fewer cell walls and therefore minimizes tolerance stack-up effects. However, it is less robust to topological imperfections than the other designs because there are fewer cell walls to compensate for a randomly missing cell wall or joint. The other designs include redundant cell walls which increase their topological robustness. Of course, if manufacturability were considered, these designs with redundant cell walls may not be feasible. The example is illustrative of the tradeoffs that are often observed between different aspects of robustness and, in general, between nominal performance and robustness.

To design these materials, we utilize the RTPDEM illustrated in Figure 7.8. The RTPDEM facilitates designing materials on mesoscopic scales by topologically and parametrically tailoring them to achieve properties that are superior to those of standard or heuristic designs, customized for large-scale applications, and less sensitive to imperfections in the material. As part of the method, robust topology design problems are formulated as cDSPs; ground structures are utilized for topology design; and local Taylor-series approximations and strategic experimentation techniques are established for evaluating the impact of dimensional and topological imperfections, respectively, on material properties.

The RTPDEM fulfills a need for robust topology design methods for materials design applications. In related topology optimization applications, the sensitivity of optimal topology to changes in prescribed loads has been investigated by considering multiple loads (e.g., (Diaz and Bendsoe 1992, Diaz, Lipton et al. 1995)), average performance under multiple loads (Christiansen, Patriksson et al. 2001), reliability (Bae, Wang et al. 2002, Maute and Frangopol 2003, Thampan and Krishnamoorthy 2001), or worst-case loads among a set of possible loads (Ben-Tal and Nemirovski 1997, Cherkaev and Cherkaeva 1999, Kocvara, Zowe et al. 2000),

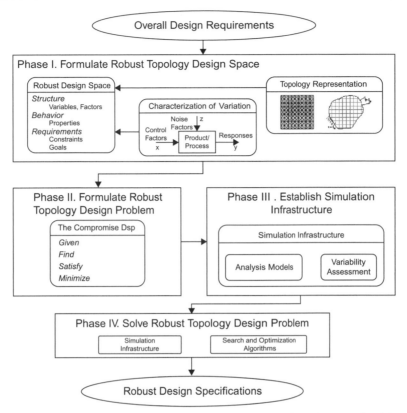

Figure 7.8 The Robust Topological Preliminary Design Exploration Method (RTPDEM) (Seepersad, Allen et al. 2006).

and Sandgren and Cameron (Sandgren and Cameron 2002) have considered the feasibility robustness of *constraints* with variations in loading and material properties. However, these examples are representative of design for mean performance or fail-safe or worst-case design, in which a structure is designed explicitly for worst-case loading, rather than robust design, in which tradeoffs are sought between preferable nominal performance values and minimal sensitivity of performance to uncontrolled variation. Furthermore, partially because topology design was originally focused on full-scale structures rather than materials, variations in the topological structure itself, such as dimensional or topological imperfections, have not been considered. The RTPDEM fills this gap.

The RTPDEM is executed in a series of four phases. In Phase I, we formulate the robust topology design space by representing the topology design space and identifying appropriate control factors (i.e., design variables), noise factors, and responses (i.e., material properties and performance parameters). As shown in Figure 7.9, a ground structure represents the topology design space of a unit cell of the doubly periodic cellular structure. The ground structure consists of a grid of regularly spaced nodes that are connected with

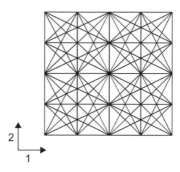

**Figure 7.9 Ground structure for cellular mesostructure design
(Seepersad, Allen et al. 2006).**

frame finite elements, with periodic boundary conditions. The design variables are the in-plane thicknesses of each of the frame finite elements. At the end of the optimization process, elements with extremely small in-plane thicknesses are typically removed in a post-processing step. Details of the specific ground structure formulation are available in Seepersad, Allen et al. (2006), and general information on the ground structure approach is available in Dorn, Gomory et al. (1964), Kirsch (1989), Ohsaki and Swan (2002), and Topping (1984).

The ground structure approach is used to design the topology and dimensions of a unit cell that achieves a set of property or performance targets as closely as possible. In this example, the properties of interest are the effective elastic compressive stiffness in the in-plane principal directions, E_{11}/E_s and E_{22}/E_s, and the effective elastic shear stiffness in the in-plane transverse direction, G_{12}/E_s, where E_s is the Young's modulus of the cell wall material. These properties are related to three of the independent elastic constants in the 2D constitutive law for a plane strain, orthotropic cellular structure with two principal axes aligned with planes of symmetry:

$$\begin{Bmatrix} \sigma_{11} \\ \sigma_{22} \\ \sigma_{12} \end{Bmatrix} = \begin{Bmatrix} \sigma_1 \\ \sigma_2 \\ \sigma_3 \end{Bmatrix} = \begin{bmatrix} C_{11} & C_{12} & 0 \\ C_{12} & C_{22} & 0 \\ 0 & 0 & C_{33} \end{bmatrix} \begin{Bmatrix} \varepsilon_1 \\ \varepsilon_2 \\ \varepsilon_3 \end{Bmatrix} \tag{7.1}$$

$$C_{11} = E_{11}/E_s \tag{7.2}$$

$$C_{22} = E_{22}/E_s \tag{7.3}$$

$$C_{33} = 2G_{12}/E_s \tag{7.4}$$

In addition to the nominal values of the properties, we are interested in the response of each stiffness component to dimensional and topological imperfections—namely, the mean value of each effective elastic stiffness component, μ_{Cij}, the range of values for each elastic stiffness component due to dimensional imperfections, ΔC_{ij}, and the standard deviation of each elastic stiffness component due to topological imperfections, σ_{Cij}. Dimensional imperfections are modeled as a range of potential dimensions, ΔX_e, for the in-plane thickness, X_e, of each cell wall, represented as a frame finite element in the ground structure. The model is based on empirical manufacturing data, as described in Seepersad, Allen et al. (2006). Topological imperfections are modeled as the probability that a joint is missing from the realized (fabricated) mesostructure, even though it is included in the designed mesostructure, according to the detailed description in Seepersad, Allen et al. (2006).

Once the design space is defined, the robust topology design problem is formulated as a cDSP in Phase II of the RTPDEM. The cDSP for the present problem is illustrated in Figure 7.10. As shown in Figure 7.10, the design variables are the in-plane thickness, X_i, of each independent frame finite element, i, in the ground structure. The challenge is to find values of these design variables—and thereby define the topology and dimensions of a unit cell—that satisfy a set of constraints and bounds and achieve a set of conflicting goals as closely as possible. The constraint is the volume fraction, v_f, of solid cell wall material in the unit cell (Eq. 7.5), which is limited to 20% in this example. The bounds include upper and lower limits for X_i of 500 and 0.1 μm (Eq. 7.9), and a set of restrictions that limit the deviation variables, d_i^- and d_i^+, to non-negative values (Eq. 7.10). The deviation variables measure the normalized difference between the target value for each goal and its actual value, as formulated in Equations 7.6 to 7.8. The cumulative deviation from goal target values is measured with an objective function, as illustrated in Equations 7.11 to 7.13. Depending on the desired set of properties, one objective function is selected and minimized during the design process. For non-robust design, only the nominal values of elastic constants are target-matched, as formulated in the objective function in Equation 7.11 and the goal in Equation 7.6. For robust design for dimensional variation, the mean values of elastic constants are target-matched while simultaneously minimizing the variation in elastic constants caused by dimensional imperfections. The corresponding objective function and goals are Equation 7.12 and Equations 7.6 and 7.7, respectively. Finally, for matching target nominal elastic constants and minimizing variation in elastic constants caused by dimensional and topological imperfections, the objective function in Equation 7.13 is utilized, along with the goals in Equations 7.6 through 7.8.

The design problem is solved with a simulation infrastructure that includes an analysis model for evaluating effective elastic constant values, models for evaluating the variation in those values due to imperfections, and an optimization algorithm for searching the design space. The analysis model for this example is a finite element-based homogenization model that evaluates the continuum level properties of the cellular structure as a function of its cellular mesostructure. The approach follows that utilized by Sigmund (Sigmund 1994, Sigmund 1995) and Neves and coauthors (Neves, Rodrigues et al. 2000); details are provided

Given
  Robust topology design space
  Simulation infrastructure
  Targets, bounds, weights
Find
  X_i In-plane element thickness, $i = 1, ..., N$
                 $N = \#\,elements$
  d_i^-, d_i^+ Deviation Variables, $i = 1, ..., 3P$
                 $P = \#\,tailored\,elastic\,constants$
Satisfy
Constraint

$$v_f \leq v_{f\text{-}limit} \qquad\qquad\qquad \text{Eq. (7.5)}$$

Goals
  Mean value of elastic constant

$$\frac{\mu_{Cij}}{\mu_{Cij-target}} + d_k^- - d_k^+ = \frac{\mu_{Cij-target}}{\mu_{Cij-target}} = 1\,;\ k = 1, ..., P \qquad \text{Eq. (7.6)}$$

  Range of elastic constant (due to dimensional variation)

$$\frac{\Delta C_{ij}}{\mu_{Cij\text{-target}}} + d_k^- - d_k^+ = \frac{\Delta C_{ij\text{-target}}}{\mu_{Cij\text{-target}}}\,;\ k = (P+1), ..., 2P \qquad \text{Eq. (7.7)}$$

  Standard deviation of elastic constant (due to topological imperfections)

$$\frac{\sigma_{Cij}}{\mu_{Cij\text{-target}}} + d_k^- - d_k^+ = \frac{\sigma_{Cij\text{-target}}}{\mu_{Cij\text{-target}}}\,;\ k = (2P+1), ..., 3P \qquad \text{Eq. (7.8)}$$

Bounds

$$X_{i,L} \leq X_i \leq X_{i,U} \qquad\qquad i = 1, ..., N \qquad \text{Eq. (7.9)}$$

$$d_i^- \bullet d_i^+ = 0\,;\ d_i^-, d_i^+ \geq 0 \qquad\qquad i = 1, ..., 3P \qquad \text{Eq. (7.10)}$$

Minimize
  Nonrobust Design:

$$Z = \sum_{k=1}^{P} W_k \left(d_k^- + d_k^+ \right),\ \sum_{k=1}^{P} W_k = 1 \qquad \text{Eq. (7.11)}$$

$$P = \#\,tailored\,elastic\,constants$$

  Robust Design for Dimensional Variation

$$Z = \sum_{k=1}^{2P} W_k \left(d_k^- + d_k^+ \right),\ \sum_{k=1}^{2P} W_k = 1 \qquad \text{Eq. (7.12)}$$

  Robust Design for Dimensional and Topological Variation

$$Z = \sum_{k=1}^{3P} W_k \left(d_k^- + d_k^+ \right),\ \sum_{k=1}^{3P} W_k = 1 \qquad \text{Eq. (7.13)}$$

**Figure 7.10 cDSP for robust topology design
(Seepersad, Allen et al. 2006).**

in Seepersad, Allen et al. (2006). The variation models include a Taylor series expansion for estimating the variation in elastic constant values, ΔC_{ij}, induced by dimensional imperfections. A series of experiments is utilized to simulate topological imperfections, by successively removing each node in the ground structure. The resulting data are used to

Table 7.1 Robust vs. nonrobust periodic cellular mesostructure for effective stiffness in both principal directions, considering topological and dimensional variation (Seepersad, Allen et al. 2006).

Robust Design for Dimensional Variation	Robust Design for Dimensional and Topological Variation	Nonrobust Design and Robust Design for Topological Variation
Material	Material	Material
Unit Cell	Unit Cell	Unit Cell
0.05 cm cell walls; 1 cm; 1 cm	0.03 cm 0.02 cm; 0.03 cm; 0.02 cm; 1 cm; 1 cm	0.02 cm cell walls; 1 cm; 1 cm
Design Performance	Design Performance	Design Performance
$\mu_{C11} = 0.10$ $\Delta C_{22} = 0.015$ $\mu_{C22} = 0.10$ $\sigma_{C11} = 0.0071$ $\Delta C_{11} = 0.015$ $\sigma_{C22} = 0.0071$	$\mu_{C11} = 0.10$ $\Delta C_{22} = 0.02$ $\mu_{C22} = 0.10$ $\sigma_{C11} = 0.0068$ $\Delta C_{11} = 0.02$ $\sigma_{C22} = 0.0068$	$\mu_{C11} = 0.10$ $\Delta C_{22} = 0.021$ $\mu_{C22} = 0.10$ $\sigma C_{11} = 0.0053$ $\Delta C_{11} = 0.021$ $\sigma_{C22} = 0.0053$

estimate the standard deviation of elastic constant values due to topological imperfections, σ_{Cij}, as detailed in Seepersad, Allen et al. (2006). For searching the design space, the analysis and variation models are paired with an optimization algorithm, namely, the Method of Moving Asymptotes (MMA) algorithm (Svanberg 1987) for this example.

In Phase IV of the RTPDEM, the design problem is solved to obtain sets of cellular structures with targeted properties. By varying targets in Equations 7.6 to 7.8, selecting a specific objective function (Equations 7.11 to 7.13), and varying the weights in the objective function, a variety of solutions can be obtained, as discussed in Seepersad (2004) and Seepersad, Allen et al. (2006).

Two representative sets of solutions are presented in Table 7.1 and Table 7.2. The first set of solutions (Table 7.1) is designed for targeted effective elastic compressive stiffness of 0.1 in each of the principal directions, C_{11} and C_{22}. The second set of solutions (Table 7.2) is designed for targeted effective elastic compressive stiffness of 0.035 and 0.09 in the first and second principal directions and effective elastic shear stiffness of 0.045. The nonrobust

Table 7.2 Robust vs. nonrobust periodic cellular mesostructure for effective stiffness in principal directions and shear, considering topological and dimensional variation (Seepersad, Allen et al. 2006).

Robust Design for Dimensional Variation	Robust Design for Dimensional and Topological Variation	Nonrobust Design
Material	**Material**	**Material**

Unit Cell

0.045 cm, cell wall

0.032 cm cell wall

1 cm

1 cm

Unit Cell

0.01 cm   0.025 cm

0.034 cm

1 cm

1 cm

Unit Cell

0.013 cm   0.025 cm

0.015 cm

0.025 cm

1 cm

1 cm

Design Performance	Design Performance	Design Performance
$\mu_{C11} = 0.032$ $\Delta C_{33} = 0.0072$	$\mu_{C11} = 0.025$ $\Delta C_{33} = 0.0064$	$\mu C_{11} = 0.029$ $\Delta C_{33} = 0.011$
$\mu_{C22} = 0.096$ $\sigma_{C11} = 0.0017$	$\mu_{C22} = 0.093$ $\sigma C_{11} = 0.0009$	$\mu_{C22} = 0.080$ $\sigma_{C11} = 0.0008$
$\mu_{C33} = 0.043$ $\sigma_{C22} = 0.0049$	$\mu_{C33} = 0.029$ $\sigma_{C22} = 0.0048$	$\mu_{C33} = 0.036$ $\sigma_{C22} = 0.0021$
$\Delta C_{11} = 0.0052$ $\sigma_{C33} = 0.0020$	$\Delta C_{11} = 0.0054$ $\sigma C_{33} = 0.0011$	$\Delta C_{11} = 0.0084$ $\sigma_{C33} = 0.0008$
$\Delta C_{22} = 0.017$	$\Delta C_{22} = 0.017$	$\Delta C_{22} = 0.022$

designs are obtained without regard for robustness, using the objective function in Equation 7.11. The robust designs for dimensional variation are obtained with the objective function in Equation 7.12 and therefore embody a balance between meeting target values for nominal properties and minimizing their sensitivity to dimensional imperfections. The robust designs for dimensional and topological variation are obtained with the objective function in Equation 7.13 to minimize not only the deviation of nominal property values from their targets but also the sensitivity of those properties to dimensional *and topological* imperfections.

These tradeoffs manifest themselves in cell morphologies and properties. As shown in Table 7.1 and Table 7.2, the nonrobust designs tend to exhibit a very dense network of cell walls. The dimensionally robust designs, on the other hand, exhibit a sparse network of thicker cell walls. The complexity of the topologically *and* dimensionally robust designs is somewhere in between those of their neighbors. This trend is to be expected. By minimizing the number of cell walls, the dimensionally robust designs reduce tolerance stack-up.

By increasing the number of cell walls, the topologically robust designs guard against degradation due to topological imperfections; the dense network of cell walls is better equipped to compensate for a missing or damaged neighbor. When both types of robustness are emphasized, we obtain compromise cell morphologies of moderate complexity.

These trends are also reflected in the quantitative properties of the cellular structures. In Table 7.1, for example, all of the designs have identical nominal properties. As dimensional robustness increases from right to left, the sensitivity of the elastic properties to dimensional variation, ΔC_{ij}, decreases. Conversely, as the topological robustness increases from left to right, the sensitivity to topological variation, σ_{Cij}, decreases, at the expense of dimensional robustness. Similar observations apply to the designs in Table 7.2.

The method and examples reported here have significant materials design implications. Imperfections are known to impact the performance of most materials significantly, but those imperfections often are not considered comprehensively in the design process. The RTPDEM has proven useful not only for building robustness into cellular structures but also for tailoring the level of complexity or simplicity of cellular structures. Analogous methods are likely to be useful for designing materials on a variety of length scales and for shortening the time to market and raising industrial confidence in the performance of designed materials. In the following section, we also explore how robust materials design methods can be used to support collaboration in the design of multifunctional materials.

7.3. Flexible, Collaborative Design of Prismatic Cellular Combustor Liner Structures

In materials design, uncertainty may stem not only from stochastic factors such as process-induced imperfections and variable usage environments, but also an inability to anticipate or predict the precise nature of any downstream changes to be made to a design. Those downstream changes may be needed for accommodating the requirements of other functional domains or spatiotemporal scales (including the design of an overall part or system). There is rarely a "master designer" who makes decisions regarding all aspects of integrating materials and systems, or even in developing materials. Decision making is a distributed process. The practical scenario involves a team of experts with differing tools and perspectives. The idea is to deliver relevant expertise at each appropriate stage of the process. In the case of prismatic cellular materials, those changes may include adjustments in the topology and dimensions of the cell walls, the properties of the base cell wall material, and the boundary conditions applied to the system, including magnitude and direction of applied mechanical loading, magnitude of applied heat flux or temperatures, and other factors. These adjustments may be suggested and undertaken by different individuals or groups in the collaborative process as part of their contribution to meeting functional requirements in iterating toward systems solutions.

We have utilized robust design techniques to build flexibility into cellular materials designs so that they can be modified subsequently for additional requirements (Seepersad 2004, Seepersad, Allen et al. 2007). The first step is to bound the set or range of potential downstream changes. Then, the changes are treated as stochastic variations during the design process. Accordingly, robust design techniques are applied, in a procedure analogous to that described in Section 7.2, to generate flexible design specifications that balance the competing objectives of (1) meeting performance targets and (2) minimizing performance variation caused by the set or range of potential downstream design changes. In other words, the robust topology design method introduced in Section 7.2 is used to design structural topologies that are robust to *intentional* downstream topological and dimensional adjustments for enhanced performance in a distinct, secondary functional domain.

7.3.1. Combustor Liner Design Example

As an example, consider the design of cellular materials as actively cooled liners for the combustion chamber of a next-generation gas turbine engine (Seepersad 2004, Seepersad, Allen et al. 2007), as illustrated in Figure 7.11. The cellular materials are required to support the internal pressure of the combustion reaction while simultaneously cooling themselves, via internal forced convection, to avoid melting and to minimize degradation of strength with increasing temperature in the base cell wall material. The actively cooled cellular materials are intended to replace standard materials, such as nickel-base superalloys, that require a barrier coating and cooling on combustion-side surfaces to prevent melting or environmental degradation, leading to increased emissions and decreased efficiency (Bailey, Intile et al. 2002, Dimiduk and Perepezko 2003). The basic design requirements are summarized in Table 7.3. The challenge is to devise a multifunctional cellular topology that balances structural properties (compliance and strength at elevated temperatures) and thermal properties (heat transfer rates and cell wall temperatures), subject to a set of constraints on cell wall thicknesses and

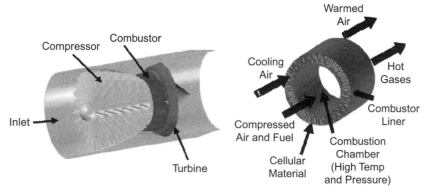

Figure 7.11 Schematic of a cellular combustor liner (Seepersad 2004, Seepersad, Allen et al. 2005, Seepersad, Allen et al. 2007).

volume fraction. The constraints are designed to accommodate the thermochemical extrusion process for fabricating the materials (Seepersad, Allen et al. 2006). The base material is a Mo-Si-B alloy, chosen for its high temperature properties and its amenability to thermochemical extrusion (Dimiduk and Perepezko 2003, Schneibel, Kramer et al. 2001).

7.3.2. Combustor Liner Design Methodology

Topology optimization techniques would be useful for determining the multifunctional topology of the material. However, topology optimization applications have been limited to customizing the thermoelastic properties of cellular materials, such as effective elastic stiffness (Sigmund 1994, Sigmund 1995), thermal expansion coefficients (Sigmund and Torquato 1997), and thermal conductivity (Hyun and Torquato 2002). In general, structural topology optimization techniques have been limited to steady state heat conduction (Hyun and Torquato 2002, Li, Steven et al. 2000, Li, Steven et al. 2004) or to coupled field problems in structural analysis that examine the interactions of deformation with temperature, electric fields, and/or magnetic fields for applications such as piezoelectric or electrothermomagnetic actuators (Sigmund 2001, Sigmund, Torquato et al. 1998) or thermoelastic materials (Sigmund and Torquato 1997). These phenomena are governed by mathematical relationships with analogous forms. In contrast, existing structural topology optimization techniques are not readily amenable to design for internal forced convection, due to the shape- and scale-dependence of the phenomenon.

Table 7.3 Design requirements and assumptions for the combustor liner application (Seepersad, Allen et al. 2007).

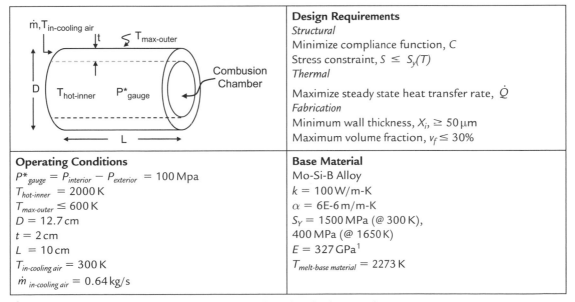

$\dot{m}, T_{in\text{-}cooling\ air}$ $\leq T_{max\text{-}outer}$, t D, $T_{hot\text{-}inner}$, P^*_{gauge} Combusion Chamber L	**Design Requirements** *Structural* Minimize compliance function, C Stress constraint, $S \leq S_y(T)$ *Thermal* Maximize steady state heat transfer rate, $\dot{Q}$ *Fabrication* Minimum wall thickness, $X_i, \geq 50\,\mu m$ Maximum volume fraction, $v_f \leq 30\%$
Operating Conditions $P^*_{gauge} = P_{interior} - P_{exterior} = 100\,\text{Mpa}$ $T_{hot\text{-}inner} = 2000\,\text{K}$ $T_{max\text{-}outer} \leq 600\,\text{K}$ $D = 12.7\,\text{cm}$ $t = 2\,\text{cm}$ $L = 10\,\text{cm}$ $T_{in\text{-}cooling\ air} = 300\,\text{K}$ $\dot{m}_{in\text{-}cooling\ air} = 0.64\,\text{kg/s}$	**Base Material** Mo-Si-B Alloy $k = 100\,\text{W/m-K}$ $\alpha = 6\text{E-}6\,\text{m/m-K}$ $S_Y = 1500\,\text{MPa}$ (@ 300 K), 400 MPa (@ 1650 K) $E = 327\,\text{GPa}$[1] $T_{melt\text{-}base\ material} = 2273\,\text{K}$

[1]The Young's Modulus is assumed to be temperature-independent for this example.

Instead of designing simultaneously for structural and thermal performance in a single step—an alternative that would force us to preselect a cell topology—a practical, two-stage robust design process is considered for facilitating topological and parametric design for multiple functional domains. As shown in Figure 7.12 and Figure 7.13, the design process is partitioned into an initial structural design stage followed by a thermal design stage.

In the first stage, the material is designed for structural performance and robustness. Specifically, it is designed to balance two competing sets of objectives: (1) minimizing compliance, and (2) minimizing sensitivity to topological or dimensional adjustments in the form of cell wall or joint removal or thickening/thinning of cell walls, respectively. The procedure is identical to that described in Section 7.2. The only difference is the initial ground structure and boundary conditions for topology design. As illustrated in Figure 7.14, the combustor liner is assumed to be axisymmetric and periodic in a basic unit wedge,

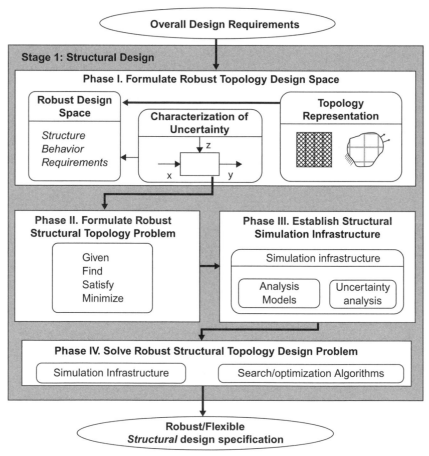

Figure 7.12(a) Outline of the multifunctional topology design method—Stage 1 (Seepersad, Allen et al. 2007).

corresponding to 1/32 of the combustor liner. The ground structure is constructed for a single-unit wedge with symmetric boundary conditions applied to either side, along with a uniform pressure of 100 MPa on the interior surface, and spring supports with stiffness 30×10^8 N/m along the outer surface to simulate the surrounding structure. Results are validated with a denser ground structure, as described in Seepersad (2004). As described in Section 7.3.3, the result of the first stage is a robust structural topology with structural properties that are relatively insensitive to a specified set of acceptable topological and dimensional adjustments.

In the thermal design stage (Stage 2 in Figure 7.12 and Figure 7.13), the robust structural design is adjusted to improve its thermal properties. Thermal adjustments are limited to the set of acceptable topological and dimensional adjustments specified by the structural designer. The thermal designer formulates a cDSP to identify the specific topology and dimensions that maximize the total rate of steady state heat transfer, $\dot{Q}$, and minimize the temperature of

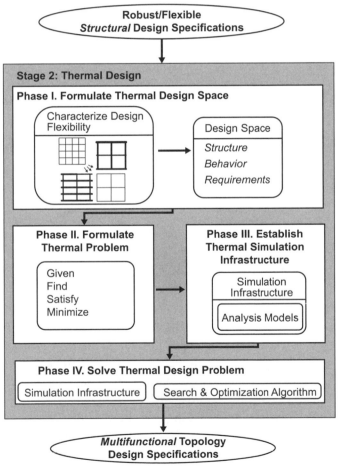

Figure 7.12(b) Outline of the multifunctional topology design method—Stage 2 (Seepersad, Allen et al. 2007).

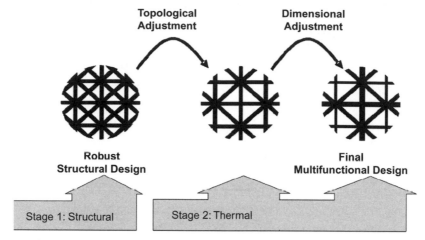

Figure 7.13 A Two-stage multifunctional topology design process (Seepersad, Allen et al. 2007).

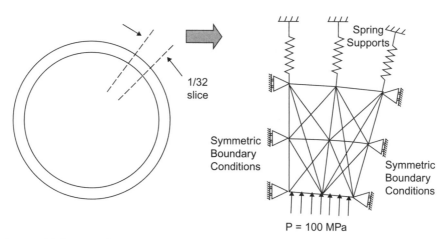

Figure 7.14 Initial ground structure and boundary conditions for structural topology design (Seepersad, Allen et al. 2007).

the solid cell wall material. The cDSP is illustrated in Figure 7.15, where $X_{i,L}$, $X_{i,U}$, and Φ_A represent the lower and upper acceptable limits for the thickness of each cell wall and the set of acceptable topologies, respectively, as specified by the structural designer. To solve the cDSP, heat transfer analysis is performed with a hybrid finite element/finite difference approach for approximating the temperature distribution in the cellular mesostructure and the total rate of steady state heat transfer from the cellular heat sink to the fluid flowing through its passageways. This model is faster than a detailed FLUENT model and is more easily reconfigurable, and it shares a common finite element representation with the structural model. These characteristics facilitate rapid design exploration. Details of the model are provided in Seepersad (2004) and Seepersad, Allen et al. (2007).

Given

Thermal design space
Simulation infrastructure
Targets, bounds, weights

Find

X_i^T In-plane element thickness, $i = 1, ..., N$

$N = \#$ *elements in final topology,* χ^T

ϕ^T Set of elements in final topology

d_k^-, d_k^+ Deviation variables $k = 1, ..., 3$

Satisfy

Constraint

$$V_f \leq V_{f\text{-limit}} \tag{7.14}$$

$$\phi^T \in \Phi^A \tag{7.15}$$

Goals

$$\frac{\dot{Q}}{\dot{Q}_{target}} + d_1^- - d_1^+ = \frac{\dot{Q}_{target}}{\dot{Q}_{target}} = 1 \tag{7.16}$$

Bounds

$$X_{i,L} \leq X_i \leq X_{i,U} \qquad i = 1, ..., N \tag{7.17}$$

$$d_k^+ \cdot d_k^+ = 0 \; ; \; d_k^-, d_k^+ \geq 0 \qquad k = 1, ..., 3; \tag{7.18}$$

Minimize

$$Z = W_1 d_1^- , \; W_1 = 1 \tag{7.19}$$

Figure 7.15 cDSP for robust structural topology design in the combustor liner example (Seepersad, Allen et al. 2007).

7.3.3. Combustor Liner Design Results

Structural results are documented in Figure 7.16 and Table 7.4. The diagrams depict a single wedge, along with a visualization of the entire combustor liner. The properties are provided for a single wedge, and robust and nonrobust designs are documented. The nonrobust design

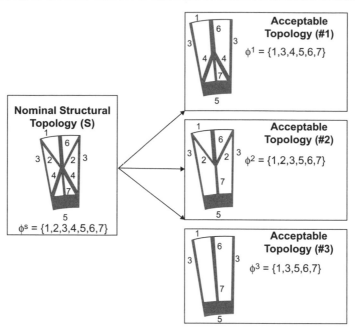

Figure 7.16 Set of acceptable structural topologies, Φ^A, derived from the robust structural topology (Seepersad, Allen et al. 2007).

is tailored only for minimum compliance. The robust design is tailored for a balance of minimum compliance and minimum sensitivity to dimensional and topological adjustments.

As shown in Table 7.4, the robust and nonrobust solutions have very different topologies. The nonrobust topology is an expected outcome for this problem. Element 5 supports the internal pressure of the combustion reaction, and elements 3 and 6 reduce the hoop stress in element 5 by transferring some of the load to element 1 and the outer spring supports. The robust topology includes additional elements 2 and 4. These elements serve as redundant pathways for transmitting loads from the inner hoop (element 5) to the outer supports, thereby increasing the topological robustness of the structure. These elements can be removed or added by the subsequent thermal designer, as needed, to improve thermal properties. The improved topological robustness of the structure is reflected in the standard deviation of compliance, σ_c, due to topological adjustments, which is significantly lower for the robust design.

The outcome of the structural design process includes not only the nominal structural topology, as illustrated in the left side of Table 7.4, but also the set of acceptable topological and dimensional adjustments for the thermal designer. Acceptable topological adjustments include removal of any combination of elements 2 and 4 from the robust design, as illustrated in Figure 7.16. Acceptable dimensional adjustments are recorded in Table 7.4, along with the dimensions of each cell wall. These adjustments are acceptable because they do not raise the

Table 7.4 Robust and nonrobust structural design results (Seepersad, Allen et al. 2007).

Robust Design				Nonrobust Design		
Cross-Section of a 1/32 Wedge	Complete Combustor Liner			Cross-Section of a 1/32 Wedge	Complete Combustor Liner	
Element	Nominal Robust Topology, ϕ^s	Dimension and Range (mm), X^s +/− ΔX^A	Stress (MPa)	Nonrobust Topology	Dimension	Stress (MPa)
1	√	0.1 +/− 0.025	430	√	0.1	430
2	√	0.5 +/− 0.07	360			
3	√	0.5 +/− 0.07	352	√	0.6	536
4	√	1.0 +/− 0.1	379			
5	√	3.7 +/− 0.37	442	√	4.0	430
6	√	1.2 +/− 0.1	525	√	1.3	433
7	√	0.6 +/− 0.08	391	√	1.3	433
Properties of Converted Topology	μ_c			116.0	112.0	
	$\Delta_\mu C$			51.0	46.7	
	σ_c			24.6	37.6	
Properties of Post-Processed Topology	Compliance, C (J)			416	337	
	Max Stress, S, (MPa)			525	536	

maximum stress in any cell wall beyond its yield strength or significantly increase the total compliance of the structure.

Thermal design results are documented in Table 7.5 and Table 7.6 for the robust and nonrobust structural designs, respectively. The robust design results reported in Table 7.5 are obtained by searching the space of acceptable topological and dimensional adjustments to the robust structural design. The specific topology and dimensions that exhibit the highest rate of steady state heat transfer and lowest cell wall temperatures are identified and recorded in the table. The nonrobust design serves as a benchmark design because its properties are representative of the thermal properties obtainable *without* dimensional or topological adjustment during the thermal design phase. Its thermal properties are recorded in Table 7.6 for comparison.

From a thermal perspective, the differences between the two designs are significant. Specifically, the ability to add cell walls and adjust cell wall thicknesses in the robust design

Table 7.5 Thermal design results for robust design from Table 7.4. Thermal design modifications are limited to acceptable topological and dimensional adjustments (Seepersad, Allen et al. 2007).

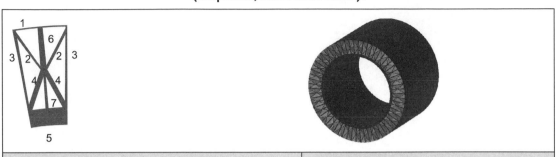

Nominal Design				Design Modified within Ranges			
Element	Dim. and Range (mm), X^s +/− ΔX^A	Avg. Temp. @ Outlet(K), T	Stress (MPa) S	Element	Final Dim. (mm), X	Avg. Temp. @ Outlet(K), T	Stress (MPa), S
1	0.1 + / − 0.025	634	427	1	0.09	656	411
2	0.5 + / − 0.07	724	356	2	0.57	750	316
3	0.5 + / − 0.07	949	355	3	0.57	962	314
4	1.0 + / − 0.1	1173	385	4	1.1	1183	356
5	3.7 + / − 0.37	1541	433	5	4.0	1527	427
6	1.2 + / − 0.1	769	543	6	1.3	795	502
7	0.6 + / − 0.08	1207	401	7	0.68	1220	362
$\dot{Q}$, Total Rate of Steady State Heat Transfer (W)			4027	$\dot{Q}$, Total Rate of Steady State Heat Transfer (W)			4095
C, Compliance			418	C, Compliance			469

increases the total rate of steady state heat transfer by over 50% relative to the nonrobust design, while simultaneously lowering the temperatures in all of the cell walls. The impact of these adjustments on structural properties is relatively small. Results are validated with detailed thermomechanical ANSYS analyses, which indicate that the proposed cellular material design should meet the design requirements. Details are available in Seepersad, Allen et al. (2007).

In summary, a practical, two-stage approach for multifunctional topology design is presented (see Figure 7.12) and successfully applied to a combustor liner example. A robust structural topology is designed to be modified topologically and dimensionally to satisfy thermal constraints and to improve thermal performance objectives without significantly degrading structural performance. The concept is applicable to many materials and product design contexts in which design and analysis activities are distributed among multiple teams or experts whose design problems are coupled (i.e., share one or more common parameters). By preserving flexibility in the values of coupled parameters, it is easier for other designers to make adjustments in order to improve system-level objectives.

Table 7.6 Thermal properties of the nonrobust benchmark design from Table 7.4 (Seepersad, Allen et al. 2007).

Nonrobust Design			
Element	Dimension, X	Avg. Temp. @ Outlet(K), T	Stress (MPa), S
1	0.1	781	430
3	0.6	1080	536
5	4.0	1642	430
6&7	1.3	1103	433
$\dot{Q}$, *Total Rate of Steady State Heat Transfer (W)*			2499
C, *Compliance*			342

It is noted that this two-stage strategy for achieving goals of target functionality in multiple domains (mechanical and thermal in this case) will not always work. For example, if there are no satisficing solutions for meeting thermal performance in the range of topologies and designs offers by the first-stage mechanical design, then the two must be coupled as a compromise. Obviously, this would be a parallel strategy, requiring more communication and coupling of the design tools and exploration of feasible design space. It should not be assumed at the outset, however, that such coupling is essential. It depends on the value-of-information that is lost in the process by decoupling the stages. Value-of-information metrics for this purpose are introduced and discussed in detail in Chapter 9.

7.4. Closing Remarks

In this chapter, we focus on illustrating the principles of concurrent design of materials and products; designing materials with functionality established by structure at *mesoscopic* length scales that are larger than microscopic features but smaller than the macroscopic characteristics of an overall part or system. The mesoscopic *topology*—or geometric arrangement of solid phases and voids within a material or product—is increasingly customizable with rapid prototyping and other manufacturing and materials processing techniques that facilitate tailoring topology with high levels of detail. Fully leveraging these capabilities requires not only computational simulation models but also a *systematic, efficient design method* for exploring, refining, and evaluating product and material topology and other design parameters in order to achieve *multifunctional* performance goals and requirements. The performance requirements for materials are typically derived from larger engineering systems in which they are embedded and often require tradeoffs among multiple criteria associated with disparate physical domains such as materials processing, heat transfer and structural mechanics. The structures and processing paths of these *multifunctional* materials must be designed to simultaneously balance these multi-physics requirements as much as possible. However, the link between preliminary design specifications and realized multifunctional performance is not deterministic. Deviation from nominal or intended performance can be caused by many sources of variability, including manufacturing processes, potential operating environments, simulation models, and adjustments in design specifications themselves during a multi-stage product development process. Topology and other preliminary specifications for materials and products should be designed to deliver performance that is *robust,* or relatively insensitive to this variability.

The RTPDEM is presented for designing complex multiscale materials and products concurrently by topologically and parametrically tailoring them for multifunctional performance that is superior to that of standard designs and less sensitive to variations. This systems-based design approach is formulated by establishing and integrating principles and techniques for robust design, multi-objective decision support, topology design, collaborative design, and design space exploration along with approximate and detailed simulation models. Robust topology design methods are used here not only to design material topologies that are relatively insensitive to manufacturing-related imperfections but also to systematically and intentionally create topological designs with built-in flexibility for subsequent modification. This flexibility is the foundation for the multi-stage, multifunctional robust topology design method introduced in Section 7.3.

Key aspects of the approach are demonstrated by designing linear cellular alloys—metallic 2D cellular materials with extended prismatic cells—for multifunctional applications. For a microprocessor application, structural heat exchangers are designed that increase rates of heat dissipation and structural load-bearing capabilities substantially relative to conventional heat sinks that occupy equivalent volumetric regions. One key characteristic of these designs is graded, rather than uniform, cell structures. Also, cellular materials are designed with structural properties that are robust to dimensional changes and topological imperfections such as missing cell walls. Although structural imperfections—or deviations from intended structural characteristics—are observed regularly in cellular materials and in other classes of materials, they have not been considered previously during the design process. Finally, cellular combustor liners are designed to increase operating temperatures and efficiencies as a case study for next-generation gas turbine engines via active cooling and load bearing within topologically and parametrically customized cellular materials.

References

Bae, K.-R., Wang, S., Choi, K. K., 2002. Reliability-based topology optimization. In: 9th AIAA/ISSMO symposium on multidisciplinary analysis and optimization. Atlanta, GA. Paper No: AIAA-2002-5542.

Bailey, J.C., Intile, J., Fric, T.F., Tolpadi, A.K., Nirmalan, N.V., Bunker, R.S., 2002. Experimental and numerical study of heat transfer in a gas turbine combustor liner Paper Number: GT-2002-30183. In: International gas turbine conference and exposition. ASME, Amsterdam. Paper Number: GT-2002-30183.

Ben-Tal, A., Nemirovski, A., 1997. Robust truss topology design via semidefinite programming. SIAM J. Optim. 7 (4), 991–1016.

Bocchini, G.F., 1986. The influence of porosity on the characteristics of sintered materials. Int. J. Powder Metall. 22 (3), 185–202.

Cherkaev, A., Cherkaeva, E., 1999. Optimal design for uncertain loading conditions. In: Berdichevsky, V., Jikov, V., Papanicolaou, G. (Eds.) Homogenization. World Scientific, Singapore, pp. 193–213.

Christiansen, S., Patriksson, M., Wynter, L., 2001. Stochastic bilevel programming in structural optimization. Struct. Multidiscip. Optim. 21, 361–371.

Diaz, A., Bendsoe, M.P., 1992. Shape optimization of structures for multiple loading situations using a homogenization method. Struct. Optimization 4 (1), 17–22.

Diaz, A., Lipton, R., Soto, C.A., 1995. A new formulation of the problem of optimum reinforcement of Reissner-Mindlin plates. Comput. Method. Appl. M. 123 (1–4), 121–139.

Dimiduk, D.M., Perepezko, J.H., 2003. Mo-Si-B alloys: Developing a revolutionary turbine-engine material. MRS Bull. September, 639–645.

Dorn, W.S., Gomory, R.E., Greenberg, H.J., 1964. Automatic design of optimal structures. J. Mec. 3, 25–52.

Evans, A.G., Hutchinson, J.W., Fleck, N.A., Ashby, M.F., Wadley, H.N.G., 2001. The topological design of multifunctional cellular materials. Prog. Mater. Sci. 46 (3–4), 309–327.

Gibson, L.J., Ashby, M.F., 1997. Cellular Solids: Structure and Properties. Cambridge University Press, Cambridge, UK.

Gu, S., Lu, T.J., Evans, A.G., 2001. On the design of two-dimensional cellular metals for combined heat dissipation and structural load capacity. Int. J. Heat Mass Tran. 44 (11), 2163–2175.

Hayes, A.M., Wang, A., Dempsey, B.M., McDowell, D.L., 2004. Mechanics of linear cellular alloys. Mech. Mater. 36 (8), 691–713.

Hyun, S., Torquato, S., 2000. Effective elastic and transport properties of regular honeycombs for all densities. J. Mater. Res. 15 (9), 1985–1993.

Hyun, S., Torquato, S., 2002. Optimal and manufacturable two-dimensional, Kagome-like cellular solids. J. Mater. Res. 17 (1), 137–144.

Incropera, F.P., DeWitt, D.P., 1996. Fundamentals of Heat and Mass Transfer, third ed.. Wiley, New York.

Kirsch, U., 1989. Optimal topologies of structures. Appl. Mech. Rev. 42 (8), 223–239.

Kocvara, M., Zowe, J., Nemirovski, A., 2000. Cascading: An approach to robust material optimization. Comput. Struct. 76 (1–3), 431–442.

Li, Q., Steven, G.P., Querin, O.M., 2000. Structural topology design with multiple thermal criteria. Eng. Computation. 17 (6), 715–734.

Li, Q., Steven, G.P., Xie, Y.M., Querin, O.M., 2004. Evolutionary topology optimization for temperature reduction of heat conducting fields. Int. J. Heat Mass Tran. 47 (23), 5071–5083.

Maute, K., Frangopol, D.M., 2003. Reliability-based design of MEMS mechanisms by topology optimization. Comput. Struct. 81 (8–11), 813–824.

Neves, M.M., Rodrigues, H., Guedes, J.M., 2000. Optimal design of periodic linear elastic microstructures. Comput. Struct. 76 (1–3), 421–429.

Ohsaki, M., Swan, C.C., 2002. Topology and geometry optimization of trusses and frames. In: Burns, S.A. (Ed.), Recent Advances in Optimal Structural Design. American Society of Civil Engineers, Reston, VA.

Sandgren, E., Cameron, T.M, 2002. Robust design optimization of structures through consideration of variation. Comput. Struct. 80 (20–21), 1605–1613.

Schneibel, J.H., Kramer, M.J., Unal, O., Wright, R.N., 2001. Processing and mechanical properties of a molybdenum silicide with the composition Mo-12Si-8.5B (at. %). Intermetallics 9 (1), 25–31.

Seepersad, C.C., 2004. A robust Topological Preliminary design Exploration Method with Materials design Applications Ph.D. dissertation. GW Woodruff School of Mechanical Engineering, Georgia Institute of Technology, Atlanta, GA.

Seepersad, C. C., Allen, J. K., McDowell, D. L., and Mistree, F. 2005. Robust design of cellular materials with topological and dimensional imperfections. Advances in Design Automation Conference. Paper No: ASME IDETC/DAC-85061.

Seepersad, C.C., Allen, J.K., McDowell, D.L., Mistree, F., 2006. Robust design of cellular materials with topological and dimensional imperfections. ASME J. Mech. Des. 128 (6), 1285–1297.

Seepersad, C.C., Allen, J.K., McDowell, D.L., Mistree, F., 2007. Multifunctional topology design of cellular materials. ASME J. Mech. Des. 130 (3), 1–13, 031404.

Seepersad, C.C., Dempsey, B.M., Allen, J.K., Mistree, F., McDowell, D.L., 2004. Design of multifunctional honeycomb materials. AIAA J. 42 (5), 1025–1033.

Seepersad, C.C., Kumar, R.S., Allen, J.K., Mistree, F., McDowell, D.L., 2005. Multifunctional design of prismatic cellular materials. J. Comput-Aided Mater. Des. 11 (2–3), 163–181.

Sigmund, O., 1994. Materials with prescribed constitutive parameters: An inverse homogenization problem. Int. J. Solids Struct. 31 (17), 2313–2329.

Sigmund, O., 1995. Tailoring materials with prescribed elastic properties. Mech. Mater. 20 (4), 351–368.

Sigmund, O., 2001. Design of multiphysics actuators using topology optimization—Part I: One-material structures. Comput. Method. Appl. Mech. Eng. 190 (49–50), 6577–6604.

Sigmund, O., Torquato, S., 1997. Design of materials with extreme thermal expansion using a three-phase topology optimization method. J. Mech. Phys. Solids 45 (6), 1037–1067.

Sigmund, O., Torquato, S., Aksay, I.A., 1998. On the design of 1-3 piezocomposites using topology optimization. J. Mater. Res. 13 (4), 1038–1048.

Silva, M.J., Hayes, W.C., Gibson, L.J., 1995. The effects of non-periodic microstructure on the elastic properties of two-dimensional cellular solids. Int. J. Mech. Sci. 37 (11), 1161–1177.

Svanberg, K., 1987. The method of moving asymptotes: A new method for structural optimization. Int. J. Numer. Meth. Eng. 24 (2), 359–373.

Thampan, C.K.P.V., Krishnamoorthy, C.S., 2001. System reliability-based configuration optimization of trusses. J. Struct. Eng. 127 (8), 947–956.

Topping, B.H.V., 1984. Shape optimization of skeletal structures: A review. J. Struct. Eng. 109 (8), 1933–1951.

Wang, A., McDowell, D.L., 2004. Effects of defects on in-plane properties of periodic metal honeycombs. Int. J. Mech. Sci. 45 (11), 1799–1813.

CHAPTER 8

Integrated Design of Materials and Products—Robust Design Methods for Multilevel Systems

Nomenclature	
$acFe$	Accumulated mass fraction of Fe
Bj	Discrete point vector on a constraint boundary
$\mathbf{B}_j^i$	Projected vector of **Bj** along i direction
$b_{j,i}$	ith component of **Bj**
$b_{j,i}^i$	ith component of **Bj**^{i}
$\beta, \hat{\beta}$	Vector of regression (estimated) parameters
DCI	Design capability index
DCI_{target}	Target DCI value
δ_{max}	Maximum deflection of a beam
E	Modulus of elasticity
ε_i	Observed random error
EMI	Error margin index
EMI_l, EMI_u	Lower and upper EMI
$EMI_{constraints}$	EMI constraint conditions
EMI_{target}	EMI target value
F	Transformation function
$f_{0,i}(x)$	Mean response (performance) functions
$f_{j,i}(x)$	Uncertainty bound functions for $f_{0,i}(x)$
$HD\text{-}EMI_i$	HD-EMI in i performance direction
$HD\text{-}EMI_{target}$	Target HD-EMI
mean	Vector of mean responses
$mean_i$	ith component of **mean**
μ_x	Means of system variables
μ_y	Mean response
P_{max}	Maximum load

P	Number of predictors
R	Radius of a circular beam section
ρ	Density of material
σ	Scale factor in conditional variance model
σ_u	Ultimate tensile strength
$t_{N-P,1-\alpha/2}$	t-value in student t-distribution with $(1-\alpha)$ confidence level
T_{ignit}	Critical temperature for chemical reaction initiation
$\theta, \hat{\theta}$	Vector of variance (estimated) parameters
U_p	Particle velocity
W	Weight
x	Volume fraction of clay
x_1	Mean radius of Al particles
x_2	Mean radius of Fe_2O_3 particles
x_3	Volume fraction of voids
x_4	Mean radius of voids
y^{tr}	Transformed response
ΔY	Response deviation
$\Delta Y_{upper}, \Delta Y_{lower}$	Upper and lower deviation from a mean response
Z	Vector of variance factors

Two main challenges for integrated design of materials and products are issued in Chapter 6. One challenge is a need for a new type (Type III) of robust design, and the other is a new method for multilevel robust design satisfying requirements that previously developed multilevel robust design methods cannot solve. In this chapter, we discuss methods for overcoming the two challenges. Readers who are not familiar with robust design and related statistical methods may wish to explore Chapter 6 as a prerequisite to this chapter.

In designing materials concurrently with products, a robust design method should satisfy two requirements: (1) computational efficiency and (2) incorporation of all types of robust design, namely, Types I, II, and III. In Section 8.1, an approach, Robust Concept Exploration Method (RCEM-EMI), that satisfies these two requirements. Following the RCEM-EMI construct, designers obtain robust ranged sets of materials design specification from the design requirements. This method incorporates a metamodel and includes bounds of uncertainty for the metamodel. The metamodels and uncertainty bounds are used in order to reduce the computational expense of uncertainty analyses, so that the design method is applicable to computationally intensive simulation or expensive experiments.

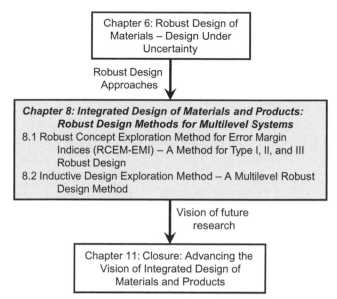

Figure 8.1 Overview of Chapter 8 and its relationship with other chapters in the book.

In Section 8.2, we discuss a multilevel robust design method, Inductive Design Exploration Method (IDEM), to facilitate top-down robust design under propagated uncertainty resulting from hierarchical information dependency in a model chain. IDEM is established based on the three solution-finding strategies (top-down design exploration, ranged sets of design solution, and parallelized function evaluation) in order to better facilitate concurrent materials and product. The techniques in IDEM include parallel discrete function evaluation, Inductive Discrete Constraint Evaluation (IDCE) based on Hyper-Dimensional Error Margin Indices (HD-EMIs). Solution strategies and techniques are discussed and demonstrated in the context of the multiscale Multifunctional Energetic and Structural Materials (MESMs) design problem.

8.1. Robust Concept Exploration Method with Error Margin Indices (RCEM-EMI)—A Method for Type I, II, and III Robust Design

The overall procedure of the RCEM-EMI is shown in Figure 8.2. The procedure consists of (a) clarification of the design task, (b) Design of Experiments (DOE) and simulation (or experiments), (c) integrated metamodel and prediction interval estimation, and (d) design space search using the cDSP for the RCEM-EMI.

In Step (a), a designer first clarifies his or her task based on the given design requirements by defining design variables, a design space, assumptions, constraints and bounds, and design goals. The designer also characterizes variability and uncertainty in design variables and the system model itself.

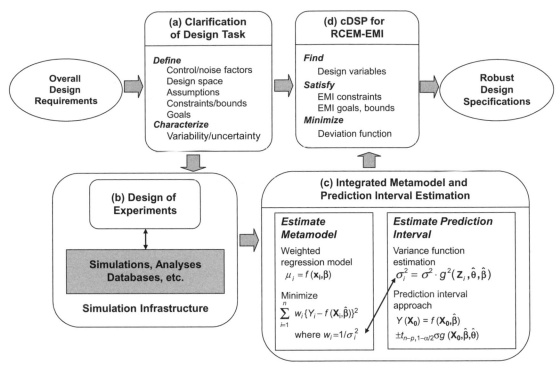

Figure 8.2 The RCEM-EMI construct.

In Step (b), simulations are designed and performed using the simulation infrastructure with given the design space and variables from Step (a). The obtained results of the simulations are transferred to Step (c).

In Step (c), a designer formulates two metamodels for rapidly mapping design space to performance based on the data obtained from Step (b). The mean response model (one metamodel) is iteratively updated based on the estimated variance function (another metamodel) until the estimation parameters in the mean response model converge. The mean response model is a function of the most probable responses at new observations. The prediction interval consists of upper and lower limits within which a new observation may be located with some confidence level. The prediction interval estimation task in this step provides upper and lower uncertainty bounds around the mean response model.

In Step (d), the mean response model and prediction interval from Step (c) alongside the variability in design variables from Step (a) are synthesized in the cDSP for RCEM-EMI. The cDSP for RCEM-EMI is solved using a search algorithm in order for a designer to find robust ranged sets of design specifications

In Sections 8.1.1 to 8.1.4, each of these steps is discussed in the context of the example of design of an energetic material based on mesoscale shock simulations, as discussed in Chapter 3.

8.1.1. Clarification of Design Task—Step (a)

In this section, a process for clarifying a design task is discussed, which is Step (a) of the construct of the RCEM-EMI shown in Figure 8.2. This task is an important step since the design goals, design parameters, uncertainty identification, and strategy for design exploration are determined in this process.

First, design goals need to be identified from design requirements. What performance criteria (responses) should be selected? For example, if designers are interested in cooling an electronic chip, they need to measure the heat transfer rate in the system. In other words, designers need to identify a parameter that indicates the system's capability for a specific purpose of design. Once performance criteria (responses) for a design task are identified, then it is necessary to identify target values. In the RCEM-EMI, requirement limits need to be identified. Depending on the type of a goal, the designer's preferences can be classified as "smaller is better," "nominal is better," or "larger is better." For each type, the requirements that need to be identified are an upper requirement limit, upper and lower requirement limits, and a lower requirement limit, respectively. For example, a requirement limit for a cooling device design is a minimum heat transfer rate (a lower requirement limit) that is necessary for proper operation. Higher heat transfer rate is desirable.

The next task is identifying control and noise factors that affect the system performances. As discussed in robust design method reviews in Chapter 6, factors that can be easily controlled by designers or manufacturers are control factors, and factors that cannot be controlled are noise factors. The number of control and noise factors determines the number of experiments (or simulations) that are required for establishing a system metamodel and variance function model in Step (c) of the RCEM-EMI. It is important to eliminate the factors that are relatively unimportant for system responses. Chen and coauthors (Chen, Allen et al. 1996) proposed an iterative screening procedure for identifying unimportant factors using the response surface method. Once control and noise factors are identified, system constraints need to be established in terms of these factors. System constraints include lower and upper bounds on the control and noise factors and models or functions that evaluate constraint conditions.

After design goals and factors are identified, the next step is identifying uncertainty in a system. As discussed in Section 6.1, uncertainty in a material system could be variability in control and noise factors (parameterizable variability), variability embedded in system behavior (unparameterizable variability), assumption and idealization of system analysis model or experiment (model structure uncertainty), and limitation of data acquisition (model parameter uncertainty).

The uncertainty classification for the energetic material design problem is discussed in Section 6.2. Please refer to the definition of uncertainty and example. In the RCEM-EMI, it is recommended to identify uncertainties in a system and classify those in terms of the types discussed in Section 6.1.

Example 1. Clarification of Energetic Material Design Task—Step (a)

To better understand this example, recall the energetic material design problem discussed in Chapters 2 and 6. The design objective is to determine the mean particle size of each constituent (Al and Fe_2O_3) and volume fraction of voids that yield robust reaction initiation satisfying given design requirements. The details of the design task are summarized in Table 8.1 where design variables, design space, and deviation of design variables are specified.

Even though the mean particle size for each constituent is accurately measured by the supplier, the mean size as sampled in a small statistic volume element (SVE) used for shock simulations could differ from the supplier's specifications. Changes in particle morphology constitute another source of variability. Some parameters, including the volume fraction of each constituent (Al and Fe_2O_3), are fixed, as shown in Table 8.1

The shock compaction process is idealized as the passage of a single 1D shock wave. Here, the plane strain assumption is invoked in order to reduce the 3D nature of microstructure to a more tractable 2D case. For the deviatoric constitutive model of Fe_2O_3, a simple elastic-plastic model was adopted, consisting of an initial linear elastic response followed by linear isotropic strain-hardening.

In this example, the performance of a reactive particle system is evaluated based on the number of sites within each statistical volume of microstructure that satisfy a microscale reaction initiation (micro-initiation) criterion during shock wave propagation. The sites that experience micro-initiation are relegated to small volumes of the SVE where the reactants are in intimate contact. The prediction of micro-initiation is based on the Merzhanov instability criterion (Merzhanov 1966). According to the Merzhanov criterion, thermal explosion of a hot spot occurs when the rate of heat generated from the chemical reaction outpaces the rate of heat conduction to the surroundings.

To evaluate the Merzhanov criterion, the following quantities must be determined at a reactant interface: (1) the temperature of the local hot spot, (2) the temperature of the hot spot surroundings, and (3) the size of the hot spot. The aforementioned quantities are calculated during shock wave propagation in the finite element simulations. This provides time histories of the estimated number of micro-initiation sites contained in the SVE. For the purposes of this example, the maximum number of micro-initiation sites during shock wave propagation within a given statistical volume element of microstructure is taken as the response of the system.

8.1.2. Design of Experiments (DOE) and Simulation—Step (b)

Based on the control and noise factors identified in the previous section, it is necessary to design experiments (plan simulation)—Step (b) in Figure 8.2—that should be performed for building an accurate mean response model and conditional variance models. There are many

Table 8.1 Clarification of energetic material design task using the shock simulation.

Clarified Items	Specifications
Control Factors	x_1 : Mean radius of Al particles
	x_2 : Mean radius of Fe_2O_3 particles
	x_3 : Volume fraction of voids
	x_4 : Mean radius of voids
Design Space	$x_1 = [0.5, 1.5]$ (μm)
	$x_2 = [0.2, 1]$ (μm)
	$x_3 = [0.02, 0.1]$
	$x_4 = [0.2, 1]$ (μm)
Deviation of Control Factors	$\Delta x_1 = \pm 0.2$ (μm)
	$\Delta x_2 = \pm 0.1$ (μm)
	$\Delta x_3 = \pm 0.01$
	$\Delta x_4 = \pm 0.1$ (μm)
Assumptions	Generalized plane strain assuming circular particles as cylinders
	Constitutive model for Fe_2O_3 : bilinear elastic-plastic model
Fixed Parameters	Volume fraction of intermetallic compound (Al + Fe_2O_3): 50% of the Total Volume
	Volume fraction of Al : volume fraction of Fe_2O_3 = 2:3 (reactants are in stoichiometric proportion)
	Standard deviation of void radius : 20% of the mean radius of Void
	Standard deviation of Al radius : 20% of the mean radius of Al particles
	Standard deviation of Fe_2O_3 radius : 20% of the mean radius of Fe_2O_3 particles
	Particle shock velocity (Up) : 1 (km/s)
	Size of SVE: 11 $\times$ 7 (μm^2)
	Particle overlapping: $-$ 0.09 (μm)
Response	Number of reaction sites with the SVE based on Merzhanov reaction initiation criterion
Design Objectives	Lower requirement limit: the number of reaction sites ≥ 1
	Achieving higher number of reaction sites

DOE techniques available for characterizing a system response in terms of the control and noise factors. Refer to Chapter 6 for the details of DOE techniques.

In the RCEM-EMI, a DOE technique using the single array (no separate array for control and noise factors) approach is incorporated. In the DOE technique, Central Composite Design

(CCD; Myers and Montgomery 1995) is recommended for building a response surface model (mean response model) since the prediction errors of the CCD are identical across the entire design space. However, designers may employ any other DOE techniques, such as two- or three-level factorial design, Latin HyperCube sampling (McKay, Conover et al. 1979), Box-Behnken (Myers and Montgomery 1995), etc., for obtaining the most accurate response surface model with the minimum number of experiments (or simulations).

Another important task is modeling the unparameterizable variability. This requires replicated or near-replicated experiments at each designed experimental point to effectively capture a conditional variance model. The number of replications for accurately capturing a conditional variance model depends on system characteristics. We discuss this step in detail in the next section.

While collecting experimental data based on DOE specifications, designers have to perform a number of simulations or experiments in a series or parallel. They may also need to connect to a remote machine where simulation software is installed. In this case, it is necessary to employ a distributed, collaborative design framework, such as Web-based Distributed Product Realization (Web-DPR) (Choi 2001, Xiao, Choi et al. 2001), eXtensible Distributed Product Realization (X-DPR) (Choi, Panchal et al. 2003), Phoenix Integration (Phoenix Integration Inc. 2001), and iSIGHT/FIPER (Engineous Inc. 2001). These design frameworks are useful for automating serialized and parallelized multiple simulation executions and collecting result data across geographically distributed computing resources. Some of the software frameworks (e.g., iSIGHT/FIPER) also provide statistical and design tools for supporting design tasks in general. We discuss distributed, collaborative design frameworks in Chapter 10.

Example 2. DOE and Shock Simulation—Step (b)

Continuing the energetic material example, CCD for four control factors (x1-x4) is employed for the DOE. Please note that there is no noise factor in this example. The CCD produces 25 sampling points, as shown in Table 8.2

As mentioned, the listed simulation points are repeated 20 times, resulting in 500 total samples to capture unparameterizable variability. The simulation time for each simulation takes approximately 30 to 60 minutes depending on the particle sizes. It is difficult to manually execute such a large number of simulations one by one. Moreover, the simulation code (RAVEN at UCSD (Benson 1995)) is located remote from the design facility. For these reasons, a distributed collaborative framework (Phoenix Integration is used for this example) is required for automated serial execution of the code. The simulation infrastructure for the shock simulation of this energetic material composed of three modules, namely the DOE module, the pre- and post-processing module, and the RAVEN simulation module.

Table 8.2 CCD for the shock simulation of energetic material.

Trials	x1 (mm)	X2 (mm)	x3	x4 (mm)
1	0.00075	0.0004	0.04	0.0004
2	0.00075	0.0004	0.08	0.0004
3	0.00125	0.0004	0.04	0.0004
4	0.00125	0.0004	0.08	0.0004
5	0.00075	0.0008	0.04	0.0004
6	0.00075	0.0008	0.08	0.0004
7	0.00125	0.0008	0.04	0.0004
8	0.00125	0.0008	0.08	0.0004
9	0.00075	0.0004	0.04	0.0008
10	0.00075	0.0004	0.08	0.0008
11	0.00125	0.0004	0.04	0.0008
12	0.00125	0.0004	0.08	0.0008
13	0.00075	0.0008	0.04	0.0008
14	0.00075	0.0008	0.08	0.0008
15	0.00125	0.0008	0.04	0.0008
16	0.00125	0.0008	0.08	0.0008
17	0.001	0.0006	0.02	0.0006
18	0.001	0.0006	0.1	0.0006
19	0.0005	0.0006	0.06	0.0006
20	0.0015	0.0006	0.06	0.0006
21	0.001	0.0002	0.06	0.0006
22	0.001	0.001	0.06	0.0006
23	0.001	0.0006	0.06	0.0002
24	0.001	0.0006	0.06	0.001
25	0.001	0.0006	0.06	0.0006

8.1.3. Estimation

Natural uncertainty is due to the inherent randomness or unpredictability of a physical system, as defined in Chapter 6. The process of quantifying variability is an important step in robust design. Variability may be quantified as variance, standard deviation, interval, integrated mean response, prediction interval, etc. However, variability due to natural randomness that cannot be parameterized in numeric form (i.e., unparameterizable variability) is difficult to quantify since it can be estimated only when the random source is parameterized.

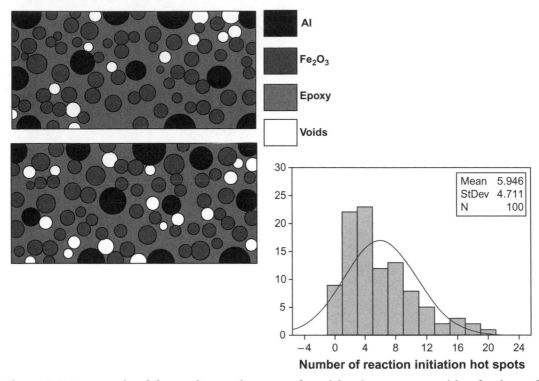

Figure 8.3 An example of the random assignment of particle microstructure with a fixed set of input parameters.

Example 3. Unparameterizable Variability in Energetic Materials

The randomness in microstructure of an SVE shown in Figure 8.3, which is generated by the pseudo-random particle placement process, cannot be parameterized but it is an important source of the response variability (i.e., variability in the number of reaction initiation hot spots per unit volume).

In Figure 8.3, the two SVEs are the instances of microstructures created by the pseudo-random particle placement process with maintaining all other conditions and parameters to be the same. The histogram is obtained out of the simulation outputs, the number of reaction initiation hot spots, of 100 SVE instances. Based on the histogram, it is identified that the randomness in microstructure within the SVE is the source of large variances in the outputs because all other conditions and parameters of the simulation model are fixed in the 100 repeated samples.

In RCEM-EMI, the unparameterizable variability is quantified as a prediction interval around the estimated mean response model for use in Step (d) of RCEM-EMI. Again, the

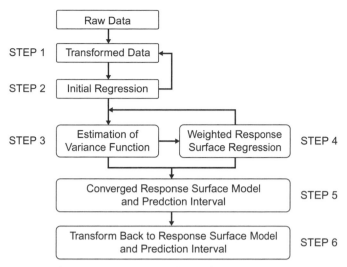

Figure 8.4 Integrated mean response model and prediction interval estimation for characterizing unparameterizable variability.

mean response model is a function of the most probable responses at new observations. The prediction interval consists of upper and lower limits within which a new observation may be located with some confidence level. These limits are schematically shown in Figure 6.8. The underlying theory of this modeling technique is rooted to the *location and dispersion modeling* approach. In this approach, the mean response and variance are iteratively remodeled in a concurrent manner within a continuous design space. For a detailed review of this approach, we recommend readers to refer to the works by Davidian and Carroll (Davidian and Carroll 1987), Vining and Myers (Vining and Myers 1990), and Engel and Huele (Engel and Huele 1996).

The procedure of Step (c) of RCEM-EMI, estimating mean response model and prediction interval, is illustrated in Figure 8.4. It consists of six substeps (Steps 1 to 6). In Step 1, raw data obtained from simulation or experiments need to be transformed if errors—the magnitude of differences between estimated mean and obtained data—of the raw data are not normally distributed. In Step 2, the transformed data are initially regressed to form a mean response model, such as response surface model. In Step 3, a conditional variance model (function of error variance) is estimated based on the initial response surface model. In Step 4, the initial response surface model is re-estimated in consideration of the variance function estimated in Step 3. In this step, the weights of the data are calculated based on the estimated variance of errors so that the regression can be robust against large random errors (such as outliers). Step 3 and Step 4 are repeated until the parameters in the response surface model are converged. In Step 5, a converged response surface model is obtained as a result of the previous iterative process. In the last step, the converged model will be transformed back to the raw data space by inverse transformation of Step 1. Each of the six substeps is explained in detail, together with an example of the energetic material design for reaction initiation sites in the following.

STEP 1: Since the obtained data often represents non-normal distribution of residuals (differences between obtained data and estimated mean), it may be necessary to transform raw data (y) into another form so that a mean response model (e.g., response surface model) can fit the transformed data (y^{tr}) accurately. Box-Cox transformation is useful for selecting a transformation function. For readers who are not familiar with this transformation, please refer to Neter, Kutner et al. (1996). The required transformation is defined by the equation

$$y^{tr} = F(y) \qquad (8.1)$$

where F is the transformation function.

In Figure 8.5 a, the normal probability plot of the residuals obtained by fitting the raw data is skewed, which means the distribution of the residuals is not balanced. The raw data are transformed using $y^{tr} = \text{In}(y + 2)$ and the transformed data are fitted again. As shown in the Figure 8.5 (b), the normal probability plot of the residuals with transformed data indicates that the residual is now balanced.

STEP 2: y^{tr} is regressed using a response surface model, $f(x_i,\beta)$, where x_i is a sample vector of design variables and β is a vector of regression parameters. This response surface model represents the means of unparameterizable variability (dispersion) as a function of x. This is why we also call this the mean response model. A quadratic model is selected for $f(x_i,\beta)$ since this is often used in engineering applications. However, a designer may select

---

Example 4. Transformation of Raw Data for Energetic Material Design—Step 1

Continuing Example 2, 500 data have been obtained from Step (b) of the RCEM-EMI. In this example, a quadratic response surface model will be used as the mean response model and an exponential function powered by a quadratic response surface model as the conditional variance model. Therefore, the estimated mean response (micro-initiation) model is given by $f(\mathbf{x},\beta) = \mathbf{x}\cdot\beta'$, where

$$\mathbf{x} = [1, x_1, x_2, x_3, x_4, x_1^2 x_2^2, x_3^2 x_4^2, x_1 x_2, x_1 x_3, x_1 x_4, x_2 x_3, x_2 x_4, x_3 x_4]$$

and

$$\beta = [\beta_0, \beta_1, \beta_2, \beta_3, \beta_4, \beta_{11}, \beta_{22}, \beta_{33}, \beta_{44}, \beta_{12}, \beta_{13}, \beta_{14}, \beta_{23}, \beta_{24}, \beta_{34}]$$

The normality of the residuals is tested by fitting a quadratic response surface model and the raw data will be transformed into transformed data if necessary.

a linear or higher-order model. The model structure uncertainty induced in this process should be identified. If the transformation in STEP 1 is successful, a residual plot should show a normal distribution, $\varepsilon_i. \sim N(0, \sigma_i^2)$. The regression is

$$y_i^{tr} = f(x_i, \beta) + \varepsilon_i \tag{8.2}$$

where i = 1, ..., N (total number of samples).

A major assumption in regression is that the variances (σ_i^2) are homoscedastic; that is, that the random error distributions are constant over the entire design space. However, in reality,

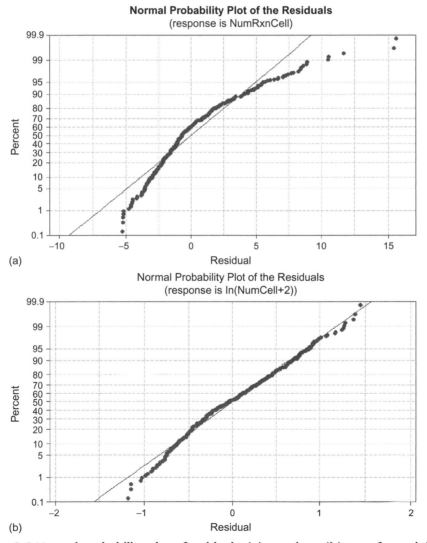

(a)

(b)

Figure 8.5 Normal probability plot of residuals: (a) raw data, (b) transformed data.

variances are not always constant across an entire design space but may vary. This is referred to as heteroscedasticity. Unparameterizable variability is one of the primary sources of heteroscedasticity. In this step, a mean response model is estimated by obtaining a vector of estimated regression parameters ($\hat{\beta}$) that is used in the next step.

STEP 3: The response variance should be modeled in terms of factors. A generalized representation of a conditional variance model, proposed by Davidian and Carroll (Davidian and Carroll 1987) is given by

$$\sigma_i^2 = \sigma^2 v^2(z_i, \theta, \hat{\beta}) \tag{8.3}$$

where $i = 1, \ldots, N$ (total number of samples), σ is a scale factor, z_i is a vector of variance factors, θ is a vector of variance parameters, and $\hat{\beta}$ is the vector of estimated regression parameters obtained in STEP 2. Many different types of conditional variance models have been proposed. In industrial contexts, Box and Meyer (Box and Meyer 1986) as well as many others (Chan and Mak 1995, Engel and Huele 1996, Grego 1993, Nair and Pregibon 1988), suggest

$$\sigma_i^2 = \exp(\theta, x_i') \tag{8.4}$$

where $i = 1, \ldots, N$ (total number of samples), σ is assumed to be unity, and z_i are assumed to be x_i with a given $\hat{\beta}$. An exponential form effectively represents heteroscedastic variance since estimated variances are always positive, and θ can be efficiently estimated via log-link conversion, as discussed later.

For estimating θ, variance model-fitting methods, such as the maximum pseudolikelihood method (Aitkin 1987, Gong and Samaniego 1981), the least squares on squared residuals estimation (Amemiya 1977, Jobson and Fuller 1980), the least squares on absolute residuals estimation (Glejser 1969, Theil 1971), or the logarithm method (Engel and Huele 1996, Harvey 1976) are required. The maximum pseudolikelihood estimation function for estimating θ in Equation 8.3 is given by

$$l(\beta, \theta, \sigma) = -N \log\sigma - \sum_{i=1}^{N} \log\{v(z_i, \beta, \theta)\} - \frac{1}{2\sigma^2} \sum_{i=1}^{N} \{\frac{y_i - f(x_i, \beta)\}^2}{v^2(z_i, \beta, \theta)}\} \tag{8.5}$$

Applying Equation 8.4 to Equation 8.5 with transformed data (y^{tr}_i), the pseudolikelihood estimation function becomes

$$l(\theta) = -\frac{1}{2} \sum_{i=1}^{N} [\theta x_i' - \frac{\{y^{tr}_i - f(x_i, \hat{\beta})\}^2}{\exp(\theta x_i')}] \tag{8.6}$$

By maximizing this function, we obtain a vector of estimated variance parameters ($\hat{\theta}$) with given $\hat{\beta}$.

The estimation function of the logarithm method for estimating θ in Equation 8.3 is

$$ls(\beta,\theta,\sigma) = \sum_{i=1}^{N}[\log\{y_i - f(\mathbf{x_i},\beta)\}]^2 - \log\{\sigma^2 v^2(\mathbf{z_i},\beta,\theta)\}^2 \qquad (8.7)$$

Applying Equation 8.4 to Equation 8.7 with transformed data (y^{tr}_i), we obtain the logarithm least-squares estimation function

$$ls(\theta) = \sum_{i=1}^{N}[\log\{y_i^{tr} - f(\mathbf{x_i},\hat{\beta})\}^2 - \theta\mathbf{x_i}']^2 \qquad (8.8)$$

By minimizing Equation 8.8, we obtain $\hat{\theta}$ with the given $\hat{\beta}$. Although two methods for estimating θ in the conditional variance model are introduced, the maximum pseudolikelihood method is employed in this chapter since it is suitable for any type of random error distribution.

STEP 4: Once we estimate the conditional variance model for heteroscadastic observations, the mean response model must be refit using the weighted regression method (Neter, Kutner et al. 1996). With this approach, the mean response model becomes more robust to the effect of adding large variance data (such as outliers). For estimating the updated $\hat{\beta}$, the following weighted squares of residuals should be minimized:

$$wls(\beta) = \sum_{i=1}^{N}w_i\{y_i - f(\mathbf{x_i},\beta)\}^2 \text{ where, } w_i = 1/\sigma_i^2 \qquad (8.9)$$

Applying the estimated conditional variance model in STEP 3 to Equation 8.9, we obtain the weighted least squares estimation function

$$wls(\beta) = \sum_{i=1}^{N}w_i\{y_i^{tr} - f(\mathbf{x_i},\beta)\}^2 \text{ where, } w_i = \exp(-\hat{\theta}\mathbf{x_i}') \qquad (8.10)$$

Based on the updated mean response model, we re-estimate the conditional variance model obtaining updated $\hat{\theta}$. STEP 3 and 4 are iterated until $\hat{\beta}$ converges.

> **Example 5. Iteratively Reweighted Mean Response Model and Conditional Variance Function Estimation—Step 2, Step 3, and Step 4**
>
> With the transformed data from Example 4, initial regression (Step 2) is performed. The initially estimated parameters ($\hat{\beta}_{init}$) of the response surface model are listed in Table 8.3. After estimating the initial mean response model, we perform iteratively reweighted regressions (Step 3 and Step 4).

The conditional variance model used here is an exponential function with a quadratic response surface model

$$\sigma^2 v^2(\mathbf{x},\beta,\theta) = \exp(\mathbf{x}\theta')$$

where

$$\mathbf{x} = [1,x_1,x_2,x_3,x_4,x_1^2,x_2^2,x_3^2,x_4^2,x_1x_2,x_1x_3,x_1x_4,x_2x_3,x_2x_4,x_3x_4]$$
$$\theta = [\theta_0,\theta_1,\theta_2,\theta_3,\theta_4,\theta_1^2,\theta_2^2,\theta_3^2,\theta_4^2,\theta_{11},\theta_{22},\theta_{33},\theta_{44},\theta_{12},\theta_{13},\theta_{14},\theta_{23},\theta_{24},\theta_{34}]$$

The maximum pseudolikelihood estimation method in Equation 8.6 is used to estimate parameters of the conditional variance model (θ), i.e.,

$$l(\theta) = -\frac{1}{2}\sum_{i=1}^{500}[\theta\mathbf{x}_i' - \{y_i^{tr} - f(\mathbf{x}_i,\hat{\beta}_{init})\}^2/\exp(\theta\mathbf{x}_i')]$$

The initially estimated response surface model, $f(\mathbf{x},\hat{\beta}_{init})$, is used for the estimation for this first iteration. By maximizing this estimator, parameters ($\hat{\theta}$) in the conditional variance model can be obtained. This is Step 3. Based on the obtained conditional variance model, weighted regression is performed by minimizing weighted least square estimator, Equation 8.10, to obtain updated response surface model, $f(\mathbf{x},\hat{\beta}_{updated})$. This is Step 4. After ten iterations of Step 3 and Step 4 in this manner, converged estimated parameter vectors ($\hat{\beta}$ and $\hat{\theta}$) are obtained, as shown in Table 8.3

STEP 5: After obtaining the converged mean response model and the conditional variance model based on the procedure (up to Step 4 iteratively), we estimate an interval around the regression function at a new observation ($\mathbf{x}_0$) using the prediction interval estimation with a 100(1-α) % confidence level given by

$$y^{tr}(\mathbf{x}_0) = f(\mathbf{x}_0,\hat{\beta}) \pm t_{N-P,1-\alpha/2}\sigma v(\mathbf{x}_0,\hat{\beta},\hat{\theta}) \tag{8.11}$$

Table 8.3 Estimated parameters in mean and conditional variance models.

Subscripts	$\hat{\beta}_{init}$	$\hat{\beta}$	$\hat{\theta}$
0	3.9704e + 000	3.9924e + 000	6.8363e − 001
1	−1.8342e + 003	−1.7704e + 003	−1.5518e + 003
2	−1.3850e + 003	−1.4896e + 003	4.4727e + 002
3	−1.3503e + 001	−1.3798e + 001	−3.2056e + 001
4	−1.2722e + 003	−1.2988e + 003	−8.4064e + 002
11	1.1298e + 006	1.1115e + 006	−1.2844e + 003
22	8.9132e + 005	9.2869e + 005	−4.4852e + 002
33	1.7331e + 002	1.7190e + 002	8.6503e + 001
44	6.8733e + 005	6.9570e + 005	7.7073e + 002
12	−1.0642e + 006	−1.0284e + 006	−1.3846e + 003
13	4.8862e + 002	4.0649e + 002	2.1635e + 004
14	−7.0772e + 005	−7.9062e + 005	−1.5430e + 003
23	−7.2651e + 003	−7.3911e + 003	−1.2607e + 004
24	1.2218e + 006	1.2735e + 006	−3.9634e + 002
34	2.3488e + 003	3.3732e + 003	1.3293e + 004

P is the number of predictors and α is the confidence level. Applying Equation 8.4 to Equation 8.11, the transformed estimated prediction interval is

$$y^{tr}(\mathbf{x}_0) = f(\mathbf{x}_0,\hat{\beta}) \pm t_{N-p,1-\alpha/2}\exp\left(\frac{\hat{\theta}\mathbf{x}_0{}'}{2}\right) \quad (8.12)$$

STEP 6: Convert the mean response model and prediction interval estimation into the original function, i.e.,

$$y(\mathbf{x}_0) = F^{-1}\left\{f(\mathbf{x}_0,\hat{\beta}) \pm t_{N-p,1-\alpha/2}\ \exp\left(\frac{\hat{\theta}\mathbf{x}_0{}'}{2}\right)\right\} \quad (8.13)$$

In this section, we discussed an integrated, iterative variability and uncertainty quantification method based on quantifying uncertainty bounds due to unparameterizable variability and limited sample size. In Section 8.1.4, we discuss a design exploration technique for searching Type I, II, and III robust design specifications using quantified uncertainty bounds. In the design exploration process, the mean response model, $F^{-1}(f(\mathbf{x}_0,\hat{\beta}))$, and the upper and lower limits of the prediction interval, Equation 8.13, are used as a mean response function and uncertainty bound functions, respectively.

Example 6. Building the Mean Response Model and Prediction Interval— Step 5 and Step 6

Continuing our energetic material design example, the transformed mean response model and prediction interval at a new observation ($\mathbf{x}_0$) are calculated according to

$$y^{tr}(\mathbf{x}_0) = \hat{\beta}\mathbf{x}_0 \pm t_{N-p,1-\alpha/2}\exp\left[\frac{\hat{\theta}\mathbf{x}_0'}{2}\right] \tag{8.14}$$

where the number of samples (N) is 500, the total number of predictors (P) is 30, and the confidence level (α) is 0.99. $\hat{\beta}$ and $\check{\theta}$ are the converged parameter in Table 8.3.

Transforming back to the original coordinate, the inverse trasformation function is $F^{-1}(y^{tr}) = \exp(y^{tr}) - 2$. Therefore, the mean response model, $f_0(\mathbf{x}_0)$, the upper limit of prediction interval $f_1(\mathbf{x}_0)$ and the lower limit of prediction interval $f_2(\mathbf{x}_0)$ are obtained as follows:

$$f_0(\mathbf{x}_0) = \exp(\hat{\beta}\mathbf{x}_0) - 2$$

$$f_1(\mathbf{x}_0) = \exp\left\{\hat{\beta}\cdot\mathbf{x}_0 + t_{N-p,1-\alpha/2}\;\exp\left(\frac{\hat{\theta}\mathbf{x}_0'}{2}\right)\right\} - 2$$

$$f_2(\mathbf{x}_0) = \exp\left\{\hat{\beta}\cdot\mathbf{x}_0 - t_{N-p,1-\alpha/2}\;\exp\left(\frac{\hat{\theta}\mathbf{x}_0'}{2}\right)\right\} - 2 \tag{8.15}$$

The mean response and prediction interval models are plotted in Figure 8.6. The models are depicted in terms of the voids' volume fraction and mean radius when the mean radius of Al is 0.75 (μm) and that of Fe_2O_3 is 0.4 (μm). As shown in Figure 8.6, the obtained samples' responses are dispersed within the prediction interval limits at fixed design parameters because of the unparameterizable variability (due to the changes in the microstructure of SVE). In this section, we identified the mean response model and functions for the prediction interval limits.

8.1.4. Compromise Decision Support Problem (cDSP) for RCEM-EMI

The last step—Step (d) in the RCEM-EMI—is to identify a robust solution based on design specifications obtained in Step (a) and the mean response models and prediction interval obtained in Step (c).

In many engineering applications, including materials design, design problems involve multifunctional systems. As discussed in Chapter 3, an energetic material can also be a multifunctional material if it is required to carry structural loads or have a certain target stiffness. Balancing those multiple performances of a system is important in

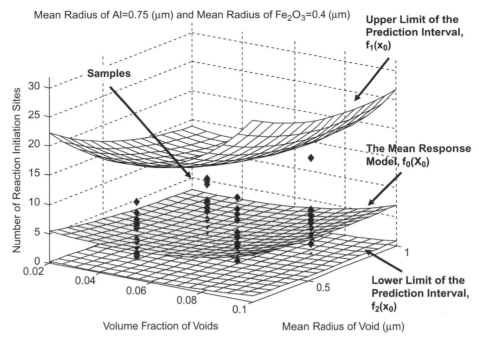

Figure 8.6 The mean response model and upper and lower limits of the prediction interval for the estimation of number of reaction sites within a given SVE.

engineering systems design under uncertainty. In Step (d), we employ the compromise Decision Support Problem (cDSP) discussed in Chapter 5, together with an index, called the Error Margin Index (EMI), in order to design multifunctional systems under system uncertainty. This is the final step of RCEM-EMI for achieving Type I, II, and III robust design specification.

8.1.4.1. Error Margin Indices (EMIs)

Before getting into the details of Step (d) of RCEM-EMI, we discuss EMIs used for searching robust design specifications. EMIs are metrics indicating the degree of reliability of a decision that satisfies system constraints or bounds. In other words, EMIs represents the amount of safety margin against system failure due to uncertainty (or errors) in a model used for the decision making. In the mathematical sense, EMIs are ratios calculated based on variability of achieved (or designed) system performance (or response) and design requirements of the system performance; therefore, it is dimensionless. The EMIs are used in search algorithms to find ranged sets of design specifications that satisfy a range of system requirements and are a key construct for Type III robust design for multilevel design problems. The procedure for obtaining the EMIs is as follows: (1) obtain the upper and/or lower deviation of a response and (2) calculate the EMIs based on this deviation.

Obtaining the Upper and Lower Deviations of a Response with Mean Response Model and the Prediction Interval Found in Step (c)

The first step in finding a design variable vector $\mathbf{x} = \{x_1, x_2, \ldots, x_n\}$ that ensures a robust response is to estimate the response variation due to variations in the design variable vector ($\mathbf{x}$) using a first-order Taylor series expansion. Assuming that variations (Δx_i) in design variables are small, the response variation (ΔY) is

$$\Delta Y = \sum_{i=1}^{n} \left| \frac{\partial f}{\partial x_i} \right| \Delta x_i \tag{8.16}$$

where n is the number of design variables. The representation of the response deviation in Equation 8.16 is close to the worst-case scenario, which assumes that all fluctuations occur simultaneously in the worst possible combination (Chen and Lewis 1999). However, ΔY in Equation 8.16 is the response variation due to variations in design variables only. *It does not include the response variation due to the variability of the model itself since it uses only the mean response model (f).* For example, the mean response model (*f*) used in Equation 8.16 is only the $f_0(x_0)$ among the three functions obtained in Example 6.

To consider the effect of unparameterizable variability in the system response variation, we use the following procedure. Assuming a system model has k uncertainty bounds (e.g., upper and lower predication interval limits), we obtain a response variation (ΔY_j) from each of them, i.e.,

$$\Delta Y_j = \sum_{i=1}^{n} \left| \frac{\partial f_j}{\partial x_i} \right| \Delta x_i \tag{8.17}$$

where $j = 1, 2, \ldots, k$ (the number of uncertainty bounds). Evaluating multiple variances in a mean response function as well as k uncertainty bound functions with the same procedure by variability in n design variables, we identify minimum and maximum responses considering variability in design variables and uncertainty bounds around a mean response functions, i.e.,

$$Y_{max} = Max[f_j(x) + \Delta Y_j] \tag{8.18}$$

$$\text{and}$$

$$Y_{min} = Min[f_j(x) - \Delta Y_j]$$

where $j = 0, 1, 2, \ldots, k$, $f_0(x)$ is the mean response function, and $f_1(\mathbf{x}) \ldots f_k(\mathbf{x})$ are uncertainty bound functions. Upper and lower deviations, which are the deviations from the mean response to the maximum and minimum responses, respectively, are represented using

$$\Delta Y_{upper} = Y_{max} - f_0(\mathbf{x}) \text{ and } \Delta Y_{lower} = f_0(\mathbf{x}) - Y_{min} \tag{8.19}$$

Example 7 Computing Upper and Lower Response Deviation Using a Mean Response Model and Two Prediction Interval Limits

As an example, the case of a mean response model, $f_0(x)$, and upper and lower prediction interval limits, $f_1(x)$ and $f_2(x)$, in terms of a single design variable, x, is illustrated in Figure 8.7. In this example, normal distributions are used for demonstration purposes. The mean response model (the most probable response model) is shown as a solid curve with the two prediction interval limits shown as dotted curves in the figure. As shown in the right-hand side of the figure, the corresponding response variations of the mean response function and the prediction interval limits within the interval of the design variable's variance are

$$\Delta Y_0 = \left| \frac{\partial f_0}{\partial x} \right| \Delta x, \ \Delta Y_1 = \left| \frac{\partial f_1}{\partial x} \right| \Delta x,$$

and

$$\Delta Y_2 = \left| \frac{\partial f_2}{\partial x} \right| \Delta x.$$

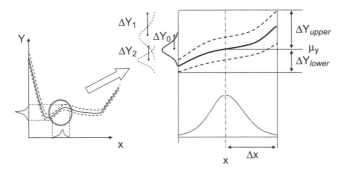

Figure 8.7 Formulation of uncertainty bounds due to variations in a design variable and a model.

Where $f_0(\mathbf{x})$ is the mean response, Y_{max} is the maximum response, Y_{min} is the minimum response, ΔY_{upper} is the upper deviation, and ΔY_{lower} is the lower deviation.

Maximum and minimum responses considering the prediction interval are given by

$$Y_{max} = Max\ [f_0(x) + \Delta Y_0, f_1(x) + \Delta Y_1, f_2(x) + \Delta Y_2]$$

and

$$Y_{min} = Min\ [f_0(x) + \Delta Y_0, f_1(x) - \Delta Y_1, f_2(x) - \Delta Y_2]$$

Finally, upper and lower deviations of the response at x are expressed as

$$\Delta Y_{upper} = Y_{max} - f_0(x) \text{ and } \Delta Y_{lower} = f_0(x) - Y_{min}, \text{ respectively.}$$

The energetic material design example that is continuously discussed also includes a mean response model and upper and lower prediction interval limits as this example. The procedure for getting upper and lower deviation of the response at $\mathbf{x}$ is similar. The only difference is the energetic material design example includes four design variables, $\mathbf{x} = [x_1, x_2, x_3, x_4]$. Therefore, the corresponding response variations are

$$\Delta Y_0 = \sum_{i=1}^{4} \left| \frac{\partial f_0}{\partial x_i} \right| \Delta x_i \Delta Y_1 = \sum_{i=1}^{4} \left| \frac{\partial f_1}{\partial x_i} \right| \Delta x_i,$$

and

$$\Delta Y_2 = \sum_{i=1}^{4} \left| \frac{\partial f_2}{\partial x_i} \right| \Delta x_i.$$

The rest of the procedure will be the same as the previous example.

Using this procedure, the response variation of a model that has different types of uncertainty bounds, including one-sided, two-sided, and crossed error bounds, can be evaluated while considering uncertainty in a model. The response deviations (ΔY_{upper} and ΔY_{lower}) can be calculated from any system model functions, including metamodels and/or engineering equations.

Calculating the Error Margin Indices (EMIs)

EMIs are calculated using the same mathematical construct underlying the Design Capability Indices (DCIs) (Chen, Simpson et al. 1999). The mathematical construct is shown in Figure 8.8. The normal distributions in this figure are given merely as examples; many material response functions do not adhere to Gaussian statistics.

The DCIs are calculated using the mean response (μ_y) obtained by the mean response function, $f_0(\mathbf{x})$, and deviation (ΔY) from a response distribution of a system model (dotted distributions in the figure); ΔY is calculated based on Equation 8.13. For example, in the "Smaller is better" item (Figure 8.8 a), the DCI is $(URL-\mu_y)/\Delta Y$, where URL is Upper Requirement Limit. On the other hand, the EMI is calculated including μ_y and upper and lower deviations (ΔY_{upper} and ΔY_{lower}) from a combined distribution of a system model and uncertainty bounds (solid distributions in the figure); the deviation is calculated based on Equations 8.17, 8.18, and 19.

In the case where "Smaller is better," the EMI is $(URL-\mu_y)/ \Delta Y_{upper}$. In other words, the DCI includes only the response deviation of a system model due to the variations of design

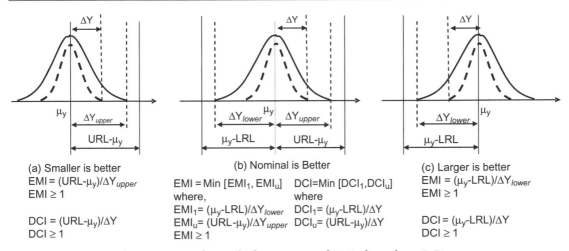

Figure 8.8 Mathematical construct of EMIs based on DCIs.

variables. The EMI, on the other hand, includes the response deviations of error bounds as well as the system model. In all cases depicted in Figure 8.8, the EMI becomes larger as decisions become more reliable. In a "Smaller is better" case, the EMI becomes larger when the location of μ_y is farther from a URL and/or ΔY_{upper} gets the smaller. An EMI of unity means that an uncertainty bound just meets a requirement limit. EMIs, smaller than unity, indicate that a requirement limit may be violated due to uncertainty in the model. The same is true when the EMI is formulated for a "Larger is better" item (Figure 8.8c). In this case, the larger EMI can be achieved by locating μ_y farther from a Lower Requirement Limit (LRL) and/or reducing ΔY_{lower}. In the "Nominal is better" case, depicted in Figure 8.8b, both the upper and lower EMI need to be calculated, and the worse case (the smaller EMI) is selected. The EMIs, formulated in this manner, are leveraged in a solution search algorithm to find solution sets that are robust to model variability as well as variations in input variables.

8.1.4.2. The Compromise Decision Support Problem (cDSP) with EMI for Design Exploration

As discussed in Chapter 5, the cDSP (Mistree, Hughes et al. 1992) is a mathematical construct for identifying design solutions in the presence of multiple conflicting goals. The generic cDSP is a hybrid formulation that incorporates concepts from both traditional mathematical programming and goal programming. Please find a detailed discussion of the generic cDSP in Chapter 5.

The mathematical instantiation of the cDSP for RCEM-EMI is shown in Table 8.4. Mean response models for multiple performances, $f_{o,i}(\mathbf{x})$, and upper and lower prediction interval limits, $f_{1,i}(\mathbf{x})$ and $f_{2,i}(\mathbf{x})$ of each mean response model are obtained by Step (c) as discussed in Section 8.1.3. System constraints and goals are formulated to capture the demands placed on the system and the requirements that a designer wishes to satisfy to the extent possible,

Table 8.4 Mathematical forms of the cDSP for RCEM-EMI.

cDSP for RCEM-EMI
Given
n, (number of system variables)
m, (number of system goals)
q, (number of inequality constraints)
$f_{0,i}(x)$, (multiple mean response models)
$f_{1,i}(x)$, (multiple upper uncertainty bound, i.e., prediction interval limit)
$f_{2,i}(x)$, (multiple lower uncertainty bound, i.e., prediction interval limit)
$g_{0,i}(x)$, (multiple mean constraint functions)
$g_{1,i}(x)$, (multiple upper constraint bound functions)
$g_{2,i}(x)$, (multiple lower constraint bound functions)
$URLi$ and $LRLi$, performance requirements
Δx, deviations of system variables.
$EMItarget,i$, EMI targets
Find
μ_x (mean of system variables)
d_i^+, d_i^-, (deviation variables)
Satisfy
System Constraints:
$EMI_{constraint,i}(x) \geq 1$ *where i = 1,...,q*
System Goals:
$EMI_i(x) / EMI_{target,i} + d_i^- - d_i^+ = 1$ *where i = 1,...,m*
Bounds:
$x_i^{min} \leq x_i \leq x_i^{max}$ *where i = 1,...,n*
$d_i^-, d_i^+ \$ 0$ and $d_i^- \cdot d_i^+ = 0$
where i = 1,...,m
Minimize
Preemptive
$Z = [f_1(d_i^-, d_i^+),...,f_k(d_i^-, d_i^+)]$
Archimedean
$Z = \Sigma W_i(d_i^-, d_i^+), \Sigma W_i = 1$

respectively. The *LRLs* and/or *URLs* represent the required lower and/or the upper system performance limits that must be met. These are the design requirements that must be obtained from customers' requirement survey. Deviations (variability) of system variables must be specified by the designer.

A solution search algorithm is used to find the location of the mean of system variables, μ_x, and the deviation variables, d_i^+, d_i^-, from EMI targets. Solutions must satisfy the system constraints considering variability and uncertainty in system variables and constraints functions. These uncertain system constraints are formed by $EMI_{constraint,i} \geq 1$ for multiple constraints. The $EMI_{constraint,i}$ are calculated based on $g_i(\mathbf{x})$, instead of $f_i(\mathbf{x})$, with the same procedure discussed in Section 8.1.4; the EMI formulation is always the "Smaller is better" case and *URL* is zero. Multiple goals are formulated in terms of multiple EMI

Example 8 Design Exploration Using cDSP for the Energetic Material Design Problem—Step (d)

We formulated the design specifications in Example 1 and the mean response model and prediction interval limit models $f_0(\mathbf{x_0})$, $f_1(\mathbf{x_0})$, and $f_2(\mathbf{x_0})$, in Example 6 for the energetic material design. Based on the results, a cDSP is formulated incorporating the EMIs, discussed in Example 7, to achieve robust system responses with ranged sets of design specifications as shown in Table 8.5 In the table, two other design formulations, a cDSP incorporating DCIs and traditional optimization, are also introduced and performed for the purpose of highlighting the difference of design results.

As shown in the right column of the CDSP for RCEM-EMI, the mean response function, $f_0(\mathbf{x})$, upper uncertainty bound function, $f_1(\mathbf{x})$, and lower uncertainty bound function, $f_2(\mathbf{x})$, are determined from the mean response model and the upper and lower prediction interval limits identified in Example 6. Other required inputs are derived from the specifications identified in Example 1.

The LRL for the response is set to 1, which means the number of reaction sites should be greater than or equal to 1, and the larger number is better in this example. This is the case in Figure 8.8c. The target EMI is set to 10. The deviations of design variables (control variables) are also identified in Example 1. The objective is finding the mean location of design variables (x_1, x_2, x_3, and x_4) for achieving the target EMI as closely as possible.

---

targets ($EMI_{target,i}$). As discussed in Section 8.1.4.1, EMIs are calculated in terms of system variables based on the mean response model and lower and upper prediction interval limits. In the cDSP for RCEM-EMI, ranged sets of design specifications for multiple targets are obtained by minimizing the objective function that is stated in terms of the deviation variables only.

The bounds of the design variable are derived from the design space identified in Example 1. By minimizing the deviation, d_1^-, underachievement to the target EMI, we obtain ranged sets of design specifications that are robust to the deviation in design variables, unparameterizable variability, and model parameter uncertainty due to the limited number of samples. We do not have to consider the deviation variable, d_1^+, in the objective function in this case, which means the EMI overachievement, exceeding the target value of EMI, will not be considered better system performance.

In the second row of the table, mathematical forms of the cDSP for RCEM-DCI and optimization for this design problem are illustrated. Comparing with the cDSP for RCEM-EMI, the cDSP for RCEM-DCI incorporates with only $f0(\mathbf{x})$. The DCI is calculated based on mathematical constructs in Figure 8.8c and response deviations (ΔY) are obtained by calculating the deviations of $f0(\mathbf{x})$ due to variations in system variables *without considering $f1(\mathbf{x})$ and $f2(\mathbf{x})$*. Solutions with the two other design exploration methods are also identified for validating usefulness of RCEM-EMI in the next section.

Table 8.5 Three algorithms for searching energetic material design specifications.

Methods	Mathematical Form
DSP for RCEM-EMI	**Given** $\mathbf{x} = \{x_1, x_2, x_3, x_4,\}$, (System variables) $f_0(x)$, (Mean response model) $f_1(x)$, $f_2(x)$, (Upper and lower prediction interval limits) $LRL = 1$ (Lower requirement limit, larger-is-better case) $EMI_{target} = 10$ (Target EMI) $\Delta\mathbf{x} = [\pm 0.2, \pm 0.1, \pm 0.01, \pm 0.1]$ (Variations in system variables) **Find** μ_x, (Mean location of system variables) d_1^+, d_1^- (Deviation variables) **Satisfy** *Goals*: $EMI(x) / EMI_{target} + d_1^- - d_1^+ = 1$ $\quad$ where $EMI(\bar{x}) = \{ f_0(x) - LRL \} / \Delta Y_{lower,}$ $\quad \Delta Y_{lower}$ is obtained at Example 7 *Bounds*: $0.5 \leq x_1 \leq 1.5, 0.2 \leq x_2 \leq 1,$ $0.02 \leq x_3 \leq 0.1,$ and $0.2 \leq x_4 \leq 1$ $d_1^-, d_1^+ \geq 0, d_1^- \cdot d_1^+ = 0$ **Minimize** $\quad Z = d_1^-$
cDSP for RCEM-DCI	**Given** $\mathbf{x} = \{x_1, x_2, x_3, x_4,\}$, $f_0(\mathbf{x})$ (Mean response model), $LRL = 1, DCI_{target} = 10$ (Target DCI), $\Delta\mathbf{x} = [\pm 0.2, \pm 0.1, \pm 0.01, \pm 0.1]$ **Find** μ_x, d_1^+, d_1^- **Satisfy** *Goals*: $DCI(\mathbf{x}) / DCI_{target} + d_1^- - d_1^+ = 1$ $\quad$ where $DCI(\mathbf{x}) = \{ f_0(\mathbf{x}) - LRL \} / \Delta Y$ $\quad$ where $\Delta Y = \sum_{i=1}^{4} \left| \dfrac{\partial f_0}{\partial f_1} \right| \cdot \Delta x_i$ *Bounds*: $0.5 \leq x_1 \leq 1.5, 0.2 \leq x_2 \leq 1, 0.02 \leq x_3 \leq 0.1,$ and $0.2 \leq x_4 \leq 1$ $d_1^-, d_1^+ \geq 0, d_1^- \cdot d_1^+ = 0$ **Minimize** $\quad Z = d_1^-$
Optimization	**Given** $f_0(\mathbf{x})$ (Mean Response Function) **Find** $\mathbf{x} = \{x_1, x_2, x_3, x_4,\}$

Table 8.5 (Continued)

Methods	Mathematical Form
	Satisfy
	$0.5 \leq x_1 \leq 1.5$, $0.2 \leq x_2 \leq 1$, $0.02 \leq x_3 \leq 0.1$, and $0.2 \leq x_4 \leq 1$
	Maximize
	$Z = f_0(\mathbf{x})$

8.1.5. Results and Discussion: Energetic Material Design

In this section, we discuss the usefulness of RCEM-EMI by comparing our example energetic material design results obtained based on three design methods: RCEM-EMI, RCEM-DCI, and traditional optimization.

8.1.5.1. Achieving Robustness Under Unparameterizable Variability

The energetic material design solutions based on the three methods are listed in Table 8.6. The values in the shaded cells, which cannot be achieved using either the optimization or RCEM-DCI method, are obtained via the RCEM-EMI approach. Clearly, all three approaches produce different design solutions.

The optimal solution is found on the upper bound of x_1 (mean radius of Al particles) and x_3 (volume fraction of void), and on the lower bound of x_2 (mean radius of Fe_2O_3 particles) and x_4 (mean radius of voids). Using traditional optimization, the maximum performance is determined using only the mean response function. We obtained the maximum mean response (the number of reaction sites within the SVE) of 21.8 at the upper bounds of x_1 and x_3 and the lower bounds of x_2 and x_4.

On the other hand, based on the RCEM-DCI approach, a robust design solution is found at the lower bounds of x_2~x_4 and 0.7 of x_1 by considering deviations in design variables. The RCEM-DCI approach leads to the robust solution using only the mean response function. At the solution, we achieved the maximum DCI, 2.74, over the entire design space.

Using the RCEM-EMI approach, we determine a robust design solution at the lower bounds of x_1, x_2, and x_4 and the upper bound of x_3 (volume fraction of voids). At the solution point, the maximum EMI of 1.10 over the entire design space is achieved. The minimum estimated numbers of reaction sites is 2.69. These are estimated by consideration of the deviation in the design variables, unparameterizable variability, and model parameter uncertainty. In other words, taking into account the deviations in x_1,x_4, unparameterizable variability and 20 replicates, we estimate the minimum number of reaction sites to be 2.69 with a 99% confidence level.

As shown in Table 8.6 the mean performance is the best at the optimal solution and the DCI is the highest at the RCEM-DCI solution. However, if we consider unparameterizable

Table 8.6 Comparison with optimal, RCEM-DCI, and RCEM-EMI design solutions.

Methods	$x_1(\mu m)$	$x_2(\mu m)$	x_3	$x_4(\mu m)$	EMI	DCI	Mean Num of Rxn Sites	Minimum Num of Rxn Sites
Optimization	1.5	0.2	0.1	0.2	1.05	1.75	21.8	1.92
RCEM-DCI	0.7 ± 0.2	0.2 ± 0.1	0.02 ± 0.01	0.2 ± 0.1	0.91	2.74	8.8	0.26
RCEM-EMI	0.5 ± 0.2	0.2 ± 0.1	0.1 ± 0.01	0.2 ± 0.1	1.10	2.27	18.9	2.69

variability, the performance deviation (the interval between the estimated minimum and maximum) at the two solution points is larger than that at the RCEM-EMI solution point as shown in the last two columns of the table. This is due to the optimization and RCEM-DCI approaches not accounting for the unparameterizable variability in the system during the design exploration. Additionally, the unparameterizable variability of the results of this shock simulation is quite large.

At the RCEM-DCI solution, it cannot be said for certain whether the system performance is satisfactory to the LRL (at least one reaction site) since the minimum estimated response is 0.26 with 99% confidence. At the optimal solution, the EMI is 1.05, which indicates the performance deviation due to uncertainty in the model is still satisfactory to the LRL. However, a smaller response variation and larger estimated minimum response (2.69) are achieved at the RCEM-EMI solution point, which produce the maximum EMI value (1.10). This result is supported by the plots shown in Figure 8.9.

These plots are the models that estimate the number of reaction sites in terms of x_1 and x_3. The mean radii of Fe_2O_3 and voids are fixed at 0.2 microns since all solutions are predicting this value as the design solution. As we expected, the optimal solution is on the peak of the mean response function, and the RCEM-DCI solution is on the flat region of the mean response function. On the other hand, the RCEM-EMI solution is on the other side of the optimal solution. At the RCEM-EMI solution point, the lower limit of the prediction interval is higher than others. Compared with the optimal solution, the response variation due to unparameterizable variability is smaller, while mean response is sacrificed somewhat. At the RCEM-DCI solution point, the lower limit of the prediction interval is very low; therefore, the response variation might violate the LRL.

All solutions are on either the edge or the vertices of the design space, which means the system performances and robustness could be improved further by examining the broader design space. In this example, we extend our design space up to the simulation capability limits, such as the limitation of particles' sizes that could be properly analyzed by the Raven simulation code. As the shock simulation capability is extended to incorporate the smaller particles, we may need to redesign our specifications in the broader design space.

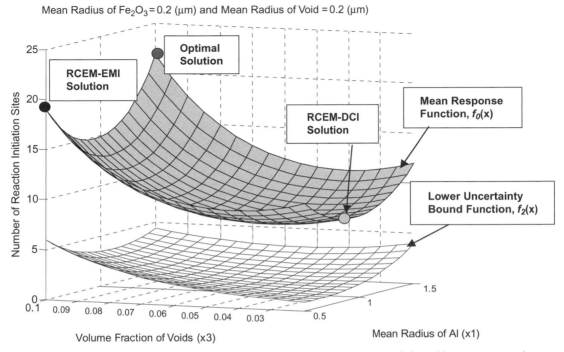

Mean Radius of $Fe_2O_3 = 0.2$ (μm) and Mean Radius of Void $= 0.2$ (μm)

Figure 8.9 The locations of the solutions in the mean response model and lower uncertainty bound functions (the upper uncertainty bound is not shown).

As discussed in this section, the RCEM-EMI solution embodies the smallest lower performance deviation (deviation from the estimated mean performance to the estimated minimum performance). Further, the mean performance is reasonably good when compared to the optimal solution. In other words, the RCEM-EMI searches for a solution to achieve not only the best mean performance but also smaller performance deviation considering unparameterizable variability. This is why the RCEM-EMI solution for the energetic material differs from the traditional optimization solution.

In this energetic material design example, the LRL is set to 1. The designer may want to increase this limit to higher number. Let's assume the LRL is set to 15. Then, the solution obtained by RCEM-EMI will not be the same as before. The RCEM-EMI will find the optimal solution that has been found by the traditional optimization since the EMI at the optimal solution will be the highest. Therefore, in RCEM-EMI, it is very important for designer to carefully set the requirement limits (LRL and/or URL) considering operating condition and problem objective.

An argument regarding the mathematical construct of the EMI is that the EMI cannot explicitly convey the designer's preferences since an EMI is calculated based on a set of combinations of mean location and performance deviation. For this reason, an EMI with a smaller distribution and a mean closer to a requirement limit could be identical to an EMI with a large distribution and a mean farther from the limit. In some cases, however, a designer may prefer the larger distribution, whose mean is located farther from the limit. In order to improve this limitation, we may have a few options. One of them is to separate the mathematical combination of performance mean and variance in order to control them individually. Two individual goals for mean and variance are created in the cDSP formulation. Formulating an objective function composed of deviation variables, it is necessary to study the tradeoff between the achievement of mean and variance. In order to extend design freedom in downstream activities, designers may find sets of design specifications at which system performances (i.e., means and variances) lie on a Pareto frontier.

Another alternative of the EMI is to use preference functions. The utility function (Myerson 1991, Otto and Antonsson 1993) and the Taguchi's Loss function discussed in Chapter 6 are useful for customizing the designer's preference. Using the preference functions, designers should consider the uncertainty in a formulated preference function itself.

The EMI goal formulation, however, is still valid and useful to identify a robust solution if a designer is not concerned about system performance once the performance satisfies a requirement limit. For example, if a number of activated cells in a SVE are good enough to initiate the reaction over the whole SVE of the energetic material, then designers will not care about the number of reacted cells. Instead, they will focus on whether or not the SVE will include at least the number of activated cells. In the energetic material design example, we believe that the EMI goal formulation is more reasonable. In addition, the simplicity of the EMI construct eliminates the backward performance check regarding performance requirement limits, which is necessary in the separate mean and variance goal formulation approach.

8.1.5.2. *Achieving Robustness Under Model Parameter Uncertainty*

In Chapter 6, different types of uncertainty are discussed. It is argued that the RCEM-EMI approach yields solutions that are robust to unparameterizable variability and model parameter uncertainty—Type III robust design. In this section, the RCEM-EMI approach is validated with respect to searching for a solution that is robust to model parameter uncertainty. For the purpose of this demonstration, we proceed from the reduced sample size (11 replicates) to the full sample size (20 replicates). With increasing sample size, design exploration results based on three different search methods are plotted in Figure 8.10.

While the trend in solution convergence of RCEM-EMI with increasing sample size is stable, those of other approaches are not. All design solutions of RCEM-EMI converge and are stabilized at 15 replicates. Using the traditional optimization approach, the solutions of x_2, x_3, and x_4 converge at 16 replicates, but the solution of x_1 is still unstable at 20 replicates. Using the RCEM-DCI approach, the solutions of x_1, x_2, and x_4 converge at 15 replicates,

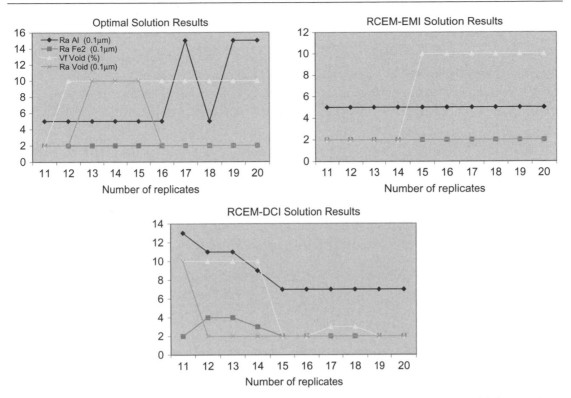

Figure 8.10 Solution convergence of optimization, RCEM-DCI, and RCEM-EMI with increasing sample size.

but x_3 is still somewhat unstable at 20 replicates. This means that more samples are necessary to achieve converged design solutions for the optimization and RCEM-DCI approaches.

The reason for this stability is that the prediction interval, employed in RCEM-EMI, is the estimation of the limits of new sample locations that will be added later. Design exploration with EMI considers this prediction interval when it searches for the solution. Since the mean response model for all methods is obtained using an iteratively reweighted regression technique, the mean response model is more robust to large random errors. With a normal regression technique, the convergence of the design solutions using the optimal and RCEM-DCI should be less stable.

The advantages of RCEM-EMI approach are (1) the design solutions are the more robust against model parameter uncertainty due to lack of data, and (2) convergence is achieved at smaller sample sizes compared with the traditional solution search algorithms. This latter characteristic renders RCEM-EMI highly desirable when complex sets of simulations are required.

8.1.6. Summary of RCEM-EMI

Various sources of uncertainty in a model, including natural, parameter, and model structure uncertainty, can be ignored in some cases. However, in most engineering applications, the uncertainty in the system model must be considered when decisions (design of system specifications) are made. Many robust design methods have been developed. None, however, have been developed to explicitly account for the inherent uncertainty embodied in models. Taking this source of uncertainty into account, a robust design method, the RCEM-EMI, is presented in this chapter. The RCEM-EMI helps a designer make decisions under a system's random variability and/or model parameter uncertainty in a model. Applying the RCEM-EMI in the shock simulation of energetic materials, we observe the following advantages over conventional optimization and robust design methods (RCEM-DCI) from the results shown in Section 8.1.5:

- Unparameterizable random changes in a material's microstructure cause large variations in material performance. A smaller performance deviation at the RCEM-EMI solution point occurs than at the solution points obtained using traditional optimization and RCEM-DCI. Further, the mean performance at the RCEM-EMI solution point is still reasonably good. Therefore, we suggest that the performance of a material system designed using RCEM-EMI is more robust to unparameterizable variability than a material system designed using either traditional optimization or RCEM-DCI.

- As shown in Figure 8.10, using the RCEM-EMI, we obtain stable design solutions with rapid convergence as the sample size is increased. This suggests that the RCEM-EMI design specifications are robust to model parameter uncertainty that arise as a result of a lack of sampling data. Hence, we suggest that a designer can make decisions that are robust even if these statistical models are uncertain due to the limitations in sample size.

- In order to minimize the computational load, RCEM-EMI embodies an integrated mean and variance metamodeling approach for estimating the amount of variability embedded in the system. A converged design solution with 375 samples using RCEM-EMI is illustrated in Figure 8.10. If traditional uncertainty analysis (e.g., Monte Carlo simulation) is invoked in a design exploration process, a large amount of sampling is needed to evaluate a single point, resulting in a method that is infeasible due to the computational requirements. Herein lies the advantage of RCEM-EMI; computationally intensive materials analyses and simulations are embodied in RCEM-EMI.

- In robust design, using traditional optimization, it is often difficult for designers to determine the tradeoff between performance and performance sensitivity. However, performance and performance sensitivity are formulated into a single index (EMI or DCI) in RCEM-EMI and RCEM-DCI, thus circumventing this problem

It is important to reduce uncertainty in a system by getting more knowledge or data of the system. However, if the system uncertainty cannot be reduced, the RCEM-EMI presented in this chapter helps designers pursue robust and reliable solutions despite remaining uncertainty in the system.

8.2. Inductive Design Exploration Method—A Multilevel Robust Design Method

As discussed in Chapter 6, uncertainties in the models are propagated through a multilevel model chain and the performance estimation at the end of the models chain may have high degree of uncertainty. We call this "propagated uncertainty." A simple analysis chain is illustrated in Figure 8.11.

Input variables ($x1$, $x2$) may be parameters related to a material's process route or structure that may be tailored by materials designers. The functions ($f1$, $f2$) may be simulation models for predicting material properties ($y1$, $y2$), such as modulus of elasticity, ultimate strength, yield strength, the Hugoniot relation, etc. The derived materials properties or responses are interfaced to a product-level model (g), such as a finite element analysis model, and then a system-level response of interest (z), such as structural integrity, thermal behavior, etc. In this simple schematic example, uncertainty may be accumulated and amplified through the sequential chain, making the variance of the final response (z) unacceptably large. Small variations or errors in input parameters may cause high levels of variability in the system response. Therefore, the following three important strategies are formulated in order to overcome the challenges.

- **Explore hierarchical design space in an inductive (top-down) manner**: The basic idea of finding a set of solutions that are robust against propagated uncertainty is to pass down the feasible solution range in an inductive (top-down) approach, i.e., *finding a feasible range in the lower-level space from a given desired range in the higher-level space*. The term *inductive* is adapted from Olson's materials design hierarchy (Olson 1997). The inductive approach is the opposite of the deductive (bottom-up) approach, in which the

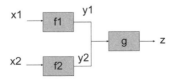

Figure 8.11 An example of information flow in a multilevel model chain.

mean and variance (or deviation) of an output is passed to the higher level subsystems in a hierarchy.

*   **Find ranged sets of design specifications:** The inductive design exploration method allows designers to find not only a single solution, but also ranged solution sets among which designers may choose the best solution based on their preferences. While the feasible solution space is passed down along a model chain, it is sometimes necessary to determine specifications of the upstream design variables[1] in the middle of the model chain. A guiding principle for this decision is to maintain as large a feasible range as possible when upstream design variables need to be determined.

*   **Parallelize multiple function evaluations**: The two processes, design exploration and uncertainty analyses, are tightly coupled in previous approaches. Considering the large amounts of resources and time required for uncertainty analysis, it is desirable to decouple those two processes so that we can parallelize the uncertainty analysis. In multilevel robust design, this decoupling involves concurrent evaluation of the means and deviations of performances, storing the results in a database, and exploring a design space by retrieving the results

Applying our strategies discussed previously, the following advantages are expected in integrated materials and product design. Product designers may input performance requirements in product level performance and the IDEM will find feasible spaces in materials properties domain. The identified feasible spaces (i.e., feasible data sets) will be passed down to material designers. The materials design will again find feasible spaces in the domain of parameters of materials structures or processing paths using the IDEM. In the IDEM, we perform parallel evaluation of the simulation or material models in the model chain before conducting the inductive feasible space search.

8.2.1. Overview of IDEM

The overall procedure of IDEM is illustrated in Figure 8.12 and described in the following section.

*   Step 1: It is necessary to define the rough search space in design variables (i.e., search space in **x**), interdependent variables (i.e., search space in **y**), and the performance variables (i.e., search space in **z**). Within the search spaces of all the variable vectors, discrete points are generated for the model evaluation in the next step with a reasonable amount of resolution.

---

[1] Design variables that may be introduced in the middle of a model chain (e.g., design variables that are inputs to model "g" but outputs from neither "f1" nor "f2" in Figure 8.10)

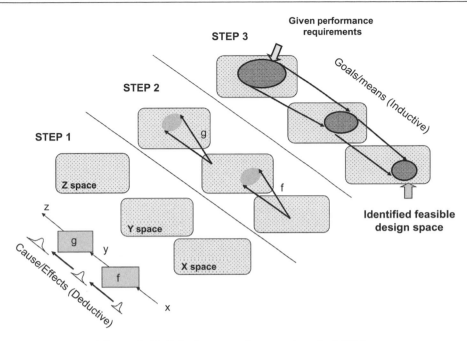

Figure 8.12 Solution search procedure in IDEM.

- Step 2: Each of the generated discrete points in the spaces is evaluated in parallel using the mapping models (models **f** and **g** in Figure 8.12). Evaluated data sets, composed of a discrete input point and output range, are stored in a database. We claim that the model will produce not only discrete output but also ranged output taking into uncertainty in input parameters and model itself. Details about input and ranged output are discussed in Section 8.2.2.

- Step 3: With a given final performance range in **z** space, feasible ranges in **y** and **x** spaces are sequentially identified as shown in Figure 8.13. We call this step Inductive Discrete Constraints Evaluation (IDCE); its details are discussed in Section 8.2.3.

In the following, we introduce a simple example, the design of a clay-filled polyethylene cantilever beam design, for better understanding the overall procedure of IDEM. This example is a representative problem of integrated materials and product design, which employ a chain of simulation models.

The analysis model chain for the cantilever beam design is depicted in Figure 8.15. First, the volume fraction of clay (x) is an input of models f_1, f_2, and f_3 for the density (ρ), modulus of elasticity (E), and tensile strength (σ_u) of the composite material, respectively. The properties of the composite material and the radius of the beam are the inputs of g_1, g_2, and g_3 that are used to estimate the beam weight (W), beam deflection (δ_{max}), and maximum load (P_{max}),

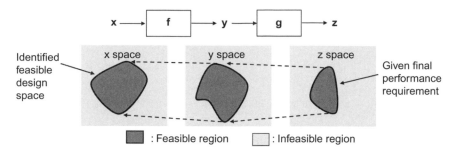

Figure 8.13 Inductive discrete constraints evaluation (IDCE).

Example 9. Circular Sectional Cantilever Beam Design with Clay-Filled Polyethylene

Many engineering polymers that contain fillers and extenders are particulate composites used for various engineering applications (Askeland 1994). As shown in Figure 8.14, designers can tailor the composite material's properties for a specific need, adding clay particles into polyethylene as an extender. The tensile strength of the composite decreases and the modulus of elasticity increases as the volume fraction of clay in polyethylene increases; therefore, it is important to tailor the volume fraction of clay in polyethylene for desirable product performance.

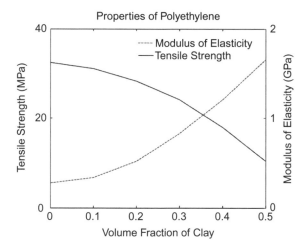

Figure 8.14 Effect of clay on the properties of polyethylene (Askeland 1994).

respectively. Let's assume that the length of the beam is fixed as 1 meter long. The details of the mapping models are presented here.

$$\rho = f_1(x) = x\rho_{clay} + (1\text{-}x)\rho_{polyethylene}(kg/m^3)$$

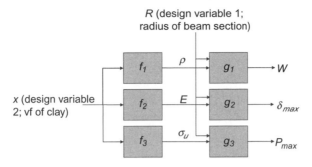

Figure 8.15 Chain of clay-filled cantilever beam analyses.

$$E = f_2(x) == 4875.6x^2 + 365.42x + 264.96 \ (MPa)$$

$$\sigma_u = f_3(x) = -80.028x^2 - 3.9152x + 32.331 \ (MPa)$$

$$W = g_1(\rho,R) = \pi R^2 L \rho \cdot 10^{-3} (kgf)$$

$$\delta_{max} = g_2(E,R) = 9.8PL^3 / 3EI \ (mm)$$

$$P_{max} = g_3(\sigma_u,R) = \sigma_u I / 9.8LR \ (kgf)$$

Where,

$$\rho_{clay} = 1150 \ (kg/m^3), \quad \rho_{polyethylene} = 940 (kg/m^3),$$
$$L = 1000 (mm), \ I = \pi R^4 / 4$$

The design objective in this problem is to determine the radius of the circular section of the cantilever beam (R) and the volume fraction of clay in polyethylene (x) for the beam material with the following requirements:

- The deflection with 150 *kgf* load at the end of the beam must not exceed 35 mm, and lower deflection is better.

- The beam must withstand at least a 500-*kgf* load without failure, and being able to withstand a larger load is better.

- The weight of the entire beam must not exceed 30 *kgf*, and a lighter beam is better.

Associating the variables and models in Figure 8.12 with the beam example discussed previously, the design variable vector, **x,** includes the volume fraction of clay, x;

therefore, $\mathbf{x} = \{x\}$. Similarly, the interdependent variable vector is $\mathbf{y} = \{ R, \rho, E, \sigma_u \}$ and the performance variable vector is $\mathbf{z} = \{W, \delta_{max}, P_{max}\}$. The material model vector is $\mathbf{f} = \{f_1, f_2, f_3\}$ and the product model vector is $\mathbf{g} = \{g_1, g_2, g_3\}$.

- Step 1: $\mathbf{x}$ space is defined as $\mathbf{x} = \{[0, 0.5]\}$ such that we search for design solutions within the space of clay volume fraction between 0 to 0.5. Similarly, $\mathbf{y}$ space is $\{ [1, 101](mm), [940, 1150](kg/m^3), [200, 1800](MPa), [0, 40](MPa) \}$ and $\mathbf{z}$ space is $\{[0, 150](kgf), [0, 100](mm), [100, 5100](kgf)\}$. The resolution of x may be 0.01; therefore, it creates 51 discrete evaluation points in $\mathbf{x}$ space. Similarly, we generate discrete points in $\mathbf{y}$ and $\mathbf{z}$ space with a resolution of each variable.

- Step 2: All discrete values of clay volume fraction (generated discrete points in the $\mathbf{x}$ space) are evaluated based on the material models, f_1, f_2, and f_3, generating ranged material properties, ρ, E, and σ_u (ranged output in $\mathbf{y}$ space). At the same time, all discrete points in the $\mathbf{y}$ space also evaluated based on the product model, g_1, g_2, and g_3, generating ranged output of W, δ_{max}, and P_{max} (ranged outputs in $\mathbf{z}$ space).

- Step 3: All feasible combinations of R, ρ, E, and σ_u (feasible points in $\mathbf{y}$ space) that satisfy the given design requirements in deflection, maximum load capacity, and total weight (given requirement ranges in $\mathbf{z}$ space) are searched. Similarly, with obtained feasible combinations of R, ρ, E, and σu (feasible points in $\mathbf{y}$ space), all feasible specifications of clay volume fraction (feasible points in $\mathbf{z}$ space) will be searched in the same manner.

- Although this beam example is simple, the procedure is valid for finding robust ranged sets of specifications in all types (sequential, parallel, and hierarchical) of subsystem network. The details of these procedures are explained in Sections 8.2.2 and 8.2.3.

8.2.2. Discretization and Parallel Discrete Function Evaluation—STEPS 1 and 2

In this section, we discuss the parallel discrete function evaluation, projecting a discrete input vector to an output space. A discrete function evaluation receives a discrete input vector as well as input deviations at the input vector and produces *a range of output, not a single deterministic output*, given the effects of quantified uncertainty. This function evaluation is defined as a "*mapping*," and the associated uncertainty measures can be calculated based on available uncertainty analysis techniques. In most cases, *the output space is multidimensional* and multiple attributes are evaluated in parallel. For example, the input space shown in Figure 8.16 is 2D $\mathbf{x} = \{x_1, x_2\}$ and the output space is 2D $\mathbf{y} = \{y_1, y_2\}$. We have two mapping models (e.g., evaluation functions for a mapping), $\mathbf{f} = \{f_1, f_2\}$. Here, it is assumed that an output is evaluated by a single mapping model. Therefore, the rank of the mapping model vector should be identical to the rank of the output vector. Even if a simulation model produces multiple outputs, the number of mapping models is identical to the number of

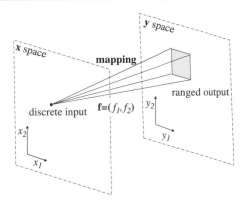

Figure 8.16 An example of mapping between input and output spaces.

multiple outputs in a systems perspective. In this case, the simulation models are the same for the outputs, but the mapping models are different from each other.

In many engineering applications, design variables are shared by multiple mapping models, as shown in Figure 8.17. The parallel discrete function evaluation process is described in detail in the context of the shared variable example. In this example, the mapping models, $f1$ and $f2$, share a design variable ($x2$), while both have independent variables. It is assumed that the space of each design variable is discretized by two discrete points. Therefore, $x1 = \{1, 2\}$, $x2 = \{a, b\}$; and $x3 = \{I, II\}$.

The process is as follows:

- **Discretizing**: all possible combinations of the discrete points of the design variables are created. In the example, discretized points in the design space of $x1, x2,$ and $x3$ are

$$\mathbf{DP} = \begin{Bmatrix} (1,a,I), & (1,a,II), \\ (1,b,I), & (1,b,II), \\ (2,a,I), & (2,a,II), \\ (2,b,I), & (2,b,II), \end{Bmatrix} \tag{8.20}$$

- **Grouping**: The discretized points created by the discretization process are grouped as input sets for mapping models. In this example, two groups are formed for the mapping models, $f1$ and $f2$. One group, $\{(1,a), (1,a), (1,b), (1,b), (2,a), (2,a), (2,b), (2,b)\}$, is formed by the first and second elements of the discretized points, which is subject to $f1$ and requires inputs $x1$ and $x2$. The other group is formed by the second and third elements of the discretized points $\{(a,I), (a,II), (b,I), (b,II), (a,I), (a,II), (b,I), (b,II)\}$, which is subject to $f2$ and requires inputs $x2$ and $x3$. Once the two groups are formulated, duplicate points

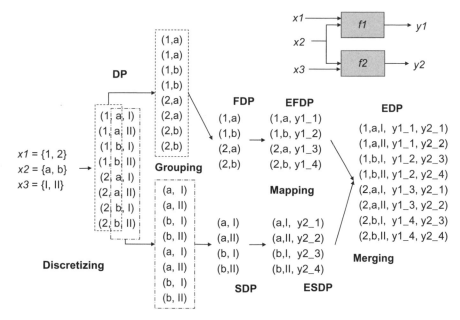

Figure 8.17 An example of function evaluation with multiple inputs and outputs with shared variables.

are eliminated in each of the groups. Thus, the first group becomes **FDP** = {(1,a), (1,b), (2,a), (2,b)}, and the second one becomes **SDP** = {(a,I), (a,II), (b,I), (b,II)}.

- *Mapping*: The points in each group are evaluated in parallel by each mapping model and the evaluation results are stored with the input points. In the example, the input points in **FDP** are evaluated based on *f1*, and the results are stored as **EFDP** = {(1, a, y1_1), (1, b, y1_2), (2, a, y1_3), (2, b, y1_4)}. Similarly, the input points in **SDP** are evaluated using *f2*, and the results are stored as **ESDP** = {(a, I, y2_1), (a, II, y2_2), (b, I, y2_3), (b, II, y2_4)}. Here, a result of *y1*, *an output range of y1*, is represented as *y1_i*, and a result of *y2* as *y2_j*, where *i* and *j* are positive integers.

- *Merging*: Sets of input points and corresponding outputs obtained in the *mapping* step are merged to form a set of evaluated original discrete points. This point set includes original discreet points obtained in the previous *discretizing* step with corresponding evaluation results. In the example, **EFDP** and **ESDP** are merged to:

$$
\mathbf{EDP} = \begin{cases}
(1, a, \ I, \ y1\_1, \ y2\_1), \ (1, a, \ II, \ y1\_1, y2\_2), \\
(1, b, \ I, \ y1\_2, \ y2\_3), \ (1, b, II, \ y1\_2, y2\_4), \\
(2, a, \ I, y1\_3, \ y2\_1), \ (2, a, \ II, \ y1\_3, y2\_2), \\
(2, b, \ I, \ y1\_4, y2\_3), \ (2, b, II, \ y1\_4, y2\_4),
\end{cases}
\tag{8.21}
$$

In Equation 8.21, the first element of **EDP,** (1, a, I, y1_1, y2_1)**,** is formed using the first element of **EFDP,** (1, a, y1_1), and the first element of **ESDP,** (a, I, y2_1) since the inputs of the elements, (1, a) and (a, I), are combined to be (1, a, I). The other **EDP** elements may be formed in the same way.

This process facilitates independent mappings associated with *f1* and *f2* that are amenable to *parallel computation* to increase efficiency, and it captures the interdependency of multiple outputs due to shared input variables.

8.2.3. *Inductive Discrete Constraint Evaluation (IDCE): STEP 3*

The third step, the IDCE process, is described in this section. In this step, designers sequentially find feasible ranges in the spaces of interdependent variables[2] and design variables, based on the data generated in Step 1. The IDCE process includes the following three substeps:

- Step (a): find satisfactory points in an input space with given constraints (feasible ranges) in an output space based on Hyper-Dimensional Error Margin Indices (HD_EMIs), to be discussed in the next subsection,

- Step (b): obtain contours for the borders of the feasible regions in an input space, creating border points between discrete satisfying points and points that are not satisfying, and

- Step (c): sequentially repeat steps (a) and (b) to find feasible regions at the lower levels using the borders of feasible regions obtained in the previous step (b) as the constraint bound in the output space.

IDCE requires feasible ranges in a hyper-dimensional (multiple outputs) output space that may be given as requirements from a design problem. However, these feasible ranges may also be identified from the previous Discrete Constraint Evaluation (DCE) process. This means that IDCE is an inductive cascading process of DCE.

Step (a) is discussed in Section 8.2.3.1, Step (b) in Section 8.2.3.2, and Step (c) in Section 8.2.3.3.

8.2.3.1. *Find Feasible Ranges (Satisfactory Discrete Points) in an Input Space Based on Hyper-Dimensional Error Margin Indices (HD_EMIs)*

Step (a) of IDCE is checking whether the mapping of each discrete point from an input space to an output space lies within given feasible ranges in output space. For this feasibility check, we use the HD_EMIs as metrics. The HD_EMIs are extensions of EMIs (Choi, Austin et al. 2005)

---

[2] Parameters that are an input to a model as well as an output from another model.

from single-dimensional (single-output) to hyper-dimensional (multiple-output) spaces. In order to calculate HD_EMIs at each discrete point of input space, we need to determine whether the mean of an output range is in the feasible range or not.

This determination is not simple since the feasible range in an output space is represented as a discrete set of points when the feasible range is passed from the previous DCE task. We consider the nearest neighbor points from the location of the mean of the output range obtained by the mapping model. If more than half of the nearest neighbor points are feasible points, then we assume that the mean is in the feasible range. For more conservative applications, all nearest neighbor points must be feasible points for the mean to be in the feasible range. When the mean vector of an output range is not in the feasible range, then HD_EMIs of all outputs are

$$HD\_EMI_{all} = -1 \qquad (8.22)$$

This means the discrete input point does not satisfy the constraint bound in the output space.

When the mean vector is inside the feasible range, we identify the value of the HD_EMI in each output direction. A direction represents one output (e.g., thermal conductivity) among multiple outputs (e.g., thermal conductivity and Young's modulus) for the measurement of HD_EMI. The HD_EMI in each output direction is

$$HD\_EMI_i = Min\left[\frac{\|(\mathbf{mean}\text{-}\mathbf{B_j})\mathbf{u_i}\|}{\|(\mathbf{mean}\text{-}\mathbf{B_j^i})\mathbf{u_i}\|}\right] \qquad (8.23)$$

where $i = 1, .. ,$ number of directions, $j = 1, .. ,$ number of discrete points on a constraint boundary, **mean** is the vector of the mean in an output range, $\mathbf{B_j}$ is a discrete point vector on a constraint boundary, $\mathbf{B_j^i}$ is the projected point vector of $\mathbf{B_j}$ onto the nearest boundary of the output range along the ith direction, and $\mathbf{u_i}$ is a unit vector along the ith direction.

In Equation 8.23, an HD_EMI in the ith direction (*HD_EMIi*), is the minimum HD_EMI among all HD_EMIs that are calculated using discrete points on a constraint boundary (B), their projected points on output boundary in the ith direction (B^i), and the mean of the output range (mean). *As HD_EMIi increases, the output range moves farther from the constraint boundary in the ith direction.* Details of the HD_EMI calculation are clarified in Figure 8.18, in which the output space is two-dimensional (two outputs) and the feasible region in the output space is bounded by a contour (constraint boundary).

A rectangular region (a HyperCube in the multidimensional case) representing an output range at a discrete input has an associated mean. All points on the constraint boundary are subjected to the mapping on the output range boundary along each independent direction (coordinate) of the space. For example, a point (B$_j$) on the constraint boundary is mapped

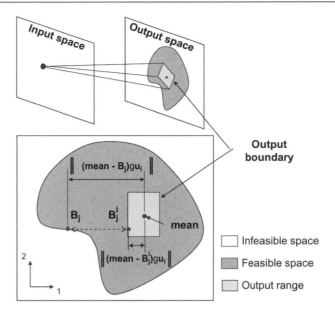

Figure 8.18 HD_EMI calculation in a given direction.

onto the boundary of the output range (rectangular) in direction 1, and a projected point (B_j^1) is created. Using these two points, B_j and B_j^1, we calculate a value of HD_EMI in direction 1, which is the ratio of the normed distance in direction 1 between the mean and B_j versus the mean and B_j^1. This HD_EMI calculation is performed for all other discrete contour points that can be likewise mapped in direction 1 onto the output boundary. Among the calculated HD_EMIs, the minimum is denoted as HD_EMI1, the HD_EMI in direction 1. The HD_EMI$_2$ is also obtained using the same technique. In a multidimensional case, HD_EMIs for all other directions are calculated similarly.

We can obtain accurate HD_EMIs even if there are isolated multiple feasible regions (even single points) since HD_EMIs are calculated based on all constraint boundary points in an output space and the minimum HD_EMI are selected for each output direction. The HD_EMIs obtained in this process are used to formulate exact discrete constraint boundary points in the input space, as discussed in the next section.

8.2.3.2. Generating Constraint Boundary Points in Input Space

A constraint evaluation with a coarse discretization can result in a rough estimate, leading to some output ranges that are considered as satisfying constraints for coarse discretization but may be unsatisfactory if much finer discretization is undertaken. Discretization errors can be alleviated by increasing resolution, which will, however, increase the amount of computation required. This problem occurs frequently in the vicinity of a border between satisfactory and unsatisfactory regions. Since an HD_EMI calculation requires accurate measurement of the distance from an output range boundary to constraint boundaries, accuracy of constraint boundaries is essential.

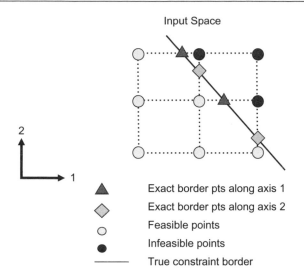

Figure 8.19 Generation of exact constraint boundary points.

As shown in Figure 8.19, an exact boundary is located between discrete points in an input space that satisfy and do not satisfy constraints, respectively. We need to get an accurate estimate of the location of the points on the border. An exact point between the feasible and infeasible points along axis 1 (the triangular points) is determined using a numerical root finding method, fixing the values constant along axis 2. An exact border point along axis 2 (the diamond points) is determined in the same way, fixing the value constant along axis 1 at the evaluation point. Since this algorithm evaluates all intervals between feasible and infeasible points, we may evaluate not only a single feasible region, but also multiple isolated feasible regions in an input space unless a feasible region is smaller than the resolution of discretization. Gradient based root finding methods, such as the Newton-Raphson Method, are not appropriate, since the HD_EMIs used for constraint evaluation are not in an explicit form. In this chapter, we use the Bisection Method (Hoffman 1992); however, False Position Methods are also applicable.

8.2.3.3. IDCE Using HD_EMIs for Multilevel Robust Design

As mentioned, the IDCE determines feasible spaces of input variables (design space) from the given final output range (performance requirements) by a recursive DCE process motivated by the inductive nature (from top to bottom) of the design process. The IDCE process, Step 3 of IDEM, is described as follows.

The evaluation procedure starts with an assumption that a required range of the final performance, **z**, is given, as shown as the gray area in **z** space in Figure 8.12. From the given required range in **z** space, we obtain feasible discrete points in **y** space. If an HD_EMI calculated in **z** space based on mapped output ranges from a discrete point in **y** space is

---

Example 10. IDCE Process of Circular Sectional Cantilever Beam Design with Clay-Filled Polyethylene

The IDEM has been implemented in MATLAB® and executed for the beam problem. The resolutions for discrete points are fixed as 5(mm), 20(kg/m^3), 100(MPa), 4(MPa), and 0.01 for R, ρ, E, σ_u, *and x*, respectively. These resolutions are reasonably small in order to be able to ignore discretization errors. In the example, it is assumed that the amount of uncertainty in the material and product models, **f** and **g**, is ±10% of the calculated output.

First, entire feasible ranges in **y** space (spaces of R, ρ, E, and σu) will be searched with given performance requirements described in Example 9. The required HD_EMIs (HD_EMI_w, HD_EMI_δ, and HD_EMIP_{max}) for mapping models (g_1, g_2, *and* g_3) are set as greater than or equal to 1, which means the quantified uncertainty (i.e., ± 10% of the calculated output) must be satisfied. Among the obtained feasible space of R, ρ, E, and σ_u, we select the value of R (radius of beam) that has the largest feasible space of the rest of the properties (i.e., ρ, E, and σ_u) because it is desirable to maintain the feasible region as large as possible until the end of the design process. As shown in Figure 8.20, the largest feasible range in the property space (i.e., the space of ρ, E, and σ_u) is achieved at $R = 81$(mm).

---

greater than or equal to a given *required* HD_EMI, then the discrete point in **y** space is a feasible point.

After identifying all feasible discrete points, we obtain exact boundary points between feasible and infeasible discrete points in **y** space. Once we obtain the feasible regions (i.e., discrete feasible points and exact boundary points) in **y** space, then we may find feasible regions in **x** space using the same procedure. The identified feasible region in **y** space becomes the range of performance requirements in this evaluation task because **y** is a vector of interdependent variables (see Figure 8.12).

Finally, the range obtained in **x** space is the solution space in which designers may select their best solution or may pass this range to other experts to further reduce the solution space obtained. However, if x and y are shared variables (i.e., a parallel chain), the feasible range identified in **y** space becomes the design space in an evaluation task for finding a feasible range in **x** space. The IDCE process can handle both cases. To demonstrate IDEM, an example is presented in the next section. This example includes mixed interdependent and shared variables with an upper-level space.

In the figure, satisfactory discrete points (circular points) and boundary points (diamond points) between the feasible and infeasible spaces are shown in the figure. With the feasible range achieved in **y** space shown in Figure 8.20, the feasible space of the volume fraction of clay (x) is identified by setting the required HD_EMIf_1, HD_EMIf_2, *and* HD_EMIf_3 as unity.

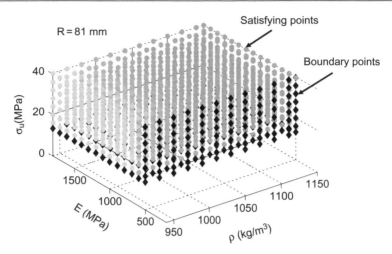

Figure 8.20 Achieved feasible discrete points in tensile strength, modulus of elasticity, and density space.

The achieved feasible space of the volume fraction of clay is between 0.19 and 0.42. Consequently, if we set radius of beam section as 81 (mm), then any clay volume fraction between 0.19 and 0.42 will guarantee satisfaction of the given design requirements, considering the uncertainties of the material and product models (i.e., $\pm 10\%$ of the calculated output) and those propagating through the beam model chain.

8.2.4. Robust Multilevel Design of an Energetic Material Using IDEM

In Chapter 2, multiscale simulation models for the analysis of exothermic reactions of energetic materials are discussed. This multiscale model chain includes heterogeneous mesoscale and homogenized continuum models. We consider this to be a multilevel model chain since the output of mesoscale model is the input of continuum scale model. Interfacing the multiscale models and propagating uncertainty through the simulation chain were also discussed. In this section, we apply IDEM to the design of energetic materials for robust reaction initiation based on the multilevel, multiscale simulation models. An overview of a multiscale model chain for reaction initiation in energetic materials is given in Figure 8.21.

The mesoscopic model for the energetic material mixture includes four design variables, mean radius of Al particles (x_1), mean radius of Fe_2O_3, (x_2), volume fraction of voids (x_3), mean radius of voids (x_4), and one output, local hot spot temperature (T_{ignit}). The Discrete Particle Mixture (DPM) model is highly heterogeneous. The continuum Non-equilibrium Thermodynamics Mixture (NTM) model, a homogenized constitutive model that reflects underlying heterogeneity, has two inputs (i.e., x_3 and T_{ignit}) and one output ($acFe$). As shown in Figure 8.21, common variables in the two models are in dotted boxes. The volume fraction

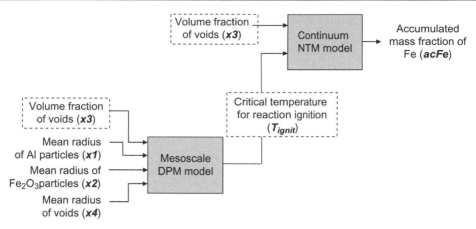

Figure 8.21 Multilevel, multiscale energetic material model chain; DPM refers to the discrete particle mixture hydrocode simulations.

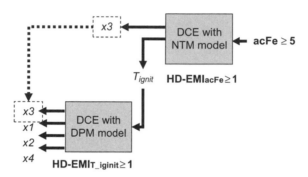

Figure 8.22 Multilevel robust energetic material design based on IDEM.

of voids is the input variable shared by the two models. The output of the DPM model, the weighted average temperature of local hot spots at initiation of the first reaction, is one of the inputs to the NTM model, and is the critical temperature for reaction initiation; this is the shared variable.

The logic for the interface between the NTM model and DPM model is to characterize local reaction initiation conditions (incipient thermal instability of chemical reactions) using the DPM model and then input the identified reaction initiation condition in the NTM model as a function of particle morphology and distribution. It is assumed that temperature is the main criterion for assessing initiation of a chemical reaction.

8.2.4.1. Clarification of Tasks in the IDEM for Multilevel Energetic Material Design

As previously described, IDEM is an inductive (top-down) design process to sequentially identify feasible design spaces from the highest-level requirements to the lowest-level design space. The IDEM for multilevel robust design of an energetic material is described in Figure 8.22.

Table 8.7 Clarification of the continuum-level design task.

Clarified Items	Specifications
Design variables and space	T_{ignit}: Critical temperature for chemical reaction initiation x_3: Volume fraction of voids $T_{ignit} = [1, 1.6]$ (1000 K) $x_3 = [0.02, 0.1]$
Response	*acFe*: Accumulated mass fraction of Fe
Uncertainty	Variability in the volume fraction of voids: $\Delta x_3 = \pm 0.01$
Fixed parameters	Volume fraction of intermetallic compound (Al + Fe2O3): 50% of the Total Volume Volume fraction of Al : Volume Fraction of $Fe_2O_3 = 2{:}3$ (reactants are in stoichiometric proportion) Left boundary velocity : 1 (km/s) Length of specimen: 4×10^{-3} (m) Time interval for calculation : $0\sim 300 \times 10^{-9}$ (seconds)
Task objectives	Achieve *acFe* ≥ 5 considering uncertainty in the NTM model Identify the feasible space of T_{ignit} and x_3

In the continuum-level design task, the first design task in IDEM is to determine the feasible design space for (x_3, T_{ignit}). The final performance requirement on accumulated mass fraction of Fe is *acFe* ≥ 5, representing the required minimum amount extent mesoscale-level design task. Details of the continuum-level design task, including initial design spaces, responses, associated uncertainty in the NTM model, fixed parameters and condition for simulation, and task objectives are listed in Table 8.7.

The mesoscale-level design task, the second design task, is summarized in Table 8.8. Deviations in the design variables ($\Delta x1 \sim \Delta x4$) represent variations from the true mean in a small statistical volume element. This variation may be present even if the supplier has measured each particle size accurately. The design space of x_3, the shared variable, is the feasible space passed down from the continuum-level design task. The response of this design task is the interdependent variable, T_{ignit}, which is the weighted average temperature of local hot spots at the first reaction initiation. This parameter is equivalent to the critical temperature in the NTM model.

8.2.4.2. Parallel Discrete Function Evaluation—Steps 1 and 2 of IDEM

The MATLAB® analysis code for the NTM model takes several minutes to complete a single calculation. Statistical uncertainty analysis requires intensive computational resources. Moreover, the solution search algorithm requires large numbers of function evaluations, including these uncertainty analyses. In this example, we adopt a response surface model, instead of using the simulation of the function evaluation directly, in order to improve computational efficiency. Here we use worst-case uncertainty propagation; however, one may apply other uncertainty analysis methods for this problem.

Table 8.8 Clarification of the mesoscale-level design task.

Clarified Items	Specifications
Design variables and design spaces	Mean Radius of Al: $x_1 = [0.0005, 0.0015]$ (mm)
	Mean Radius of Fe_2O_3: $x_2 = [0.0002, 0.001]$ (mm)
	Volume Fraction of Voids: x_3 space = the feasible range obtained from the continuum level task
	Mean Radius of Voids: $x_4 = [0.0002, 0.001]$ (mm)
Response	T_{ignit}: Weighted average temperature of local hot spots at the first reaction initiation (equivalent to the critical temperature in NTM model)
Uncertainty	$\Delta x_1 = \pm 0.0002$ (mm)
	$\Delta x_2 = \pm 0.0001$ (mm)
	$\Delta x_3 = \pm 0.01$
	$\Delta x_4 = \pm 0.0001$ (mm)
	Random noise in the results due to the randomness of the simulated microstructure.
Fixed parameters	Volume fraction of intermetallic compound (Al + Fe_2O_3): 50% of the Total Volume
	Volume fraction of Al : Volume Fraction of $Fe_2O_3 = 2:3$ (reactants are in stoichiometric proportion)
	Standard Deviation of Void Radius : 20% of the Mean Radius of Void
	Standard Deviation of Al Radius : 20% of the Mean Radius of Al Particles
	Standard Deviation of Fe_2O_3 Radius : 20% of the Mean Radius of Fe_2O_3 Particles
	Particle Shock Velocity (Up) : 1 (km/s)
	Size of SVE: 22 x 11 (10^{-6} mm^2)
Task objectives	Satisfy the feasible range of T_{ignit} identified in the continuum-level energetic material design task considering uncertainty in the DPM model
	Identify feasible space of x_1, x_4

The estimated response surface model for the mean of the response ($acFe$) is

$$acFe_{mean}(x_3, T_{ignit}) = 1.73 - 267 \cdot x_3 + 66.6 \cdot T_{ignit}$$
$$+ 237.5 \cdot x^2_3 - 46.5 \cdot T^2_{ignit} + 284. x_3 T_{ignit} \qquad (8.24)$$

In order to construct this response surface model, a central composite design of experiments is used in the initial design space. The minimum and maximum responses due to the variations in control factors at an input set of (x_3, T_{ignit}) are

$$acFe_{max} = acFe_{mean} - \Delta acFe$$

$$acFe_{min} = acFe_{mean} - \Delta acFe \qquad (8.25)$$

where

$$\Delta acFe(x_3, T_{ignit}) = \left| \frac{\partial f_0}{\partial x_3} \right| \Delta x_3 + \left| \frac{\partial f_0}{\partial T_{ignit}} \right| \Delta T_{ignit}, \tag{8.26}$$

Δx_3 is 0.01, as shown in Table 8.7 and ΔT_{ignit} is chosen to be 0.005 (5 K). Discretization resolution of T_{ignit} is 0.01 (10 K). T_{ignit} is *not* the deviation from T_{ignit} obtained as a result of the DPM model uncertainty analysis but instead is a tolerance that must be considered in NTM model uncertainty analysis. Therefore, a point in T_{ignit} represents not only a single point, but also some range (± 0.005) around that point. The amount of uncertainty (output deviation) in T_{ignit} as a result of the DPM model uncertainty analysis is represented by a set of multiple discrete points. It is recommended that ΔT_{ignit} is larger than half the amount of the discretization resolution in order to cover all of the continuous interdependent variable space with discrete points. The larger ΔT_{ignit} results in the more conservative design solutions.

The DPM model is a computationally intensive, nondeterministic simulation with many of random errors. In order to capture efficiently the minimum and maximum bounds of performance considering all sources of variability, we adapt Step 3 in RCEM-EMI, which is the integrated mean model and prediction interval estimation. Using this technique, we estimate an output deviation that uses uncertainty in the model as well as in input parameters in a computationally efficient way. The estimated mean T_{ignit} model (y_{mean}) and the upper and lower bounds of the T_{ignit} prediction interval (y_{upper} and y_{lower}) are

$$T_{ignit\_mean} = (\hat{\boldsymbol{\beta}}_{converged} \mathbf{x}_0{}')^{+1/3} - 2 \tag{8.27}$$

$$T_{ignit-upper} = \left\{ \hat{\boldsymbol{\beta}}_{converged} \mathbf{x}_0{}' - t_{N-P,1-\alpha/2} \, \exp\left(\frac{\hat{\boldsymbol{\theta}}_{converged} \mathbf{x}_0{}'}{2} \right) \right\}^{-1/3} - 2 \tag{8.28}$$

$$T_{ignit-lower} = \left\{ \hat{\boldsymbol{\beta}}_{converged} \mathbf{x}_0{}' - t_{N-P,1-\alpha/2} \, \exp\left(\frac{\hat{\boldsymbol{\theta}}_{converged} \mathbf{x}_0{}'}{2} \right) \right\}^{-1/3} - 2 \tag{8.29}$$

where

$$\mathbf{x}_0 = [1, x_1, x_2, x_3, x_4, x_1{}^2, x_2{}^2, x_3{}^2, x_4{}^2, x_1 x_2, x_1 x_3, x_1 x_4, x_2 x_3, x_2 x_4, x_3 x_4],$$

$$\boldsymbol{\beta} = [\beta_0, \beta_1, \beta_2, \beta_3, \beta_4, \beta_{11}, \beta_{22}, \beta_{33}, \beta_{44}, \beta_{12}, \beta_{13}, \beta_{14}, \beta_{23}, \beta_{24}, \beta_{34}],$$

$$\boldsymbol{\theta} = [\theta_0, \theta_1, \theta_2, \theta_3, \theta_4, \theta_1{}^2, \theta_2{}^2, \theta_3{}^2, \theta_4{}^2, \theta_{11}, \theta_{22}, \theta_{33}, \theta_{44}, \theta_{12}, \theta_{13}, \theta_{14}, \theta_{23}, \theta_{24}, \theta_{34}],$$
$$N \text{ (the number of samples)} = 360,$$

P (the total number of predictors) $= 30$, the confidence level $(1-\alpha) = 0.99$; $\check{\beta}_{converged}$ and $\check{\theta}_{converged}$ are obtained as shown in Table 8.9.

The maximum and minimum responses at point ($\mathbf{x}$) in the design space are:

$$T_{ignit-min} = T_{ignit-lower}(\mathbf{x}) - \sum_{i=1}^{4} \left| \frac{\partial T_{ignit-lower}}{\partial x_i} \right| \Delta x_i \tag{8.30}$$

$$T_{ignit-max} = T_{ignit-upper}(\mathbf{x}) + \sum_{i=1}^{4} \left| \frac{\partial T_{ignit-upper}}{\partial x_i} \right| \Delta x_i \tag{8.31}$$

Equations 8.24, 8.25, 8.26, 8.27, 8.30, and 8.31 are used to evaluate the mean, minimum, and maximum of responses at all discrete points generated in design space (x_1, x_2, x_3, x_4) and interdependent space (x_3, T_{ignit}). The resolution of discrete seeds of interdependent variables (T_{ignit}) is set to 10 K, which is a reasonably small number. The resolution of discrete seeds of design variables ($x_1 \sim x_4$) reflects the deviation in the design variables listed in Table 8.9. This is also reasonable as the entire continuous design space will be covered. The discrete points which are evaluated are stored as a database file. This database is stored with the format of (input vector, output minimum, output mean, output maximum), and for this case, it is stored as ($x_1, x_2, x_3, x_4, T_{ignit\_min}, T_{ignit\_mean}, T_{ignit\_max}$) and ($x_3, T_{ignit}, acFe_{min}, acFe_{mean}, acFe_{max}$). This evaluation process could be quite computationally intensive if the evaluation models are

Table 8.9 Converged regression parameters for the mean response model and the conditional variance model.

Subscripts	$\hat{\beta}_{converged}$	$\hat{\theta}_{converged}$
0	5.7632E − 02	−1.2886E + 01
1	1.0566E + 00	9.8022E + 01
2	−4.1796E + 01	1.9870E + 03
3	−2.8438E − 01	3.4052E + 01
4	−3.3785E + 01	5.9154E + 01
11	9.8621E + 02	5.0938E + 00
22	2.9929E + 04	2.9957E + 00
33	1.9563E + 00	−2.3105E + 02
44	1.3711E + 04	3.6701E + 00
12	2.2700E + 03	4.9126E + 00
13	−7.4761E + 01	6.0629E + 01
14	1.3515E + 03	3.8467E + 00
23	−5.5384E + 01	1.0199E + 02
24	1.0091E + 04	4.6436E + 00
34	1.9550E + 02	−3.0840E + 01

actual simulations; therefore, parallel computations for the large number of evaluations of the discrete points are valuable.

8.2.4.3. Inductive Discrete Constraint Evaluation (IDCE)—Step 3 of IDEM

The first task of IDCE, the continuum-level design task, is identifying feasible discrete points in the (x_3, T_{ignit}) space that satisfy the condition of HD_EMI$acFe \geq 1$. This condition ensures that all maximum and minimum performance limits are inside the required performance range ($acFe \geq 5$). The results of the first task are described in Figure 8.23. In the figure, feasible points and the exact boundary points between the feasible and the infeasible spaces are shown.

The identified feasible discrete points in T_{ignit} and x_3 space are passed down for the second evaluation of the IDCE, the mesoscale-level design task. The feasible discrete points of x_3 are the design space of x_3 in this task. The feasible discrete points of T_{ignit} are determined by the required performance range for this task.

In Figure 8.24, the feasible points are illustrated as filled circles and the exact boundary points as diamond points. The space is shown at $x_2 = 0.0002$ (mm). From the boundary points and feasible points, we estimate the approximate feasible space. By increasing the required minimum HD_EMI, the smaller feasible region is obtained. As shown in Figure 8.25, the number of feasible points decreases as the required HD_EMI of T_{ignit} increases, leaving only the more reliable (i.e., higher HD_EMI) design solutions.

By further reducing our feasible solution space in Figure 8.25, the 10 solutions with the highest HD_EMI for T_{ignit} are found (see Table 8.10). The solution that is most robust to

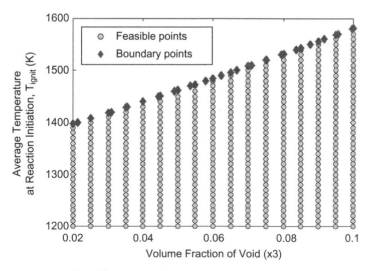

Figure 8.23 The feasible range obtained in T_{ignit} and x3 space using IDCE.

uncertainty in the DPM model is obtained at $x_1 = 0.0005$ mm, $x_2 = 0.0002$ mm, $x_3 = 0.1$, and $x_4 = 0.001$. The corresponding HD_EMI for T_{ignit} is 1.716.

To clarify the utility of IDEM, we compare IDEM solutions with solutions obtained by a traditional robust optimization as shown in Table 8.11. The DPM and NTM models for the robust calculations are the same models that are used in IDEM. In this formulation, there is a tradeoff between the mean and lower deviation of the final performance, *acFe*.

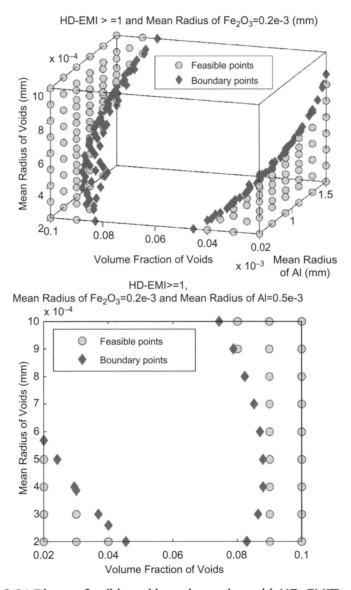

Figure 8.24 Discrete feasible and boundary points with HD_EMIT$_{ignit}$ ≥ 1.

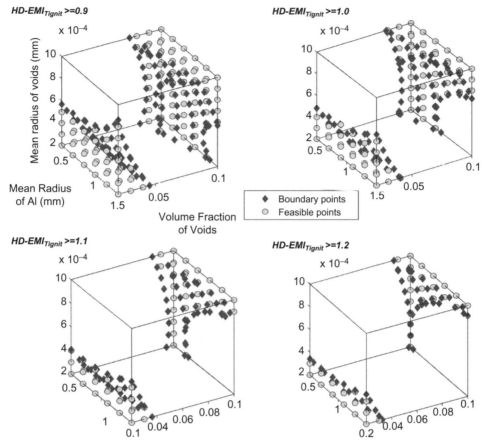

Figure 8.25 Reducing the feasible region by increasing the required HD_EMI for T_{ignit}.

Table 8.10 The 10 highest $HD\_EMI\_T_{ignit}$ solutions.

Rank	x_1 (mm)	x_2 (mm)	x_3	x_4 (mm)	$HD\_EMI\_T_{ignit}$
1	0.0005	0.0002	0.1	0.001	1.716
2	0.0015	0.0002	0.02	0.0002	1.640
3	0.0007	0.0002	0.1	0.001	1.601
4	0.0013	0.0002	0.02	0.0002	1.571
5	0.0005	0.0002	0.1	0.0009	1.563
6	0.0011	0.0002	0.02	0.0002	1.519
7	0.0009	0.0002	0.1	0.001	1.505
8	0.0009	0.0002	0.02	0.0002	1.484
9	0.0005	0.0002	0.02	0.0002	1.467
10	0.0007	0.0002	0.1	0.0009	1.451

Table 8.11 Mathematical form for robust optimization.

Given
$acFe(x_3, T_{ignit}), T_{ignit}(\overline{\mathbf{x}}), T_{ignit\_min}(\overline{\mathbf{x}}, \Delta\mathbf{x}), T_{ignit\_max}(\overline{\mathbf{x}}, \Delta\mathbf{x})$

Find
$\overline{\mathbf{x}} = \{\overline{x}_1, \overline{x}_2, \overline{x}_3, \overline{x}_4, \}$ //Find the means of design variables

Satisfy
$acFe_{mean} = acFe(\overline{x}_3, \overline{T}_{ignit})$ // Mean of the final performance $\qquad$ where $\overline{T}_{ignit} = T_{ignit}(\overline{\mathbf{x}})$ $acFe_{min} = Min\left[acFe(\tilde{x}_3, \tilde{T}_{ignit})\right]$ // Minimum of the final performance $\qquad$ where $\overline{x}_3 - \Delta x_3 \leq \tilde{x}_3 \leq \overline{x}_3 + \Delta x_3,$ $\qquad\qquad T_{ignit\_min} \leq \tilde{T}_{ignit} \leq T_{ignit\_max}$ $\qquad\qquad T_{ignit\_min} = T_{ignit\_min}(\overline{\mathbf{x}}, \Delta\mathbf{x}),$ $\qquad\qquad T_{ignit\_max} = T_{ignit\_max}(\overline{\mathbf{x}}, \Delta\mathbf{x})$ $\Delta acFe_{lower} = acFe_{mean} - acFe_{min}$ //Lower deviation of the final performance $\overline{\mathbf{x}} = [\mathbf{x_{min}}, \mathbf{x_{max}}]$ //Satisfy design space

Minimize
$z = -w_1 \cdot acFe_{mean} + w_2 \cdot \Delta acFe_{lower}$ //Trade off between mean and lower deviation

The mean and lower deviation of the propagated uncertainty are estimated as follows:

1. Identify the minimum and maximum of T_{ignit} in a range of design variables, $[\overline{\mathbf{x}} - \Delta\mathbf{x}, \overline{\mathbf{x}} + \Delta\mathbf{x}]$.

2. Identify the minimum of *acFe* using the previously identified minimum and maximum of T_{ignit}.

3. Identify the mean of T_{ignit} at $\overline{\mathbf{x}}$ using the mean model, $T_{ignit}(\overline{\mathbf{x}})$ and the mean of *acFe* using the mean of T_{ignit} and the mean model, $acFe(\overline{x}_3, \overline{T}_{ignit})$.

4. Identify the lower deviation of *acFe* using the mean and minimum of *acFe*

In this formulation, the lower deviation of the final performance is used in the objective function instead of standard deviation since the higher final performance is preferred here and the distribution of final performance is highly asymmetric. The robust optimization solutions are listed in Table 8.12.

Table 8.12 Robust optimization solutions.

w_1	w_2	x_1 (mm)	x_2 (mm)	x_3	x_4 (mm)	Rank in Table 8.10
0.1	0.9	0.00066	0.0002	0.0995	0.00094	10
0.2	0.8	0.00066	0.0002	0.0995	0.00094	10
0.3	0.7	0.00066	0.0002	0.0995	0.00094	10
0.4	0.6	0.00050	0.0002	0.1	0.00087	5
0.5	0.5	0.00150	0.0002	0.02	0.0002	2
0.6	0.4	0.00150	0.0002	0.02	0.0002	2
0.7	0.3	0.00150	0.0002	0.02	0.0002	2
0.8	0.2	0.00150	0.0002	0.02	0.0002	2
0.9	0.1	0.00150	0.0002	0.02	0.0002	2

As shown in, Table 8.12 robust optimization with different combinations of w_1 and w_2 result in only three solutions, and these are on the list of the best 10 solutions in the table. However, the other solutions obtained with IDEM cannot be obtained with traditional robust optimization. The best IDEM solution was not found by traditional robust optimization.

We performed 10,000 Monte Carlo simulations, propagating uncertainty through the multiscale model chain, at the two solution points, the best IDEM solution, $\bar{x} = [0.0005, 0.0002, 0.1, 0.001]$, and the traditional robust optimization solution with $w_1 = 0.4$ and $w_2 = 0.6$, $\bar{x} = [0.0005, 0.0002, 0.1, 0.00087]$. The probability plot of the simulation result is given in Figure 8.26.

As shown in the figure, the simulation result at the robust optimization solution yields a better mean and lower deviation in the final performance. However, it shows a longer tail in the density plot. This can be seen more clearly in Figure 8.26c. In the probability plot, all of the sampling results for the IDEM solution are above 13; however, the sampling results for the robust optimization solution are scattered from 13 down to 6 although their probability is less than 0.1 %. This indicates the sampling result for the IDEM solution is not as good as those for the robust optimization solution in terms of mean and lower deviation; however, the performance at the IDEM solutions is better if a high level of confidence is required.

The reason for the difference in the distribution of the simulation results is that the IDEM solution is obtained by emphasizing the uncertainty associated with the DPM model. The DPM model is highly uncertain (i.e., it has a large performance variation) due to random material morphology changes. We could find more robust solutions to the DPM model by selecting an IDEM solution with the highest HD_EMIT$_{iginit}$. In this way, we may find more robust solutions specifically for a highly uncertain model in a model chain.

Earlier in this chapter, we argue that IDEM is designed to facilitate parallel computation to reduce design exploration time due to discrete evaluation points. Table 8.13 compares the number of function evaluations called by IDEM and robust optimization to qualitatively

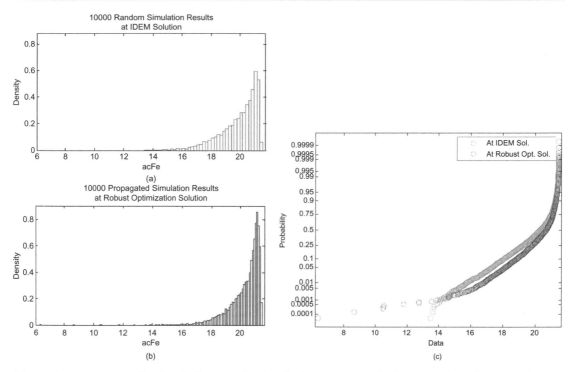

Figure 8.26 Monte carlo simulation results for the best IDEM solution and the robust optimum solution. (a) Density Plot for the IDEM solution, (b) density plot for the robust optimum solution, (c) probability plot for the two solutions.

estimate the possibility of parallelizing computation using IDEM. As discussed, the number of discrete points is the number of function evaluations in IDEM. In this example problem, the numbers of discrete points at each design variable, $[x_1, x_2, x_3, x_4]$, is [6, 9, 9, 9]; therefore, the number of discrete points in the design space is $6 \times 9 \times 9 \times 9 = 4374$. This means the DPM model is called 4374 times. Similarly, the number of discrete points in interdependent space, $[x_3, T_{ignit}]$, is 9 x 101 = 909, and the NTM model is called 909 times. On the other hand, the number of objective function evaluations called by the robust optimization with $w_1 = 0.4$ and $w_2 = 0.6$ is 403. In this optimization, each objective function evaluation calls both DPM and NTM models once. Sequential quadratic programming is used for optimization.

In this chapter, metamodels have been used instead of simulation models for the purposes of demonstration; however, the number of evaluations will be the same whether metamodels or simulations are used. We compare the computation times assuming that each model evaluation takes 0.5 hours. As shown in, Table 8.13, total evaluation time for IDEM solution is 2641.5 hours and that of robust optimization is 403 hours. However, if we use 20 CPUs for the IDEM solution process, then the total evaluation time is 132.1 hours (5.5 days). This is

Table 8.13 Comparison of the computation time for IDEM and robust optimization for both DPM and NTM models.

Design Methods	IDEM		Robust Opt.	
Simulation Models	DPM Model	NTM Model	DPM Model	NTM Model
Number of calls	4374	909	403	403
Time for an evaluation (hours)	0.5	0.5	0.5	0.5
Evaluation time (hours)	2187	454.5	201.5	201.5
Total evaluation time (hours)	2641.5		403	
Total evaluation time with 20 CPUs (hours)	132.1		403	

possible since all evaluations of DPM and NTM models are completely independent of each other. As we increase the number of CPUs, then the time for the IDEM solution process will proportionally decrease. The advantages to be gained via parallelization offered by IDEM are apparent. On the other hand, all calls to the DPM and NTM models are serialized in robust optimization. That is, each model evaluation in optimization calls the DPM model and then calls the NTM model with the results of the DPM evaluation. Moreover, these model evaluation processes are sequential due to the nature of the optimization technique. Therefore, the time for finding an optimal solution cannot be decreased simply by incorporating more CPUs in optimization schemes.

In this section, we demonstrated the solution strategies discussed at the beginning of Section 8.2. Sets of design specifications are obtained instead of an optimal solution. The inductive (top-down) design exploration procedure for multilevel materials design is established. In this example, we compare the two solution results, IDEM solution sets and a traditional robust optimization solution. Comparing the solution results in Table 8.12 and Table 8.13, the IDEM solution sets include all solutions identified by traditional robust optimization with varying weight factors for mean and deviation. However, we demonstrate that the IDEM solution process can find other robust solution sets to the uncertainty in a specific model of a multiscale simulation chain. Comparing the efficiency of the two solution finding processes in Table 8.13, the IDEM requires a longer time to identify the solution sets; however, we may significantly reduce the time by using multiple CPUs and in this situation the efficiency of IDEM will be greater than that of traditional robust optimization and decoupling the solution process and the uncertainty analysis process in IDEM is beneficial for improving the efficiency. In fact, it is an ideal application to exploit massively parallel computing. We conclude in the next section by evaluating whether IDEM addresses the challenges outlined for robust materials design.

8.3. Summary of IDEM

In this chapter, a multilevel robust design method is applied to the design of hierarchical materials and products in order to mitigate the effect of propagated uncertainty in a model chain. We have presented the IDEM based on IDCE and the HD_EMI. IDEM provides the following state-of-the-art capabilities to facilitate integrated materials and product design.

- Using IDEM, designers may identify robust solution ranges, including isolated multiple feasible solution ranges, with the consideration of propagated uncertainty in a simulation chain. By providing ranged solutions, designers can use these ranged solutions until it is necessary to make a final decision, which could significantly reduce design iteration. None of the multilevel robust design methods reviewed in Chapter 6 yield ranged solutions.

- In IDEM, the sequential uncertainty propagation analyses necessary to estimate final performance deviation are decoupled as individual uncertainty analyses at each step in a simulation chain. There is no need to establish computational interfaces between models in the simulation chain. In addition, when an analysis model in an analysis chain is changed, only the altered model needs to be reevaluated.

- Designers can easily use parallel computing techniques since sequential uncertainty analysis processes are decoupled in this method and design exploration and function evaluation processes are also decoupled in IDEM. In the energetic material design example in Section 8.2.4, response surface method is used to shorten computation time. However, if a parallel computing infrastructure is available, we can use direct simulation for the mappings. The feasibility of using direct simulation is given in Table 8.13.

- Large random errors in the model are quantified by employing the integrated mean model and prediction interval estimation in RCEM-EMI (Choi, Austin et al. 2005). Interval models to quantify those errors are used in parallel discrete function evaluations.

- Although IDEM is designed to be generally applicable to hierarchical design synthesis problems, it has some limitations that need to be addressed in future work.

- One challenge that cannot be overcome by IDEM is the management of uncertainty in material models due to idealization associated with a lack of complete knowledge of a given phenomenon and its description. For example, there is uncertainty in the constitutive models which are used to represent behavior of individual phases and the Merzhanov criteria, in which it is assumed that the probability of reaction initiation can be assessed using only local temperature and not stress state, for example.

- IDEM may be computationally intensive if the number of design variables is large or the simulation model is expensive. For use of this method, we recommend that readers reduce the number of design variables by a screening procedure (eliminating the design variables that do not significantly affect system performance) and also use parallel computation for function evaluations.

- Since IDEM evaluates discrete points, it is impossible to avoid discretization errors. We have included the exact constraint boundary generation technique to reduce discretization errors. However, errors still exist in the constraint boundary representation and the feasibility checking due to resolution size. If resolution (or tolerance) in an interdependent variable space (e.g., ΔT_{ignit}) is larger than output deviation (e.g., $[T_{ignit\_min}, T_{ignit\_max}]$, feasible points can be estimated as infeasible points. Authors recommend designers to investigate rough estimation of output deviation in the interdependent variable space so that the tolerance and resolution of interdependent variable space is sufficiently smaller than the output deviations. We also recommend the tolerance should be larger than half the amount of a discretization resolution. The larger tolerance in an interdependent variable space will produce the more conservative design solutions. Additional future work is required for overcoming these difficulties in setting the amount of tolerance and resolution in interdependent variable spaces. Clearly, there is a tradeoff between resolution refinement and computational expenses. A systematic approach for this tradeoff study is another important future work for improving IDEM

With these considerations, it is clear that the IDEM method can play an important role in realizing the concurrent design of products and materials by providing a pathway to explore and conduct top-down design based on bottom-up modeling and simulation.

References

Aitkin, M., 1987. Modelling variance heterogeneity in normal regression using GLIM. Appl. Stat. 36 (3), 332–339.

Amemiya, T., 1977. A note on a heteroscedastic model. J. Econometrics 6, 365–370.

Askeland, D.R., 1994. The Science and Engineering of Materials, third ed. PWS Publishing Company, Boston.

Benson, D.J., 1995. A multi-material Eulerian formulation for the efficient solution of impact and penetration problems. Comput. Mech. 15 (6), 558–571.

Box, G.E.P., Meyer, R.D., 1986. Dispersion effects from fractional design. Technometrics 28 (1), 19–27.

Chan, L.K., Mak, T.K., 1995. A regression approach for discovering small variation around a target, Applied Statistics. J. R. Stat. Soc. Ser. C, 44 (3), 369–377.

Chen, W., Allen, J.K., Mavris, D., Mistree, F., 1996. A concept exploration method for determining robust top-level specifications. Eng. Optimiz. 26 (2), 137–158.

Chen, W., Lewis, K., 1999. A robust design approach for achieving flexibility in multidisciplinary design. AIAA J. 37 (8), 982–989.

Chen, W., Simpson, T.W., Allen, J.K., Mistree, F., 1999. Satisfying ranged sets of design requirements using design capability indices as metrics. Eng. Optimiz. 31 (5), 615–639.

Choi, H.-J., 2001, A framework for distributed product realization, M.S., thesis. G. W. Woodruff School of Mechanical Engineering, Georgia Institute of Technology, Atlanta, GA.

Choi, H.-J., Austin, R., Allen, J.K., McDowell, D.L., Mistree, F., Benson, D.J., 2005. An approach for robust design of reactive powder metal mixtures based on non-deterministic micro-scale shock simulation. J. Comput.-Aided Mater. Des. 12 (1), 57–85.

Choi, H.-J., Panchal, J. H., Allen, K. J., Rosen, D., and Mistree, F. 2003, Towards a standardized engineering framework for distributed, collaborative product realization. *ASME Computers and Information in Engineering Conference*, Chicago, IL. Paper number: DETC2003/CIE-48279.

Davidian, M., Carroll, R.J., 1987. Variance function estimation. J. Am. Stat. Assoc. 82 (400), 1079–1091.

Engel, J., Huele, A.F., 1996. Generalized linear modeling approach to robust design. Technometrics 38 (4), 365–373.

Engineous Inc. 2001. *Product Overview: iSIGHT.* < http://www.engineous.com/overview.html > .

Glejser, H., 1969. A new test for heteroscedasticity. J. Am. Stat. Assoc. 64 (325), 316–323.

Gong, G., Samaniego, F.J., 1981. Pseudo-maximum likelihood estimation: Theory and application. Ann. Stat. 9 (4), 861–869.

Grego, J.M., 1993. Generalized linear models and process variation. J. Qual. Technol. 25 (4), 288–295.

Harvey, A.C., 1976. Estimation regression models with multiplicative heteroscedasticity. Econometrica 44 (3), 461–465.

Hoffman, J.D., 1992. Numerical Methods for Engineers and Scientists. New York: McGraw-Hill.

Jobson, J.D., Fuller, W.A., 1980. Least squares estimation when the convariance matrix and parameter vector are functionally related. J. Am. Stat. Assoc. 75 (369), 176–181.

McKay, M.D., Conover, W.J., Beckman, R.J., 1979. A comparison of three methods for selecting values of input variables in the analysis of output from a computer code. Technometrics 21 (2), 239–245.

Merzhanov, A.G., 1966. On critical conditions for thermal explosion of a hot spot. Combust. Flame 10 (4), 341–348.

Mistree, F., Hughes, O.F., Bras, B.A., 1992. The compromise decision support problem and the adaptive linear programming algorithm. In: Kamat, M.P. (Ed.), Structural optimization: Status and promise. AIAA, Washington, DC, pp. 251–290.

Myers, R.H., Montgomery, D.C., 1995. Response surface Methodology, Wiley Series in Probability and Statistics. A Wiley Interscience Publication, New York.

Myerson, R.B., 1991. Game Theory: Analysis of Conflict. Harvard University Press, Cambridge, MA.

Nair, V.N., Pregibon, D., 1988. Analyzing dispersion effects from replicated factorial experiments. Technometrics 30 (3), 247–256.

Neter, J., Kutner, M.H., Nachtsheim, C.J., Wasserman, W., 1996. Applied Linear Statistical Models, fourth ed. IRWIN, Chicago, IL.

Olson, G.B., 1997. Computational design of hierarchically structured materials. Science 277 (5330), 1237–1242.

Otto, K.N., Antonsson, E.K., 1993. The method of imprecision compared to utility theory for design selection problems. In: Hight, T.K., Stauffer, L.A. (Eds.) Design Theory and Methodology—DTM '93. ASME, New York, pp. 167–173.

Phoenix Integration Inc. 2001. *Phoenix Integration: Products.* < http://www.phoenix-int.com/products/index.html > .

Theil, H., 1971. Principles of Econometrics. New York: John Wiley.

Vining, G.G., Myers, R.H., 1990. Combining Taguchi and response surface philosophies: A dual-response approach. J. Qual. Technol. 22 (1), 38–45.

Xiao, A., Choi, H.-J., Kulkani, R., Allen, K. J., Rosen, D., and Mistree, F. 2001. A Web-based distributed product realization environment. In *Computers and Information in Engineering Conference*, Pittsburgh, PA, ASME, DETC01/CIE-21766.

CHAPTER 9

Concurrent Design of Materials and Products—Managing Design Complexity

Nomenclature	
H	Hurwicz utility
α	Coefficient of pessimism, measuring decision maker's aversion to risk
v	Value of a decision
π	Payoff function
a_0	Action taken by a decision maker in the absence of additional information
a_y	Action taken by a decision maker in the presence of additional information, y
U_{min}	Lower bound on expected utility
U_{max}	Upper bound on expected utility
X_1	Decision made using the actual system behavior
X_2	Decision made using predicted system behavior
U_1	Payoff (utility) achieved if the actual system behavior is used for decision making
U_2	Payoff (utility) achieved if the predicted system behavior is used
P_I	Improvement potential
$(U_{min})^*$	Lower bound on expected payoff at decision point
$(U_{max})^*$	Upper bound on expected payoff at decision point
$max(U_{max})$	Maximum of the upper bound on utility throughout the design space
H^*	Hurwicz utility at decision point
$U_{overall}$	Overall utility—calculate by combining the individual utilities
Z	Deviation function
V_p	Particle velocity in the materials design problem
V_s	Shock speed in the materials design problem
R_{Al}	Radius of aluminum particles in the materials design problem
VF_{Voids}	Volume fraction of voids in the material
S_C	Size of the cells in the shock simulation model
S_w	Window size in the shock simulation model
G_1	Preference for "deformation" goal
G_2	Preference for "accumulated iron" goal
G_3	Preference for "variation in deformation" goal

DOI: 10.1016/B978-1-85617-662-0.00009-0

241

G_4	Preference for "variation in accumulated iron" goal
A_i	Area of the ith hotspot
T_i	Temperature at the ith hotspot

As stated in Chapter 1, concurrent design of materials and products is increasingly being pursued as a means to deliver performance characteristics that are not possible with selection of existing materials (Ashby 1999) and to accelerate insertion of new and improved materials into products. The distinguishing features of materials design are that the models at each scale (1) have a high degree of uncertainty, (2) may be immature or incomplete, and (3) are often complex, requiring significant computational run times (Panchal, Choi 2005). Challenges in the design of multilevel, multiscale systems involving materials modeling and simulation are highlighted in (Panchal et al. 2005). The primary challenges in the design of such systems include (1) uncertainty in the simulation models, (2) complexity of design processes, and (3) complexity of models.

The uncertainty in simulation models is addressed in Chapter 8 by developing a robust design approach for making decisions that are insensitive to the uncertainties in simulation models (Choi, Allen et al. 2005, Choi, Austin et al. 2004, Choi, Austin et al. 2005). In this chapter, we address challenges related to complexity in multilevel design of materials and products. An overview of this chapter and its relationship to other chapters are provided in Figure 9.1.

9.1. Managing Design Complexity

9.1.1. Complexity of Design Processes

The concurrent design of materials and products provides designers with flexibility to achieve design objectives that were not previously attainable with existing materials or to achieve enhanced performance. However, the improved flexibility comes at a cost of increased complexity of the *design processes* and *the materials simulation models* used for executing the design processes. A design process is a sequence (often a network) of activities carried out to design materials and products that satisfy the overall design objectives. Such activities include execution of models, making decisions (selection or compromise), choosing experimental points, conducting experiments, etc. The activities can be sequenced in many different ways, resulting in different configurations of the design process. For example, in a simple design process consisting of two models, the models may either be executed sequentially or concurrently, which corresponds to two different design processes. The structure of the design process has a significant impact on both the final design and the efficiency with which design objectives are achieved. Here, the efficiency is related to the time and cost involved in developing the design. In the case of multilevel design, the complexity of design processes

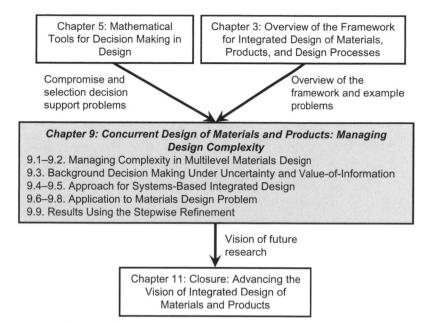

Figure 9.1 Overview of Chapter 9 and its relationship with other chapters in the book.

is generally high compared to traditional design because of the greater interdependencies between activities, resulting in significantly greater design iterations.

To illustrate the concept, consider an example of an integrated material and product design problem discussed in Chapter 3. The problem consists of two types of decisions—product-level decisions and material-level decisions (see Figure 9.2). In the product-level decision, appropriate values of the geometric variables (cross-sectional shape, outer radius, inner radius, length, etc.) must be identified. In material-level decisions, the design variables are volume fractions of constituents, particle sizes, processing parameters, etc. The information flow between these two types of decisions can be structured in a various ways to form different design processes. One example of a structure for a design process involves making the geometric decision first (i.e., fixing the values of geometric parameters) and then using those values as input deciding on the material parameters. The second structure for the process would result by reversing the sequence of the decisions. A very different design process structure would result if the two decisions are made concurrently. This design process is more complex than the process obtained by sequential decisions. It is shown later in this chapter that the designs obtained in each of these three design processes will be different, and the computational effort in executing each is also different.

Similarly, the execution of models can be carried out sequentially or concurrently. The number of design process options increases significantly as the number of models and decisions are increased. Efforts to reduce the complexity may result in increased uncertainty.

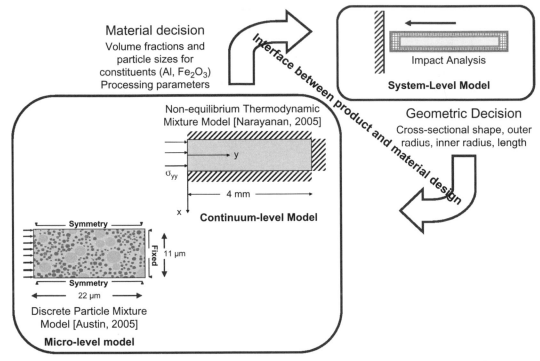

Figure 9.2 Coupling between decisions at the product level and at the material level.

For example, decoupling models may result in increased uncertainty in the model outputs. A systems-based approach is essential for managing both the complexity and the uncertainty in design processes and simulation models in concurrent material and product design. Such an approach, presented in this chapter, is based on systematically *simplifying the design processes* such that the resulting uncertainty does not significantly affect the overall system performance. Similarly, instead of striving for accurate models for multiscale systems (that are inherently complex), focus is placed on making *design decisions that are robust to uncertainties* in the models and their sequencing of execution, information flow between models, and so forth. Accordingly, hierarchical modeling is preferred for multilevel design of multiscale systems rather than concurrent multiscale models, as discussed in Chapter 1.

9.1.2. Complexity of Computational Material Models

Clearly, simulation models for multilevel design are complex in nature. The complexity of models is due to various factors such as different interrelated components, multiple coupled physical phenomena, concurrent consideration of behavior at multiple scales, consideration of a large domain, etc. The complexity of models results in potentially enormous execution time and hence enormous time required for executing the design process. Efforts to reduce model complexity may result in reduced computational expense. At the same time, these efforts may

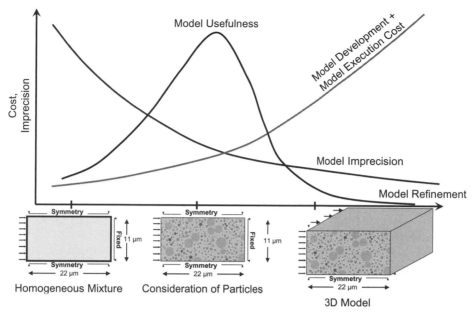

Figure 9.3 The tradeoff between model accuracy and the cost for developing and executing a model for material microstructure-property relations.

also cause reduction in the accuracy and/or predictive quality. In materials design, predictive capability of physically based models is of particular importance in informing design decisions. The degree of computational complexity of models for material structure-property relations, in general, is not necessarily related to predictive capability; for example, multiscale models that are not based on specific mechanisms and verified relations at each scale can be computationally intensive yet offer little predictive capability to support materials design exploration. The objective of material design is not just structural design at the highest length scale considered, as would be typical of product design, but rather, the design of various scales of hierarchy of the material itself. In other words, material microstructure attributes are design degrees of freedom. Hence, concurrent material and product designers are faced with the tradeoff between accuracy and computational cost associated with the development and refinement of models with more sophisticated kinds of considerations regarding physics, chemistry, and materials science than would be manifested in product design based only on materials selection.

The tradeoff between accuracy and cost in model development is conceptually shown in Figure 9.3. Three different models for predicting the behavior of a material mixture (heterogeneous multiphase microstructure) under shock loading are shown in the figure. The first model, a homogeneous approximation, is the simplest one. Here, the material is modeled as a homogeneous mixture and a 2D domain is studied under the propagation of shock wave induced by an imposed dynamic particle velocity on the left face of the statistical volume element. In the second model, the mixture is assumed to consist of prismatic cylindrical phases. Distinct, material-specific constitutive relations are used for each phase in the mixture. The third model is the most refined; the mixture is modeled in 3D and the particles are considered as spheres. Since the first model is the simplest one, the cost of developing it is the lowest and its imprecision (i.e., inaccuracy) is the highest. The usefulness of the model in design may be low because of its high degree of imprecision. At the other end, the third model is the most accurate but most complex. Again, the usefulness of the model may be low because it is costly to develop and execute; hence, the designer may not be able to explore a large number of design alternatives using this model and may be unnecessarily biased toward his/her initial concept. The intermediate model in which the domain is modeled in 2D and the particles are considered separately may offer the best compromise with high enough accuracy and low enough computational cost. The "best" compromise depends on the problem at hand.

In spite of the significant progress in developing accurate, physically based simulation models for materials and the availability of high-performance computing facilities, modeling all aspects of a complex product-material system in a simulation model is not possible using currently available (or even foreseeable) computational resources. Model approximation and idealization is a limiting factor in addition to computation. This is further emphasized by a quote by George Box: "All models are wrong, some models are useful" (Box 1979). According to this statement, all models have some approximations and they can be refined through various means.

The lack of *"perfect"* models is not entirely limiting in multilevel design because the role of simulation models in engineering design is *not to exactly* predict the behavior of a system, but to support design decisions. Despite the fact that the accuracy of simulation models influences the quality of design decisions, designers do not need (and cannot have) *perfect models* for making design decisions. A simulation model is good (i.e., useful) in the context of design if it helps designers make good design decisions. Although simulation models can be refined indefinitely, after a certain level of refinement of simulation models, simplified models can be used to make "satisficing" (good enough) decisions (Simon 1996). Hence, the fundamental question faced by a designer is—*"How much refinement of a simulation model is appropriate for a particular design problem?"* We note that this differs from the more idealized notion of optimization, which presumes that existence of precise models and accurate assignment of initial and boundary conditions.

Model refinement can be carried out in various different ways depending on the type of model. In Figure 9.3, only two pathways of refinement are shown. Other possible refinements of the shock simulation model are shown in Figure 9.4. These refinements include accounting for the statistical variation in particle sizes, improving the model physics, improving the accuracy of the parameters for constitutive equations, accounting for coupling between different physical phenomena (thermal, mechanical and chemical), etc. They are not related to geometric representation or mesh refinement per se. Given that there are many different ways in which a model can be refined, reliance on individual expert guidance is not always effective since

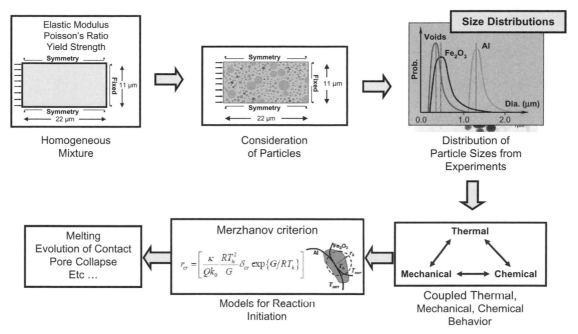

Figure 9.4 Various possible refinements in the shock simulation model.

it is usually not linked to the systems perspective, but to the expert's priorities, quality, and completeness of understanding. Domain experts will demand continued investment in resources for refinement of a given model irrespective of value added in design. Hence, a systematic quantitative approach for refinement of models is needed. In this chapter, we present such a quantitative approach anchored in *information economics* (Lawrence 1999) to decide on the level of refinement of models. Specifically, value-of-information metrics are used decide whether to make a decision using the available information or to gather more information before making a decision (Aughenbaugh, Ling 2005, Bradley and Agogino 1994, Lawrence 1999).

We present an approach based on *value-of-information* for determining the appropriate extent of refinement of simulation models. The approach consists of (1) a metric called the improvement potential for quantifying the value-of-information obtained via refinement of simulation models, and (2) a method in which this metric is utilized for supporting model refinement decisions. The improvement potential quantifies the maximum possible improvement in a designer's decision that can be achieved by refining a simulation model. The method involves starting from a simple simulation model and gradually refining it until the value of further refinement on design decisions is small. The approach is presented using a materials design problem where a complex finite element model is gradually refined. The approach presented in this chapter can be utilized by designers and analysts in developing effective simulation models for specific design problems while efficiently utilizing their model development resources.

9.2. Frame of Reference: Multilevel Materials Design

9.2.1. Design of Energetic Material Containment Systems

Consider an example involving the design of energetic materials or reactive powder metal mixtures discussed in Chapter 3. Safe handling of these materials during transportation or storage to guard against initiation of reaction demands certain multifunctional requirements of strength and reaction initiation, as shown in Figure 9.5.

In the design of energetic materials, multiscale analytical, experimental, and computational tools are employed ranging from nanoscopic- to mesoscopic-length scales:

- Ab initio and molecular dynamics calculations are conducted to estimate the equation of state for the individual constituents and to explore transition states (activation energies) for initiation of reactions in energetic materials.

- These calculations provide information to mesoscale continuum dynamic simulations of shock wave propagation through energetic material mixtures composed of discrete particles that are characteristic of actual mixtures.

- In addition to providing statistics regarding probability of reaction initiation, results of these continuum hydrocode calculations are then homogenized into equivalent

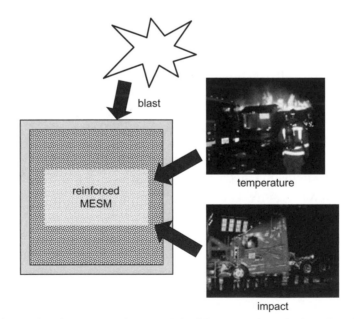

Figure 9.5 Schematic of an energetic material with a container designed to resist reaction initiation at elevated temperature and/or under impact or blast loading for purposes of safe transportation.

nonequilibrium (e.g., extended irreversible thermodynamics) models for pressure- and temperature-dependent mechanical behavior, as well as temperature-induced reaction initiation probability.

Effectively, bridges between models at multiple length and time scales are created to analyze both energetic and structural aspects of energetic materials, in close connection with companion experiments.

One of the primary advantages of concurrently designing materials and products at multiple scales is *increased design freedom* (Simpson, Rosen et al. 1998, Wood 2000) (i.e., greater flexibility in configuring the system to achieve desired behavior), which enables designers to achieve better performance. For example, by designing energetic materials concurrently with the containment systems, designers have more options for customizing the containment/reinforcement concept, as well as the weight, geometry, and dimensions of containers, by employing custom-designed energetic materials. In the energetic material design problem, the objective is to achieve desired performance at the system level (e.g., a reinforced or encapsulated energetic material). Overall system performance is a function of the reinforced energetic material behavior. Material properties in turn depend on reinforcement strategy, micro-scale interfaces between the particle mixture and reinforcement, as well as interfaces between constituent particles in the mixture. Behaviors at the mesoscale interfaces depend on nanoscale lattices and reactant interface structures and transition states for chemical reactions. This hierarchy must be considered when designing energetic material systems at multiple scales, resulting in a significant complexity due to interactions between both the simulation models and the design decisions.

Since the energetic material design scenario involves assigning material design variables (such as volume fractions and size distributions of constituent particles and matrix/binder phase) and the product design variables (such as mass density of energetic material, container dimensions, etc.), the problem involves the integrated design of materials and products (see Figure 9.6). In this problem, the product is the container along with encapsulated energetic material. Decisions regarding materials and product are coupled because they both influence

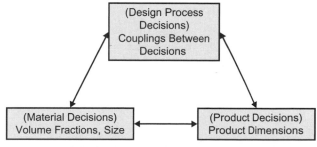

Figure 9.6 Integrated design of materials, product, and design processes.

the deformation and reaction behaviors of the system. Each decision relies on the knowledge of design variables from the other decision. Both material and product decisions require multiple simulation models that exchange information with each other. Further, the simulation models are also coupled at different scales of material and product hierarchy. These couplings between decisions and simulation models increase the complexity of the complete design problem. Since the complexity of problem severely limits the rate and scope of design exploration, it is important to consider ways to simplify design processes in a manner that does not affect the overall system performance. The degree of complexity of this example is in many ways quite representative of real-world applications of a materials design framework.

9.2.2. Addressing the Complexity of Energetic Material Design Problem by Simplification of Interactions

The strategy for addressing the complexity in the energetic material design problem is the systematic simplification of information flows between models and decisions in the design processes. In this chapter, the information flows are also referred to as *interactions*. Multiscale design processes are first modeled on the basis of interactions between models and decisions. The interactions are characterized using standard *interaction patterns* with clearly defined inputs and outputs. Explicit consideration of interactions between simulation models and decisions is an essential component in the strategy for designing complex multilevel systems, particularly for systems with high degree of nonlinearity, uncertainty, and dependence upon initial conditions, as is the case with design of materials at multiple scales. These patterns facilitate hierarchical modeling of design processes.

After modeling the design processes using interaction patterns, the simplification of design processes is carried out by identifying the interactions that do not have a significant effect on the overall outcome. Not all interactions have equal impact from a decision-making standpoint; some couplings have a significant effect on a designer's decisions, whereas others only have minor impact. Hence, the proper level of simplification of a complex design process is the one that reduces the design effort significantly without having a marked effect on the design outcome. For example, we can convert a coupled interaction (two-way information flow) between product and material decisions into a sequential interaction (one-way information flow) between material and product decisions, as discussed in Section 9.1.1. This simplification is warranted only if the final design outcome does not suffer substantially. Designs obtained by simplification that are clearly inferior in terms of functionality and performance compared to those with strong coupling between models and data at different levels of hierarchy represent those with high value-of-information since much is lost by simplification.

Hence, there is a need to *quantify the impact of simplification on the overall system performance*. We utilize a class of metrics referred to as "value-of-information" metrics to measure the difference between the design outcome using nonsimplified design processes and the outcome achieved using simplified design processes. A specific value-of-information based metric, the *improvement potential,* is used in this chapter to quantify the impact on overall performance (Panchal, Choi et al. 2007).

Considering design process simplification from a robust design perspective, the objective is to identify design processes that are robust to simplification, as opposed to design processes whose simplification results in significant degradation in the outcome. This is a novel application of the robust design philosophy initially proposed by Taguchi (Taguchi 1986) and later extended by various researchers (Box 1979, Chen 1995, Choi 2005, Choi, Austin 2004, Choi, Austin 2005, Mistree, Seepersad 2002).

9.2.3. Simulation Model Refinement

The value-of-information based approach is also extended to the problem of systematic *refinement* of simulation models from the standpoint of multi-objective design decision making. These decisions are modeled using the compromise Decision Support Problem (cDSP) (Mistree, Hughes 1993, Seepersad 2001) construct discussed in detail in Chapter 5. The approach for model refinement consists of the *improvement potential* metric to quantify the improvement in decision making gained by refining the simulation model, serving as a method for guiding stepwise model refinement. In the context of simulation-based design, a *simulation model is a source of information.* Refining a simulation model is analogous to adding more information for decision making so that the model, once calibrated, more accurately predicts the behavior of the system for conditions for which is it not calibrated. In the context of model refinement, the improvement potential quantifies the maximum possible improvement in a design decision (in terms of utility) that can be achieved by refining a simulation model.

For a given model, the improvement potential can be evaluated. If the potential is high, the model is refined further; otherwise it is used as is for making design decision. The refinement is carried out in a stepwise manner, with improvement potential calculated at each step, until the model is appropriate for the design decision. Moreover, the effects of different sequences of simplification can be considered. Using this approach, designers and analysts can develop and configure effective simulation models for specific design problems in a resource efficient manner. Details are provided along with a materials design example in the rest of the chapter.

In this chapter we focus on two items, namely, (1) the notion of *designing the design process* in concurrent design of materials and products, and (2) integration of model decisions (such as to what level should the models be refined or coupled) with the design of materials and products. The approach consists of two key components: (1) the use of interaction patterns for modeling design processes, and (2) the use of value-of-information-based metrics for making design-process-level decisions. These two components of the approach are discussed in Sections 9.4.1 and 9.4.2, respectively. In this chapter, we embody the approach into a design method that is outlined in Section 9.5.2. The method is illustrated using the energetic material design problem in Section 9.7. Finally, the results are discussed in Section 9.9, and closing thoughts are presented in Section 9.10. Before discussing the approach, however, it is important to review the different types of uncertainties and their impact on design decisions.

9.3. Background—Decision Making Under Uncertainty and Value-of-Information

In this section, we discuss different metrics for value-of-information (Section 9.3.2) and the application of value-of-information concepts in design decision making (Section 9.3.3). Recently developed concepts on applying value-of-information concepts in design are discussed in Section 9.3.3.1, and the limitations of existing efforts are discussed in Section 9.3.3.2. In order to provide a background for the existing efforts, we first present an overview of decision making under different types of uncertainty (Section 9.3.1).

9.3.1. Decision Making Under Different Types of Uncertainty

Uncertainty in simulation models is broadly classified into two types—*variability* and *imprecision*. Variability is the type of uncertainty that is inherent in the system being modeled and is represented using probability distributions. It is also referred to as *stochastic uncertainty*. In contrast to variability, imprecision is generally modeled using intervals (Moore 1966, Ward, Lozano-Perez et al. 1990) or fuzzy sets (Antonsson 2001, Antonsson and Otto 1995, Wood and Antonsson 1989). Antonsson and Otto (Antonsson and Otto 1995) define an imprecise variable as "a variable that may potentially assume any value within the possible range." Fuzzy sets are used to model a designer's preference for specific values of design parameters in addition to the imprecision in the design variables. The key difference between imprecision and variability is that imprecision is a type of epistemic uncertainty (i.e., lack of knowledge), whereas variability is a type of aleatory uncertainty (i.e., inherent randomness in the system). From a value-of-information standpoint, imprecision can be reduced by incorporating more knowledge via refinement of simulation models but the variability that is inherent in the system behavior cannot be reduced by gathering more information.

Decision making under variability and imprecision in simulation models is illustrated separately in Figure 9.7(a) and Figure 9.7(b), respectively. Consider the scenario shown in Figure 9.7(a), where a designer needs to select between design alternatives A and B. Due

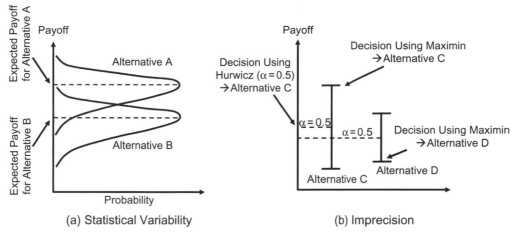

Figure 9.7 Decision making under (a) statistical variability and (b) imprecision.

to the system variability, the uncertain payoff values for both these alternatives are best represented as probability distributions. The decision criterion in such a scenario is to select the alternative that results in the maximum expected payoff when taking risk attitudes into account (Keeney and Raiffa 1976).

In the context of decision making in materials design, the payoff is generally defined in terms of the performance parameters of the design, such as heat transfer coefficient or thermal conductivity, energy release capability, stiffness, creep resistance, etc. In a multifunctional design problem, multiple performance parameters are typically of interest. We use utility functions to measure the payoff so that the preference for multiple performance parameters can be accounted for. Utility functions are used to quantify a designer's preferences for outcomes achieved for different alternatives (Von Neumann and Morgenstern 1947). Details of utility functions are provided in Chapter 4. For simplicity, we assume that the utility of an ideal outcome is 1 and the utility of an unacceptable outcome is 0. The shape of a utility function represents a designer's attitude towards risk; a convex utility function implies risk aversion, whereas a concave utility function implies risk proneness. For details on the characteristics of utility functions, please refer to (Keeney and Raiffa 1976). According to utility theory, "if an appropriate utility is assigned to each possible outcome and the expected utility of the outcome for each alternative is calculated, then the best course of action is to select the alternative whose outcome has the largest expected utility" (Keeney and Raiffa 1976). Hence in Figure 9.7(a), a designer would select alternative A because the expected utility (payoff) is greater than for alternative B.

An approach for selection of alternatives based on maximization of expected utility is presented by Fernández and coauthors (Fernández, Seepersad 2005). The decision criterion of maximization of expected utility is extended to multi-objective decisions by Seepersad and coauthors (Seepersad 2001, Seepersad, Mistree 2005) in the Utility-based cDSP. In this formulation, individual goals are formulated as single-attribute utility functions, and multiple goals are combined in the objective function using Archimedean weightings (Seepersad, Mistree 2005). The Utility-based cDSP is used in this chapter to formulate multi-objective design decisions. The mathematical formulation is provided in Chapter 5.

In the scenario shown in Figure 9.7(b), the payoffs for alternatives C and D are imprecise and are best represented by intervals. Since the representation is non-probabilistic, maximization of expected utility cannot be used as a decision criterion. In this scenario, one possible decision criterion is to select the alternative with the highest upper bound on payoff (maximax criterion). Such a criterion reflects a designer's complete optimism. An alternative decision criterion involves selecting an alternative that maximizes the lower bound on payoff (maximin criterion), reflecting a designer's complete pessimism. Using the maximax criterion, a designer would select alternative C, whereas alternative D is chosen when the maximin criterion is considered. Different decision criteria are appropriate in different design scenarios. A third type of decision criterion, called the Hurwicz criterion (Arrow and Hurwicz 1972), is based on the combination of optimistic and pessimistic criteria. The Hurwicz decision criterion involves maximizing a weighted average (H) of lower and upper bound:

$$H = (\alpha)\,(Lower\ Bound) + (1 - \alpha)\,(Upper\ Bound) \tag{9.1}$$

The weighted average is calculated using a coefficient of pessimism (α) that is a measure of a decision maker's aversion to risk under imprecision. An $\alpha = 1$ implies complete pessimism (maximin) and $\alpha = 0$ implies complete optimism (maximax). Using a pessimism index of $\alpha = 0.5$ in Figure 9.7, a designer would select alternative C.

In a design scenario, the model can have both statistical variability and imprecision. In such a scenario, the Hurwicz criterion can be applied on the *intervals of expected payoff*. For each alternative, the lower and upper bounds on expected payoff are determined. Using the pessimism index, a weighted average of expected utility is determined for each alternative. The alternative with maximum weighted average is selected. Throughout this chapter, we use this decision criterion to select an alternative or a point in the design space. A pessimism index of $\alpha = 0.5$ is used in this chapter unless explicitly stated otherwise.

9.3.2. Value-of-Information in Decision Making

At any stage in the decision-making process, designers possess some amount of information that can be used for selecting the best course of action. Designers have an option of either (1) making a decision using the available information, or (2) gathering more information and then making a decision using the updated information. In this context of decision making, the value of this added information refers to the *improvement in a designer's decision-making ability*. The refinement of simulation models is analogous to the acquisition of a new source of information, and the set of results from the execution of this refined model is analogous to additional information for decision making. Hence, value-of-information-based metrics can be used by decision makers to make metalevel decisions involving a tradeoff between reduced uncertainty and increased cost as a result of refinement of simulation models. The increase in cost is due to both refinement and execution of a more complex model.

Consider, for example, a designer who has a simulation model for predicting the system behavior and is interested in making a decision using the model. Before making the decision, he/she has the option of increasing the fidelity of the model by including additional physical phenomena. For example, a structural designer may improve the fidelity of a statics model by adding dynamic behavior or time- and temperature-dependent diffusional creep. Inclusion of a physical phenomenon in the simulation model is equivalent to an information source that generates information regarding the system behavior. The output of the simulation (i.e., the predicted system behavior) is then used to assess the added information generated by the information source.

Howard (Howard 1966) proposed value-of-information as a metric for determining whether or not to additional information is needed to make a judicious decision. The expected value-of-information, as defined by Howard, is the difference between the expected value of the objective for the option selected with the benefit of the information and that without this information.

Lawrence (Lawrence 1999) provides a comprehensive overview of metrics for value-of-information. He argues that the value-of-information for decision making can be measured at different stages in the decision-making process: (1) prior to consideration of incorporation of information, (2) after considering a message source but prior to receiving a message (*ex-ante* value), (3) after receiving additional information and making the decision but before realizing the environmental state (*conditional* value), or (4) after the addition of information and observing the outcome of the decision based on acquired information (*ex-post* value). Referring back to the structural decision maker, the decision maker can evaluate the expected value-of-information before even considering the incorporation of any additional physical phenomena. The second option (*ex-ante* value) is to decide which physical phenomena to model (i.e., which information source to choose) and to evaluate the value-of-information before executing the simulation code. The third option (*conditional* value) is to evaluate the value after executing the simulation code and making decisions about the system but before manufacturing and testing the system. The fourth option (*ex-post* value) is to evaluate the value of this additional information after making decisions and also manufacturing and testing the system.

Mathematically, the *ex-ante* and *ex-post* value-of-information-based metrics $v(x,y)$ are represented as follows:

Ex-ante value:

$$v(x, y) = E_{x|y}\pi(x, a_y) - E_x\pi(x, a_0) \qquad (9.2)$$

Ex-post value:

$$v(x, y) = \pi(x, a_y) - \pi(x, a_0) \qquad (9.3)$$

where $E_x f(x)$ is the expected value of $f(x)$ and $E_{x|y} f(x)$ is the expected value of $f(x)$ given y. The symbols a_o and a_y, respectively, represent the actions taken by the decision maker in the absence and presence of information y, and $\pi(x,a)$ represents the payoff achieved by selecting an action a when the state realized by the environment after the decision is x. It is important to realize that the key difference between *ex-ante* and *ex-post* values is that for the latter, the realization of the state x is known. However, the realization of the state x is not known in the *ex-ante* value and the expected value of the payoff is taken over the uncertain range of state x.

Ideally, designers prefer the *ex-post* value-of-information because it truly reflects the value-of-information for a decision based on the actual behavior of the system, for example, prototype behavior. At that stage, the realization of the state x is known and the system behavior can be measured exactly. However, it is not possible to predict the *ex-post* value of a decision before making the decision itself. Due to the *ex-ante* nature of decision making, decisions based on the information have to be made before the state is actually realized. Hence, the designers must use *ex-ante* value-of-information which considers uncertainties in the system.

9.3.3. Value-of-Information Metric to Support Design Decisions

9.3.3.1. Existing Research Advances Regarding Value-of-Information

A summary of the existing efforts on applying value-of-information in engineering is provided next. In the context of engineering design, Agogino and coauthors (Agogino 1997, Bradley and Agogino 1994, Bradley, Agogino and Wood 1994, Wood and Agogino 2005) present a metric called the Expected Value-of-Information (EVI) and use it for a catalog selection problem (Bradley and Agogino 1994), in which a designer is faced with the task of choosing components from a catalog in order to satisfy some functional requirements. During the conceptual design phase, selection decisions need to be made under significant uncertainty, for example, due to a limited understanding of requirements and constraints, an inability to specify part dimensions, or uncertainty in the environmental conditions. Before making the decision about selecting the right component, a designer is faced with another higher-level decision—whether to go ahead and make the decision using the available information or to spend resources and gather more information before making the selection decision. This is a process-level decision, for which Bradley and Agogino (Bradley and Agogino 1994) use EVI to quantify the expected benefit from additional information. The EVI is equal to the expectation of a value of the objective function defined in terms of the value of the uncertain variables with new information minus the expectation of the objective function based on the current state of information. Evaluation of the EVI requires the availability of probability distributions over uncertain variables with new information that is not available before the refinement of the simulation model. Hence, the EVI is not suitable for model refinement decisions. Further, the focus of research by Agogino and coauthors is on the *refinement of information regarding design concepts* and the evaluation function in the conceptual design phase (Pahl and Beitz 1996), whereas our focus in this chapter is on *refinement of simulation models* to be used in the embodiment design phase (Pahl and Beitz 1996).

Poh and Horvitz (Poh and Horvitz 1993) use a value-of-information metric for refining *decisions*. The authors present three dimensions along which decisions can be refined—quantitative, conceptual, and structural. Quantitative refinement of a decision can be carried out by reducing the uncertainty in the decision problem or by refining the preference models. Conceptual refinement is carried out by refining the definition of alternatives and design variables, whereas structural refinement requires addition of dependencies in the simulation model. Poh and Horvitz use the value-of-information metric to determine which dimension is critical for refinement of the decision problem.

In order to model variability for evaluating value-of-information, it is generally assumed that the probability distributions are available. However, if these probability distributions are not available, they are typically generated through an *educated guess* that is based on prior knowledge of a designer (or analyst otherwise involved in the design process). In order to address the problem of lack of knowledge about the probability distributions, Aughenbaugh and coauthors (Aughenbaugh, Ling 2005) present an approach of measuring the value-of-information based on probability bounds. They assume that although the exact probability distributions are unavailable, the lower and upper bounds on these probability distributions are available in terms of p-boxes (Ferson and Donald 1998). Using this p-box approach, they evaluate the value of added information that reduces the size of the interval for probability distributions (i.e., tightens the probability bounds).

9.3.3.2. Limitations of Existing Approaches in the Context of Simulation Model Refinement

Howard (Howard 1966), Bradley and Agogino (Bradley and Agogino 1994), and Lawrence (Lawrence 1999), calculate the value-of-information by considering only the stochastic variability, with the decision being made by maximizing the expected value of the objective function. They do not address the case of decision making under imprecise information. For example, consider a scenario where a designer has an option of making a decision using only one of two available simulation models. One of the simulation models has a higher-fidelity representation of physics than the other. The differences in the results from simulation models are best represented as imprecision rather than statistical variability. Such a scenario is extremely common in multilevel design problems of hierarchical materials and products. The consideration of imprecision while measuring the value-of-information in addition to variability is very important from the standpoint of simulation model refinement. It forms a basis for determining the extent of refinement of simulation models. Hence, the primary need for a new value-of-information based metric for model refinement is *quantifying the impact of imprecision* in simulation models. Except for Aughenbaugh and coauthors (Aughenbaugh, Ling 2005), imprecision in data, which cannot be modeled in terms of probability distribution functions, is often not considered at all in existing approaches. Aughenbaugh and coauthors (Aughenbaugh, Ling 2005) focus on modeling the effect of imprecision in probability distribution functions on data collected from physical experiments, whereas in this chapter, we focus on modeling the imprecision in the outputs parameters of simulation models and the effect of reducing this imprecision by refining the models. A metric is proposed that accounts for both imprecision and variability in simulation models.

9.4. Approach for Systems-based Integrated Design of Materials, Products, and Design Processes

In this section, we present two key components of the approach: (1) interaction patterns for modeling design processes, and (2) value-of-information based metrics for making design process decisions. These two aspects of the approach are discussed in Sections 9.4.1 and 9.4.2, respectively, and are utilized in the method presented in Section 9.5.2.

9.4.1. Modeling Design Processes Using Interaction Patterns

9.4.1.1. Overview of Interaction Patterns

At any level of abstraction, design processes can be modeled using repeating *patterns*. Any complex network can be broken down into a common set of patterns. One classification of patterns for simulation-based multilevel design processes for hierarchical multiscale systems is presented in Figure 9.8, where nine patterns are identified based on the type of information flow between different simulation models and/or design decisions. The patterns are organized in a matrix, whose three rows are:

(1) Information flow between simulation models,

(2) Information flow between decisions, and

(3) Multifunctional decisions.

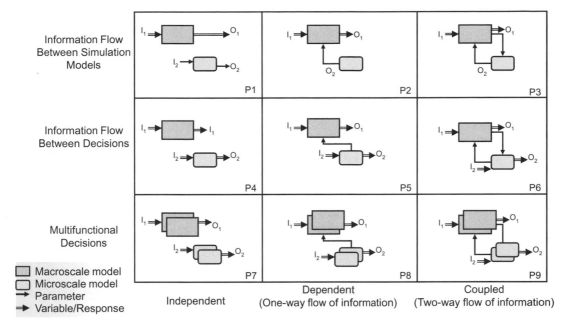

Figure 9.8 Interaction patterns in multilevel design.

The three columns of the matrix are labeled:

(1) Independent interaction,

(2) Dependent interaction, and

(3) Coupled interaction.

The interaction patterns are labeled from P1 through P9. In pattern P1, there are two independent models with separate inputs (I1 and I2) and outputs (O1 and O2). For example, consider two models for simulating the thermal and structural responses of a linear cellular alloy (LCA; discussed in Chapters 3 and 7). If the geometry and material are fixed, the inputs to the models are different boundary conditions and the outputs are the overall heat transfer rate and elastic compliance, respectively. The models can be executed independently. In pattern P2, there is a sequential information flow between models. The output of one model (O2) is an input to the other model. For example, consider a scenario involving two models—a product-level model and a material-level model. The material model is used to predict the material properties such as stiffness, heat transfer coefficient, etc. which are then used in the product model. If the product model does not feed information back to the material model, then it is modeled as a sequential pattern P2. In pattern P3, the information flow between models is bidirectional. This kind of an interaction is common in concurrent multiscale models where one model is used to capture the material behavior at lower scales which is used in the higher scale model. At the same time, the higher-scale model provides the boundary conditions for the lower-scale model.

Patterns P4, P5, and P6 capture the interactions between design decisions. These patterns refer to independent, sequential, and coupled decisions respectively. Pattern P4 is used when two decisions can be made independently. For example, in the pressure vessel problem from Chapter 4, if the dimensions of a pressure vessel can be determined independent of the orientation of the fibers in the composite, the two decisions (about these design variables) can be made concurrently. If the decisions can't be made concurrently and the knowledge of one set of design parameters is required before the values of other parameters can be determined, this scenario can be modeled using pattern P5, the sequential pattern. Pattern P6 is used when both the decisions must be made concurrently because the output from one decision serves as an input of the other decision and vice versa. Pattern P6 has a more parallel, iterative character than P5. For example, in the two decisions shown in Figure 9.2, the outputs of the material decision are the material variables (volume fractions, particle sizes, processing parameters, etc.). These variables are used as inputs to a decision regarding the product geometry. Similarly, the outputs from the geometry decision (cross-sectional shape, outer radius, inner radius, and length) are used as the inputs to the material decision. The details of this problem are provided later in this chapter.

Finally, patterns P7, P8, and P9 refer to the information flow between multifunctional decisions (e.g., thermal, structural, etc.) at multiple scales. These patterns are not used in this chapter, but they are important in practice since different models, expertise, and stakeholders in the design process are often used in supporting decisions regarding performance in different physics domains. The field of concurrent multiphysics modeling, for example, would be represented by P9 on this chart. Although the interaction patterns are defined for two interacting decisions/models, the same principles extend to processes with more than two decisions/models. For example, the problem discussed in this chapter involves three models that exchange information with each other. The use of interaction patterns for the energetic material design problem is discussed in Section 9.8.5.

9.4.1.2. Design Process Simplification Using Interaction Patterns

The complexity of design patterns increases as we move from left to right in the matrix in Figure 9.9. For example, pattern P3 is more complex than pattern P2, which is more complex than the pattern P1. Similarly the complexity increases as we move from the top to bottom (i.e., pattern P8 is more complex than pattern P5, which is more complex than pattern P2). The increased complexity in moving from top to bottom owes to the notion that modeling of unit processes (top) embeds into informing design decisions (middle), which further embeds into consideration of performance attributes in multiple physical domains (bottom).

Based on the nine interaction patterns, designers can simplify the design processes in the following three ways: *scale decoupling*, *decision decoupling,* and *functional decoupling*. These three types of simplification are illustrated in Figure 9.9 and discussed next.

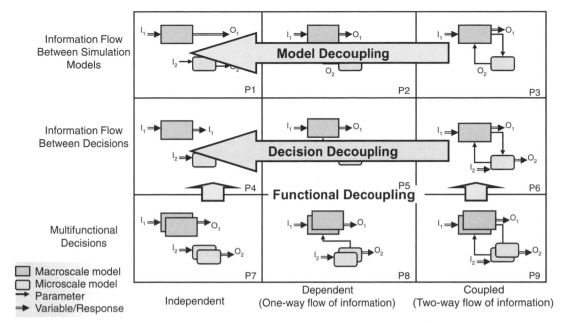

Figure 9.9 Simplification of design processes via scale, decision, and functional decoupling.

Scale decoupling refers to the simplification of interactions from pattern P3 to pattern P2, and from pattern P2 to pattern P1. It refers to the simplification of information flow between two simulation models at different scales of the multiscale hierarchy used for making a single decision. In scale decoupling, there is a single set of design variables and a set of objectives related to a single function that can modeled using a cDSP. The information needed to make that decision is generated from (at least) two separate simulation codes (generally at different scales) that may need to be executed in a coupled fashion. The task in scale simplification is to determine whether the coupled nature of simulation models (pattern P3) is important for making the decision or it can be simplified into a sequential information flow (pattern P2) or into an independent execution (pattern P1). For example, in a multiscale model, two models at different scales can be coupled with each other via bidirectional information links. Through scale decoupling, the information links are systematically removed and replaced with *uncertain* parameters; uncertainty is introduced as a result of simplification. The simplification is carried out only if the impact on the overall design decision is insignificant. Uncertainty introduced by decoupling the scales is accounted for, and robust design decisions are made using the simplified model interactions.

Decision decoupling refers to the simplification of interaction patterns from pattern P6 to pattern P5, and pattern P5 to pattern P4. It refers to the simplification of information flow between decisions from a coupled decision making to an independent decision making. A decision decoupling scenario is characterized by multiple decisions—each associated with a set of design variables that need to be decided upon. Each of the set of design variables affects a common set of objectives. The task in decision decoupling is to determine an appropriate interaction level between the decisions such that the design objectives are satisfied with the minimum complexity in the design process.

For example, in the case of coupled product and material decisions shown in Figure 9.2, the information flow is represented by coupled pattern P6. However, the coupled pattern can be simplified to a sequential scenario by breaking one of the information links. If the information link from the product decision to the material decision is broken, then the design process becomes a sequential process where the material variables are fixed first; the material variables are passed on to the product decision and, using the knowledge of the material variables, a decision is made regarding an appropriate geometry. A different sequential process would result if the information link from the material decision to the product decision is broken. Decision decoupling can be further carried out to convert the sequential decisions into independent decisions. In that case, the product and material decisions are carried out independently, as might be typical of conventional materials selection.

Functional decoupling refers to the simplification from pattern P9 to pattern P6, pattern P8 to P5, and pattern P7 to P4. This is important in the case of multifunctional design, where the product is designed to satisfy multiple functional requirements that drive the design into different directions. Such design scenarios are characterized by multiple sets of design variables (possibly overlapping), whose values can be selected for satisfying multiple objectives. For example, in the design of LCAs, thermal and structural aspects are coupled with each other (Seepersad, Allen 2008). The same set of topological and geometric parameters determines both the thermal and structural behavior. The task of functional decoupling is to determine which functional requirements can be satisfied independently and which should be designed for in a concurrent fashion. Functional decoupling also depends on how the design variables are partitioned for satisfying different functional requirements. Hence, the task in functional decoupling is also to determine the appropriate design space partitioning. It is often highly desirable in practice to functionally decouple since different experts, codes, and databases are commonly employed for each class of functionality. Functional decoupling is not discussed in this chapter.

In order to determine whether a particular type of simplification of the design process is appropriate, it is important to quantify the impact of these simplifications on the design outcome. This leads to the second requirement—the ability to quantify the impact of simplification on the overall performance. A value-of-information-based metric called the *improvement potential* is presented in the next section to address this requirement.

9.4.2. Meta-Design Decisions Using the Improvement Potential Metric

9.4.2.1. Measuring the Value of "Perfect" Information

As discussed in Section 9.1.2, there is no such thing as a "perfect" model. All models have some assumptions built into them. Designers can only try to refine their models so that the predictions from the model are *closer* to the actual behavior of a system. Perfect information can be achieved by building the real system and using it in the actual operating conditions, which is generally not possible during the design phase. However, the notion of a perfect model serves an important baseline to quantify the usefulness of different models. In this section, we discuss a value-of-information metric assuming that perfect information is available. This metric is helpful in gaining an understanding of the concept of value-of-information. The assumption is relaxed in the next section to determine a more usable metric called *improvement potential*.

Consider the scenario shown in Figure 9.10, where the horizontal axis is the design variable and the vertical axis is the corresponding payoff that is achieved by selecting a particular numerical value of the design variable. The design variable can be some physical dimension that a designer has control over, whereas the payoff represents profit, in some sense, which depends on system behavior such as performance, strength, and cost. A designer's objective is to maximize the expected payoff through selection of an appropriate design variable value. The solid line represents the expected payoff evaluated using *actual* system behavior and the dashed line represents the payoff evaluated from system behavior *predicted* by the simulation model. The difference in actual and predicted behavior is attributed to the imprecision and is due to simplifications or idealizations in the model.

In this scenario, the statistical variability in the actual system is accounted for by taking the expected payoff *for any particular value of the design variable*, where the expected value is the

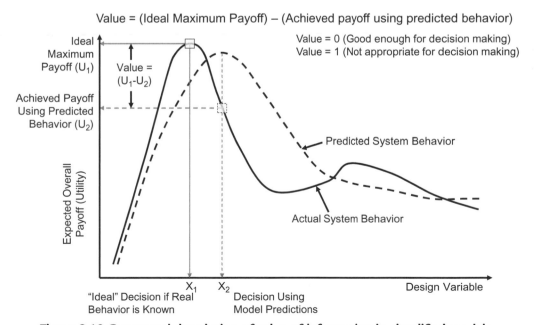

Figure 9.10 Conceptual description of value-of-information in simplified models.

sum of possible payoff (ui) weighted by the probability of achieving that payoff, $p(ui)$. In the equation here, U denotes the expected payoff at a particular value of the design variable, i.e.,

$$U = E(u) = \sum_i u_i p(u_i) \qquad (9.4)$$

For example, there may be variability in the physical system due to inherent randomness. Due to this inherent randomness, the actual system behavior may have some variation. The focus of model refinement in this chapter is only on reducing the imprecision resulting from the simplified nature of simulation models.

Considering Figure 9.10, if a designer makes a decision using the simulation model only, the decision point is X_2, because it maximizes the expected payoff based on the predicted behavior. By selecting the decision point X_2, in reality the designer would achieve the expected payoff equal to U_2 because the system's actual behavior is given by the solid line. A designer would have selected decision point X_1 if the actual (real) behavior of the system were known (by using a *perfect* model). For the decision point X_1, the actual payoff achieved is the maximum ($=U_1$). Hence, the value of using the perfect model over the simpler model is the difference in expected payoff *actually achieved*. It is important to note that the value-of-information is evaluated by measuring the difference in the expected payoff using the actual system behavior (U_1-U_2). It can only be evaluated if (a) the actual system behavior is known or (b) it is calculated after the decision is made and the outcome is realized. These requirements on evaluating the value-of-information metric are similar to those of the *ex-post* value used in the literature.

The value-of-information shown in Figure 9.10 captures the benefit (the improvement in expected payoff) of using the actual system behavior over the predicted system behavior based on the simplified model. Hence, it refers to the *value of perfect information*. As mentioned in Section 9.3.1, the payoff is quantified using utility functions whose lower and upper bounds are 0 and 1, respectively. Since the value of perfect information is measured as a difference between two utility functions, it also lies between 0 and 1. It is 0 if the decision made using the simulation model is the same as the decision made using the actual system behavior (i.e., $X_1 = X_2$). This scenario is shown in Figure 9.11(a). If this value of perfect information is 0, the benefit from using the actual system behavior over the predicted behavior is 0. This implies that the simulation model is perfect for decision making. On the contrary, if the value of perfect information is high, one should consider refining the simulation model. In other words, this metric quantifies the value of refining the simulation model. The worst-case scenario, shown in Figure 9.11(b), occurs when the maximization of expected utility using the predicted model results in U_2 being equal to 0, while the ideal maximum payoff, $U1$, equals 1. In this scenario, the value of perfect information is equal to 1 (because $U_1-U_2 = 1$). Further, the value of perfect information is always non-negative.

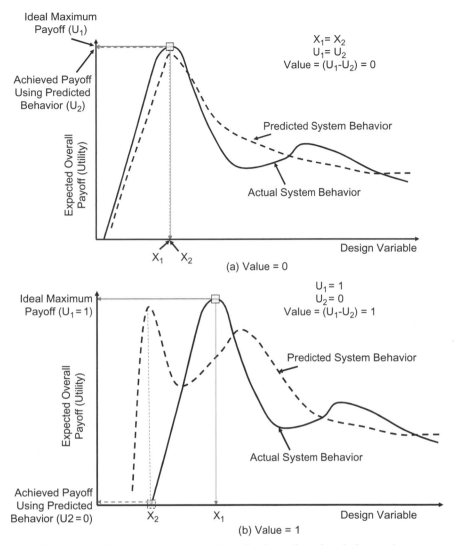

Figure 9.11 Two extreme scenarios of value of perfect information.

It is important to realize that the value-of-information metric does not only depend on the accuracy of the model. It also depends on the complete decision formulation that includes constraints, preferences, domain of the design space considered, etc. The same concept extends to higher-dimensional problems for which there are many design variables and the payoff is determined by multiple conflicting criteria. In the case of multiple design variables, the curve corresponds to a multidimensional surface. In the case of multiple design criteria that affect the payoff, the criteria are combined into an overall payoff function based on designers' preferences.

The value of perfect information can be augmented to measure the value-of-information from model refinement, which is the increase in the expected utility achieved when a refined model is used as compared to a simplified model. For calculating the value of perfect information (as shown conceptually in Figure 9.10) before making a decision, a designer needs to know the actual system behavior. However, in most design cases, the difficulty is that the exact system behavior is unknown. If the exact system behavior were available, there would be no need to use the simulation model to predict the behavior. Hence, the notion of value of perfect information shown in Figure 9.10 is impractical in a real design scenario. To overcome this difficulty in practical use, a variation of the value of perfect information approach is presented next.

9.4.2.2. Measuring the Improvement Potential

Although the *exact* system behavior is unknown throughout the design space in most design scenarios, in many cases, it is possible to determine an upper and a lower bound on the behavior predicted by a simulation model. Designers may be able to generate lower and upper bounds through physical experiments, or through analysts' insights into the system-level behavior. For example, consider a finite element model that utilizes different material parameters as inputs. The imprecision in the output of the model is in part due to the imprecision in the values of material properties such as the yield strength, Poisson's ratio, Young's Modulus, Equation-of-State parameters, etc. If the imprecision associated with these parameters can be experimentally determined in terms of lower and upper bounds, the imprecision in the outputs of the finite element model (i.e., the behavior predicted by the simulation model) can be evaluated. Similarly, for various other scenarios where the bounds on inputs can be evaluated, the bounds on the behavior predicted by the simulation model can be determined. In other situations where the model imprecision is not due to imprecise parameters, but rather due to the model structure, physical experiments can be carried out to determine the actual system behavior at targeted experimental points. The behavior information at these experimental points can be used to fit a response surface, which can be compared with the outputs of the simulation model. The difference between the output of the simulation model and the response surface developed from the experimental data can be used to determine the imprecision bounds. If it is not possible to carry out physical experiments before the design, a more accurate model (which is generally more expensive to build and execute) can be used instead of physical experiments to determine the bounds on the system behavior.

If the imprecision bounds on the output of the model can be evaluated using the approaches described previously, these bounds translate to bounds on the overall utility function, as shown in Figure 9.12. In this figure, the lower and upper bounds on the utility function (U_{min} and U_{max}, respectively) represent the range within which the expected utility calculated from actual system behavior (i.e., the solid curve shown in Figure 9.10) lies. The value-of-information metric developed in this chapter is based on the assumption that bounds on imprecision of simulation models are available.

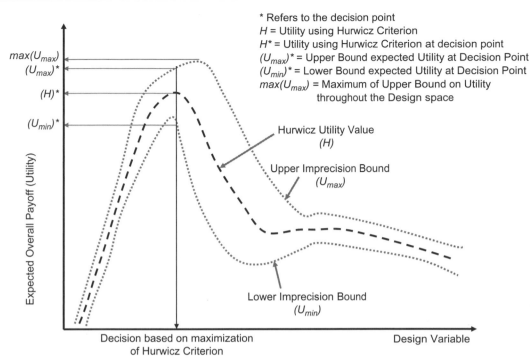

Figure 9.12 Decision made using bounds on payoff.

With the available information about lower and upper bounds on payoff, the decision maker can select a decision rule based on which he or she selects numerical values for the design variables. As discussed earlier in this section, the decision rule can be (1) maximize the lower bound on achievable payoff (i.e., the worst-case scenario), (2) maximize the upper bound on achievable payoff (i.e., the best-possible scenario), or (3) maximize a weighted combination of payoff (using the Hurwicz criterion). For the selected value of the design variable, there is a range of achievable payoffs as a result of imprecision in the simulation model. The lower bound on expected payoff is denoted by U_{min}, the upper bound by U_{max}, and the payoff evaluated using the Hurwicz criterion by H. The lower and upper bounds on expected payoff at the decision point are denoted as $(U_{min})^*$ and $(U_{max})^*$, respectively. The maximum payoff that can possibly be achieved by any value of the design space is $max(U_{max})$, and is evaluated by maximizing the upper imprecision bound on payoff. Since the exact value of the payoff is not known at different values of design variables, it is not possible to calculate the exact value-of-information as illustrated in Figure 9.10. However, since the lower and upper bounds on payoff are known throughout the design space, we can determine the *maximum possible value-of-information*. This upper bound on the value-of-information (maximum possible value) is referred to as the *improvement potential (PI)* and is given by

$$P_I = \max(U_{\max}) - (U_{\min})^* \tag{9.5}$$

where $max(U_{max})$ is the maximum expected payoff that can be achieved by any point in the design space and $(U_{min})^*$ is the lowest expected payoff value achieved by the selected point in the design space (after making the decision without added information). We propose the *improvement potential* as a value-of-information metric for deciding whether further refinement in the simulation model is necessary or not. This metric is used in the method for stepwise refinement discussed in Section 9.5.1.

In the context of model refinement, a higher improvement potential *(P_I)* indicates that there is a higher scope for improvement in the decision through the addition of more information, whereas values close to zero indicate that little benefit can be achieved in the quality of decision by adding more information. The value-of-information-based approach has various applications in model development and refinement for materials design. For example, the metric can be used to determine whether more experimental data are needed to refine the constitutive model parameters to be used in the simulation of material behavior. Further, the approach can be used to determine whether there is a need to refine the mesh size in finite element analyses. Such an example is presented in Section 9.6.

Similarly, in the context of selecting appropriate patterns, addition of information is equivalent to moving from decoupled patterns to coupled patterns. Comparing two patterns (see Figure 9.8) such as P2 (sequential interaction) and P3 (coupled interaction) by including the coupling in P3, we add information about the system that is not accounted for in P2. If the expected value of this added information is greater than certain threshold value, pattern P3 should be used instead of P2. The notion of using the improvement potential for determining whether coupled models can be converted into sequential or concurrent models has significant applications in multilevel materials design. For example, it can be used in multiscale simulation models to determine whether the lower-scale model should be tightly coupled with the higher scale model or whether a the output of the lower-scale model can be used to build a metamodel, which can be integrated with the higher-scale model. Use of fully coupled, concurrent multiscale models must be justified based on value-of-information. The application of the improvement potential metric in an energetic structural material design example is presented in Section 9.7.

To illustrate how the improvement potential can be used in deciding whether to refine the models, consider an example of the shock simulation model of energetic materials discussed in Chapter 3. The details of the model, the use of the metric, and the results are presented later in Section 9.6. The example is presented here only for the purpose of understanding the process of determining the improvement potential. The process is shown in Figure 9.13.

The first step is to identify the parameters resulting statistical variability and imprecision. The examples of parameters associated with statistical variability include the shapes and sizes of particles. These parameters are modeled using probability functions. Examples of parameters resulting in imprecision in the model include uncertain model parameters (e.g., Equation-of-State parameters), assumptions in the model (such as 2D domain, symmetry, etc.), and numerical discretization errors. These parameters are modeled using lower and upper bounds on the model outputs. The design variables (which the designers can control) and the responses (which depend on the design variables) are identified. A response is synonymous to a model output. In this example, the design variables are volume fractions of constituents, particle sizes, etc. The responses are the number density of contact sites, number density of reaction initiation sites, and Hugoniot data for the mixture.

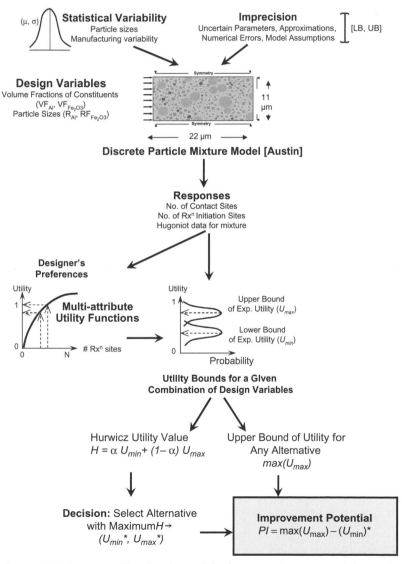

Figure 9.13 Process of evaluation of the improvement potential metric.

Using the model, a response surface is developed that relates the design variables and the responses. Variability of these responses is associated with uncertainty (both statistical variability and imprecision). The uncertainty in the response is quantified by executing the model at several combinations of the uncertain parameters. Statistical variability in the inputs results in a probability function on the response values, whereas imprecision results in an interval of response values. A combination of both these types of uncertainty in the inputs results in two probability distributions associated with the lower bound and the upper bound on each response. Independent of the model, the designers' preferences for the responses are determined in terms of utility functions. The responses and associated uncertainties are mapped on to the utility space, resulting in two probability functions—one corresponding to the upper bound on overall utility, and the other corresponding to the lower bound. The expected value of the lower and upper bounds correspond to U_{min} and U_{max} in Figure 9.12. These bounds are use to determine the Hurwicz utility value and the decision point, which is then used to determine the improvement potential, P_I.

9.4.2.3. Properties of the Improvement Potential Metric

The first important property of the improvement potential metric discussed in Section 9.4.2.2 is that it also captures the effect of imprecision in simulation models in the evaluation of value-of-information. Statistical variability is accounted for by using the *expected* utility and imprecision is accounted for by using the *lower and upper bounds* on expected utility. Hence, it overcomes the limitations of existing metrics discussed in Section 9.3.3.2. Although the metric can be used in the cases where both imprecision and variability are present, in this chapter, we only focus on the imprecision in simulation models.

The second important property of the value-of-information metric is the ability to *quantify the opportunity for improving the design solution by adding more information*. That is, the value-of-information metric quantifies the upper bound on the benefit that can be achieved by obtaining perfect information. The opportunity for improving the design solution is quantified in the literature using the Expected Value of Perfect Information (EVPI) (Bradley and Agogino 1994), which is calculated as the expected value-of-information based on setting the numerical values of uncertain parameters equal to their actual realization. If there are multiple uncertain parameters, the expected value of perfect information corresponding to each parameter is evaluated for individual parameters by setting their actual values. The greater the expected value of perfect information, the greater is the *opportunity of improving the design solution through information gathering*. The limitation of this EVPI, however, is that the exact values of parameters are generally not available before gathering the information. In contrast, the advantage of the improvement potential is that it provides an indication of the opportunity without necessitating the use of perfect information.

9.4.2.4. Scope of the Improvement Potential Metric for Simulation Model Refinement

The value-of-information metric presented in this chapter (improvement potential) is based on the assumption that information about error bounds (i.e., the lower and upper bounds on the imprecise variables) is available. The metric is not applicable if this information is not available.

For example, if results from different fidelities of simulation models are available but there is no information about the bounds within which the actual behavior lies, then the metric cannot be used. Hence, experiments and limited prototyping are necessary accompanying features of the design process. The metric is ineffective if one of the bounds is incorrect (i.e., the real value lies outside the bounds). For example, consider a model with a set of parameters that are calibrated using experimental data points. If there are not enough experimental points to estimate the bounds on these parameters, it is not possible to correctly identify the bounds on the output of the simulation model. In this case, the specific metric presented in Section 9.4.2.2 is not effective. There is a need to develop new metrics that address such scenarios. Messer (Messer 2008) has developed one such metric to account for this limitation of the improvement potential metric.

The improvement potential is developed considering only the information about the improvement in payoff of the decision, which is defined using the utility functions. It does not include the cost of gathering the additional information (the cost of reducing the range of imprecise variables in the case of simulation model refinement). Specifically, costs may be associated with performing additional experiments, better calibration of the model, or developing a lower-scale model to better predict the parameters used in the model under consideration. It is assumed that for a given step in the series of refinement steps, a designer evaluates the estimated cost of gathering information and the value of this added information. Using these two indicators, a designer makes the decision on whether additional information is worth pursuing. Strictly speaking, there is a tradeoff between the improvement potential and the cost (of refining a simulation model and executing the refined model). In this chapter, we limit the scope of the improvement potential to the benefit achieved by refining the simulation model. The cost of model refinement will be considered in a future work.

Finally, the improvement potential can only be used to determine whether a given level of refinement of simulation model is appropriate for making a particular decision or not and it is done after the computations have been made. It does not help designers in determining *how much* further refinement is required in the simulation model. The metric can be used for any type of model refinement as long as it is possible to quantify the error bounds on the output. Examples include refinement of the finite element mesh, refinement of the model of physical behavior, refinement of parameters used in the model, etc.

9.5. Utilizing the Improvement Potential for Management of Complexity in Design Processes

9.5.1. *Method for Stepwise Model Refinement Using Value-of-Information*

A method for stepwise refinement of simulation models is shown in Figure 9.14. The method consists of seven steps, outlined in the following:

Step 1: Formulate a design decision using a cDSP. The mathematical decision formulations for the design problems are presented in Section 9.6.2.1.

Step 2: In order to make this decision, a designer starts with a simple simulation model that is associated with imprecision in outputs. The strategy adopted in this method is to start with a simple model (that is computationally inexpensive) and to refine it gradually until it is appropriate for the decision. For each of the refined models being considered, the designer performs steps 3 through 7.

Step 3: Information about the lower and upper imprecision bounds on the output of a simulation model is gathered. Using this information, two optimization problems are solved in Steps 4 and 5, respectively.

Step 4: In this step, the first optimization problem is solved, in which the designer determines a point in the design space that maximizes the Hurwicz weighted average of expected utility (H). This point is called the *decision point.* The lower bound on the expected utility at this decision point is determined $(U_{min})^*$.

Step 5: In the second optimization problem, the designer determines $max(U_{max})$, which is the maximum of the upper bound on the expected utility (maximax criterion).

Step 6: The improvement potential is determined using $P_I = max(U_{max}) - (U_{min})^*$.

Step 7: If the improvement potential evaluated in Step 6 is high (implying that gathering more information via model refinement is beneficial), the model is refined and an updated decision is made. The model is systematically refined in this manner until the value of additional information is low (i.e., the model is good enough for decision making). The magnitude of the improvement potential directly relates to the achievement of the

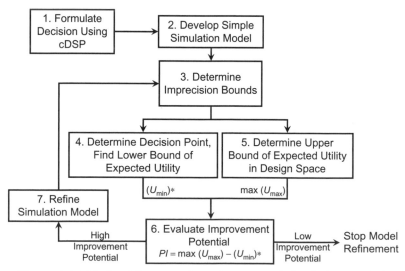

Figure 9.14 Method for stepwise refinement of simulation model.

designers' goals. Hence, deciding whether the improvement potential is low enough depends on the design problem and the cost of improving the model to achieve that improvement.

This process of starting with simplified models and gradually increasing complexity is enabled through the introduction of the improvement potential, defined in Equation 9.5. Otherwise, strongly coupled models would serve as a logical starting point if value-of-perfect-information was used as the metric.

The approach discussed in this section is illustrated using a materials design example. In the material design example, a finite element model is used to predict the behavior of a material by changing multiple design variables. The imprecision in this example is due to the small size of the domain considered and the discretization errors. The materials design example represents a typical simulation-based design problem.

9.5.2. Method for Systems-based Integrated Design of Materials, Products, and Design Processes

The method presented in this section consists of nine steps shown in Figure 9.15. We start with a simple interaction pattern to design the product and material; for this simple interaction pattern, the design is carried out and the improvement potential is evaluated. If the improvement potential is high, then there is a need to refine the interaction patterns, whereas if the improvement potential is low, then the selected interaction patterns are appropriate. Hence, the outcome of the decisions is used as the final design. The steps are discussed next.

Step 1: Formulate Decisions Using the cDSP—The first step in the method is to mathematically formulate the design decisions using the cDSP construct (Mistree, Hughes 1993, Seepersad 2001), which is a multi-objective decision model with a hybrid formulation based on mathematical programming and goal programming. The cDSP consists of four key blocks—(1) given, (2) find, (3) satisfy, and (4) minimize. The product and material decisions for the energetic material design scenario modeled using cDSP are presented in detail in Section 9.7. In the case of coupled decisions, it is important to identify the design variables specific to different decisions and constraints and goals that are common to the decisions. In this chapter, we use the utility-based formulation of cDSP (Seepersad 2001) where the goals are formulated as utility functions, which are used to quantify the designers' preferences toward achievement of goals on a scale of 0 to 1. In a multi-objective design problem, separate utility functions are formulated for different goals and these utility functions are combined into a single *overall* utility function. The objective in the cDSP is to minimize the deviation of overall utility from the maximum possible value of 1.

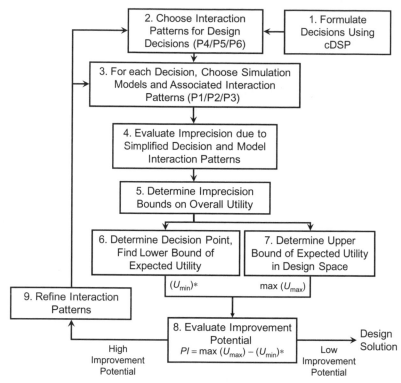

Figure 9.15 Method for designing design processes for integrated material and product realization.

Step 2: Choose Interaction Patterns for Design Decisions—After modeling the decisions and identifying the shared and independent design variables and goals, the next step is to model the interactions between decisions using interaction patterns P4, P5, and P6 presented in Section 9.4.1. Initially, the designers choose a simple interaction pattern P4 and make the decisions independently. Instead of considering the coupling variables while making the two decisions, fixed nominal values of coupled variables are used for making decisions independently.

Step 3: For Each Decision, Choose Simulation Models and Associated Interaction Patterns—For each decision identified in Step 1, the required simulation models and the interactions between them are modeled using patterns P1, P2, and P3. The interaction patterns for the energetic material design scenario are shown in Section 9.8.5. In accordance to the general strategy, the designers start with simple interaction patterns for models and sequentially refine them based on the improvement potential.

Step 4: Evaluate Imprecision due to Simplified Decision and Model Interaction Patterns—Since some of the information flow between decisions and models is ignored, the resulting expected utility (i.e., payoff) is imprecise (see Figure 9.12 for lower and upper

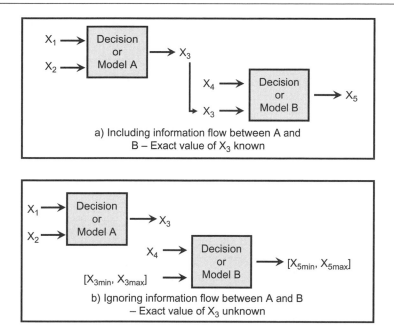

Figure 9.16 Evaluation of imprecision bounds as a result of the simplification of a pattern from sequential to independent.

bounds on utility). This imprecision in overall utility needs to be quantified in terms of lower and upper bounds in order to evaluate the improvement potential. Consider the simplification of the interaction pattern shown in Figure 9.16(a), where two models/decisions A and B are linked with a single information flow of variable X_3. In Figure 9.16(b), the flow of information is ignored and the information flow is replaced by the range of possible values taken by the variable X_3, i.e., $[X_{3min}, X_{3max}]$. Due to this range of inputs for decision/model B, the output is a range of values for X_5, i.e., $[X_{5min}, X_{5max}]$. This range is due to the simplification of the interaction pattern. The range for X_5 is evaluated by selecting various points in the range of X_3 and recording the corresponding results for X_5 after execution of decision/model B. If A and B are decisions, the possible range of X_3 is directly known from the bounds on design variables. However, if they represent simulation models, the ranges of intermediate coupling variables need to be explicitly evaluated by executing model A for various combinations of X_1 and X_2. Coupled interaction patterns P3 and P6 can be similarly simplified into sequential or independent patterns by replacing the simplified information flows by corresponding ranges.

Step 5: Determine Imprecision Bounds on Expected Utility—Information regarding imprecision in the output of models and decisions translates to the imprecision range for the overall utility (see Figure 9.12). Using this information, two variations of the cDSPs with different objective functions are solved in Steps 6 and 7, respectively.

Step 6: Determine Decision Point and Find Lower Bound of Expected Utility—In this step, the first variation of the cDSP is solved, in which the designer determines a point in the design space that maximizes the Hurwicz weighted average of expected utility (H). This point is called the *decision point.* The lower bound on the expected utility at this decision point is determined $(U_{min})^*$.

Step 7: Determine Upper Bound on Expected Utility in Design Space—In the second variation of cDSP, the designer maximizes the upper bound on the expected utility, which is max(U_{max}).

Step 8: Evaluate Improvement Potential—The improvement potential is determined using $PI = \max(U_{max}) - (U_{min})^*$.

Step 9: Refine Interaction Patterns—If the improvement potential evaluated in Step 8 is high (implying that gathering more information via model refinement is beneficial), the model is refined and an updated decision is made. There are two options for refining the interaction patterns—refining the patterns for simulation models and refining the decision patterns. As mentioned earlier in this section, the strategy is to start with a simple interaction pattern and sequentially refine it until the improvement potential is low.

The method is applied to the energetic material design problem in Section 9.7.

9.6. Application of the Simulation Model Refinement Approach to a Materials Design Problem

In this section, we present an example of simulation model refinement from the materials design domain. In this design example, the objective is to design an energetic material to achieve the desired probability of exothermic reaction initiation under shock loading. A mixture of aluminum and iron oxide particles in an epoxy matrix (as binder) is considered. A designer of such mesostructured materials can control material parameters such as constituent volume fractions and particle sizes to obtain desired properties. The performance of the material is defined in terms of its strength, reaction energy release capability, and other characteristics, and is simulated computationally using an explicit Eulerian hydrocode. The model is imprecise and can be refined through various modes of refinement. As the model is refined, the complexity of the model and the associated runtime increases. In order to make decisions efficiently, an appropriate level of refinement of the model is desired. In this section, we show how the approach can be used to determine the appropriate level of refinement of the model. Different modes of refinement of the numerical model are explicitly considered in this section. In order to understand the details of the materials design problem, the simulation model used to predict the material behavior is described in Section 9.6.1. The design problem is discussed in Section 9.6.2, and the results from refinement are discussed in Section 9.6.3. For complete details of the problem, refer to (Panchal 2005).

9.6.1. Shock Simulation Model

9.6.1.1. Model Description

The design problem in this section is to choose appropriate values for the control parameters of the energetic material, in order to achieve target values of mixture properties as a shock wave propagates through the material. The model discussed in this section is used to support such a design problem by simulating the propagation of a shock wave through an energetic material. The inputs to the model include the shock speed, the dimensions of material under consideration, the volume fractions of constituents, and the size of constituent particles. Using information regarding the material properties of the constituents (aluminum, iron-oxide and epoxy) as input, the model generates the overall material properties of the mixture as the shock propagates. The simulation model is developed by Austin (Austin 2005). Only the relevant details of the model are discussed here. The shock simulation consists of two steps—(1) the generation of a synthetic microstructure based on experimental information, and (2) the simulation of shock propagation using an Eulerian hydrocode. In the first step, information obtained from microscopy of fabricated energetic materials is used to generate information about the distribution of particles in the mixture. This information is quantified in terms of the probability distributions for sizes and nearest-neighborhood distances of the particles and voids. It is then used to randomly generate statistic volume elements (SVEs) of microscale particles (i.e., aluminum particles, iron oxide agglomerates, and voids). The SVE represents a small volume of the material through which a shock is propagated, with sufficient size to capture essential aspects of heterogeneity and particle interactions in terms of statistical sample or realization of the mesostructure, without being large enough to reflect complete (i.e., statistically homogeneous) statistics of responses. Use of a SVE is necessitated by computational tractability for such highly nonlinear problems. Hence, response statistics must be built up from analyses of a number of SVEs. The sizes and spatial locations of the particles in the SVE are modeled such that they closely approximate experimentally generated probability distributions.

After the particle morphology is generated, the next step is to perform a numerical simulation using finite element techniques. In this model, the properties of individual particles are used as inputs for the model. Based on the individual constituent properties, the effective properties of the mixture are determined. Specifically, the propagation of a shock wave through the reactive particle system is simulated to understand the effect of material properties and morphology on the hydrostatic behavior of the overall mixture. The simulation is performed using an Eulerian hydrocode (RAVEN) (Benson 1995). The boundary conditions on the SVE are shown in Figure 9.17. The shock propagation phenomenon is idealized as a 1D shock wave. A compressive shock wave is propagated through the mixture by applying a Lagrangian velocity boundary condition to the left surface of the SVE. The boundary on the left-hand side is provided an initial velocity, i.e., particle velocity (Vp). Symmetry planes serve as Lagrangian boundary conditions for the top and bottom surface of the model. A fixed Lagrangian boundary condition is imposed on the right hand side surface. Based on the results of the simulation model, the speed of the shock wave (V_s) is determined. The relationship between the V_p and V_s is used to model the dilatational behavior of the mixture by employing the Gruneisen Equation of State (for details of the calculation, please refer to Austin (2005). Since the hydrostatic behavior of the mixture is directly dependent on the shock wave speed, a designer can specify the target performance of the material in terms of the shock wave velocity when the imposed particle velocity is fixed.

9.6.1.2. Imprecision in the Shock Simulation Model

Although there are various sources of imprecision in the shock simulation model, we focus only on the imprecision that is due to the fact that (1) only a small portion of the domain (SVE) is modeled, and (2) the domain is discretized using a finite element mesh. The material morphology in the SVE is randomly generated based on the statistical properties of the distribution of both the size of the particles and the distance between them. Since the particles are randomly distributed, the material morphology is different every time a new set of particles is generated, even for the same set of parameters. The chosen *size of the SVE* is one of the main factors determining the imprecision in response. Smaller SVEs have greater imprecision than that of larger SVEs in terms of effects of symmetry planes and approximations of statistical behavior of the ensemble. As the size of the SVE increases, these sources of imprecision reduce in magnitude. Hence, the model can be refined by *increasing the size of the SVE*. Multiple SVEs are necessary to achieve a statistically representative response in any case, unless the SVE becomes large enough to exhibit statistically invariant results for number density of reaction initiation sites (response in this case) with further increase of size. In this case, the SVE would constitute a representative volume element (RVE). For rare-event, heterogeneous phenomena such as reaction initiation, the RVE is too large to be computationally tractable.

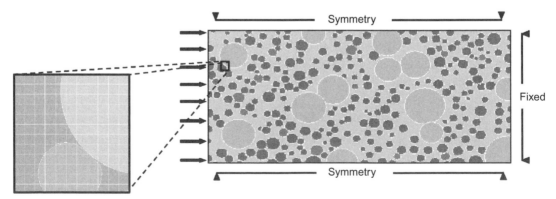

Figure 9.17 Boundary conditions of the discrete particle shock simulation for energetic material design (Austin 2005).

After the microstructure morphology is generated, the simulation is deterministic, i.e., the same morphological instantiation with the same boundary conditions will result in the same values for the output parameters. However, since the particle shock simulation is based on a model, there is imprecision in the outputs due to discretization as well. The parameter that can be used to control imprecision due to discretization is the *size of the individual cells* in the 2D Eulerian mesh. Hence, the simulation model can be sequentially refined by (1) increasing the size of statistical volume element (SVE) and (2) reducing the size of individual cells in the finite element mesh. Hence, the refinement question faced by the material designer is "What are the appropriate values of these two *model refinement parameters* for designing the material?"

9.6.2. Description of the Materials Design Problem

From a decision-making perspective, the design problem can be viewed as two decisions: a decision about the material representation and a decision about model refinement. In the first decision, a designer needs to choose appropriate values of the material parameters such as volume fractions of individual constituents, particle sizes, etc., whereas in the second decision, a designer is interested in choosing appropriate values of the two model parameters (SVE size and simulation cell size). Both these decisions depend on each other. Depending on the target requirements for material properties and/or responses, the appropriateness of the level of refinement of the model may change. Similarly, depending on the level of model refinement chosen, the decision about material properties or responses may change.

9.6.2.1. Materials Design Decision

The materials design decision is presented in Table 9.1. In this problem, we assume that the following two material parameters are variable: the size (radius) of aluminum particles and the volume fraction of voids. The objective is to achieve a shock wave speed of 4.5 km/sec at which the reaction initiates. Any shock wave speed below 3.5 km/sec has a utility of 0 and a shock wave speed greater than 4.5 km/sec has the same preference. A utility of 1 is assigned to shock wave speed equal to 4.0 km/sec. A quadratic utility function is used between 3.5 km/sec and 4.5 km/sec.

The range of radius of aluminum particles considered is [0.5, 1.5] μm, and the range of volume fraction of voids is [0.02, 0.10]. All other material parameters are assumed to be constant. The input particle velocity is fixed at 1 km/sec. The objective is to achieve the target shock wave velocity. Shock speed is chosen as an objective because the hydrostatic behavior of the mixture is influenced directly by the shock speed. As discussed previously in this section, the relevant material properties are associated with the Gruneisen Equation of State (EOS), which is determined by fitting a straight line to the particle velocity versus shock front velocity data. Hence, the material properties are dependent on the shock speed achieved for a given particle speed.

Table 9.1 Materials design decision.

Given

Cell Size (S_c) = [0.035, 0.140] microns
Window Size (S_w) = [0.014, 0.028] mm
Particle Speed (V_p) = 1km/sec
Particle Shock Simulation Model
    $V_s = f(R_{Al}, VF_{Voids}, S_c, S_w)$
    (that predicts shock speed as a function of Radius of Aluminum, Volume Fraction of Voids, Cell Size, and Window Size)

Find
Values of design variables Size of Aluminum Particles (R_{Al}) Volume Fraction of Voids (VF_{voids})

Satisfy
Bounds on design variables Size of Aluminum Particles = [0.5, 1.5]μm Volume Fraction of Voids = [0.02, 0.10] *Preference for shock wave speed* $$U_{Overall} = U_{V_s} = \begin{cases} 0, & V_s \leq 3.5 \\ 1.4(V_s - 3.5) - 0.4(V_s - 3.5)^2, & 3.5 < V_s < 4.5 \\ 1, & V_s \geq 4.5 \end{cases}$$

Minimize
Deviation from Maximum Utility: $Z = (1 - UOverall)$

9.6.2.2. Refinement Decision

As mentioned before, the refinement decision involves choosing the size of the SVE and the cell size. The size of SVE (S_w) is defined in terms of its length scale, which lies in the interval [0.014, 0.028] mm. The width is taken as half of the length in order to maintain the same aspect ratio for all scenarios. The cell size (S_c) is similarly defined in terms of its length. The length of a cell in the mesh lies within the range [0.035, 0.140] μm. The width of the cell is half of its length. The size of SVE (also referred to as *window size*) and the cell size can be varied continuously between the lower and upper bounds, which provides an infinite set of options of simulation model refinements. However, exploring all those options is not effective from a decision-making perspective. In order to reduce the computational load, we just explore nine different simulation model refinement options. These options are generated by taking all combinations of three levels each of the size of the SVE (with lengths of 0.014, 0.021, and 0.028 mm) and the cell size (with lengths of 0.035, 0.0875, and 0.14 μm).

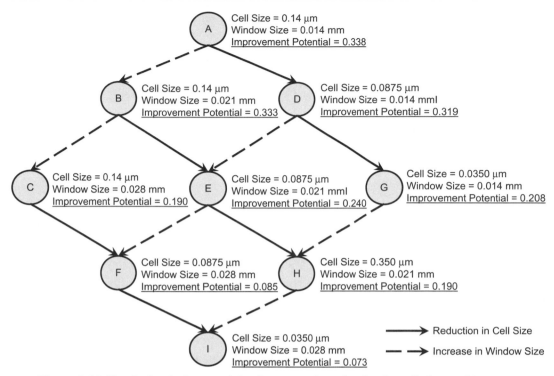

Figure 9.18 Shock simulation model refinement via reduction in cell size and increase in window size.

The nine options are labeled from A through I and are shown in Figure 9.18. The approach is to select the simplest model (window size = 0.014 mm, and cell size = 0.14 μm) and sequentially refine it to a level that is just appropriate for making decisions regarding the material properties. The refinement of the model is shown as a graph in Figure 9.18. In this figure, all possible paths of refinement are shown for illustrative purposes. Only one path will be followed by a designer. The dashed lines represent refinements through increases in window size and the solid lines represent refinements through reductions in cell size. The model that is most accurate is labeled as I. A designer starts with the simplest model, A, and refines it sequentially using the method presented in Section 9.5.1. The decisions are made using the Hurwicz criterion with a pessimism index of 0.5. At each stage, the improvement potential is evaluated, and a decision is made as to whether there is a need to refine further or not.

9.6.3. Results from the Energetic Material Design Example

The decisions made using the different levels of refinement of the simulation model are presented in Table 9.2. The table contains columns for the cell size in the mesh and the

Table 9.2 Decision outcomes for different refinement scenarios.

	Refinement Case			Design Variables		Value Metric
Label	Cell Size (microns)	Window Size(mm)	Utility Using Hurwicz Criterion	Mean Al Size (microns)	VF Void	Improvement Potential
A	0.1400	0.014	0.826	1.40	0.02	0.338
B	0.1400	0.021	0.790	1.50	0.02	0.333
C	0.1400	0.028	0.777	1.50	0.02	0.190
D	0.0875	0.014	0.830	1.44	0.02	0.319
E	0.0875	0.021	0.794	1.50	0.02	0.240
F	0.0875	0.028	0.779	1.50	0.02	0.085
G	0.0350	0.014	0.850	1.50	0.02	0.208
H	0.0350	0.021	0.797	1.50	0.02	0.190
I	0.0350	0.028	0.779	1.50	0.02	0.073

window size that determine the level of refinement of the shock simulation model. The fourth column is for the maximum overall utility achieved at the decision point. The following two columns are for design variables—mean size of aluminum and volume fraction of voids. Finally, the last column indicates the improvement potential, which is the difference between the maximum overall utility that can be achieved at any point in the design space and the minimum utility achieved at the decision point. This improvement potential is used as a metric for value-of-information that can be achieved by refining the simulation model further. If the improvement potential is low enough, the model does not need to be refined, whereas if the improvement potential is high, there is an opportunity for improving the decision via refinement of the simulation model.

The improvement potential for different model refinement scenarios is also presented in graphical form in Figure 9.18. It is observed from the graph that as the simulation model is refined, the numerical value of the metric reduces because it means that the possible improvement in the solution by refining the model also reduces. This is observed in Figure 9.18 while going from A→B (increasing the window size) or from A→D (reduction in cell size). The improvement potential reduces from 0.338 to 0.333 by refining the model from A→B, and the value reduces from 0.338 to 0.319 by refining the model from A→D. Similarly, the improvement potential reduces as the model is further refined either by increasing the window size or reducing the cell size. Note that the improvement potential is dependent only on the current level of refinement of the simulation model. It does not depend on the path followed for refinement. For example, whether the refinement is carried out as A→B→E or A→D→E, the improvement potential of E is the same.

Ideally, a designer would like to select the model with the minimum possible improvement potential. Hence, Model I may appear to be the best option. However, it is important to note that as the cell size is reduced or the window size is increased, the computational cost also increases. As a designer refines the simulation model from A through I in Figure 9.18, the cost of computation also increases. Accordingly, there is a tradeoff between the improvement potential and the associated costs. At each stage of refinement in Figure 9.18, a designer must compare the improvement potential against the increase in computational cost. If the added benefit apparent from the improvement potential exceeds the cost, then further refinement is appropriate. For example, Model F has an improvement potential of 0.085 (which is very low). If this improvement potential is smaller compared to the increase in cost by using Model I, further refinement of the Model F is not essential. Hence, in that situation, the answer to the refinement question posed in Section 9.6.1.2 is that a cell size of 0.0875 μm and a window size of 0.028 mm results in a simulation model that is appropriate for designing the material.

Reflecting on this example, one very important point arises. Convergence of the numerical solutions has not been mentioned as a function of cell size refinement. In fact, this is not essential with regard to informing the materials design decision for systems having monotonic convergence characteristics. The selection of the interval or range of cell (solution mesh) sizes is presumably such that the solutions are not highly sensitive to variation of cell size, but the effects of refinement are considered in terms of the value-of-information delivered. The point is that fully converged numerical solutions are not essential to the process of informing design decisions. Of course, results from unstable and/or incorrect numerical solutions will have no utility in this regard; it is the purview of the analyst to understand and specify the range of cell sizes such that weak convergence is achieved (relaxed tolerance for convergence). This differs substantially from conventional notions of optimization in simulation-based design in which numerical convergence must be emphasized prior to engaging in iterative "optimization" looping.

In summary, by using the value-of-information based approach presented in this chapter, designers can sequentially refine the simulation models and calculate the potential for improvement in the decision. If the improvement potential is low, then model refinement can stop. It is important to note that the level of refinement depends not only on the accuracy of simulation models, but also depends significantly on the decisions to be made, problem constraints, and a designer's preferences. By using the approach presented in this chapter, designers can exploit such situations and make good design decisions efficiently (with less resource utilization). In the materials design problem, the metric helps designers in gaining insights into the model refinement process. Using the metric, designers are able to identify that the two approaches of model refinement (i.e., increasing the window size and reducing the cell size) are not independent. Designers are also able to determine the appropriate level of refinement of the computational model. In other words, the metric helps designers in making informed simulation model decisions and utilize resources in an appropriate manner. Hence, the metric serves as a guide for designers in determining when to stop refining simulation models.

9.7. Applying the Method for Integrated Design of Materials, Products, and Design Processes to the Energetic Material Design Problem

The energetic material design problem discussed in this section is summarized as follows—"Design an energetic material system concurrently with a storage container such that the overall system has sufficient strength to remain intact under dynamic loading. The energetic material is a mixture of aluminum and iron oxide particles. The material should not react at a specified impact velocity." The details of the problem formulation are provided in Section 9.7.1.

9.7.1. Energetic Material Design Problem Formulation Using cDSPs (Step 1)

The first step in the design method is the formulation of design decisions. The integrated energetic material-container design problem is formulated in terms of coupled material-product design decisions using the cDSP construct. The "Given" block captures the information available to designers for decision making, which includes the available simulation models that generate information regarding the system behavior at different scales and a designer's preference for container performance. The preference for the objectives is expressed in terms of utility functions. Different objectives can be considered. For example, we might wish to maximize the strength of the storage container and minimize the probability of reaction initiation as a lower bound on the loading conditions. On the other hand, we might also ask the complementary question of how severe the loading conditions must be in order to promote full reaction of the energetic material. These two scenarios effectively provide bounds on behavior. The former problem is not as strongly multiscale as it primarily requires assessment of conditions for reaction initiation based on discrete particle simulations. Reaction propagation is not considered. Of more direct interest to this work is the latter scenario, in which we set an objective of complete reaction, since it requires both discrete particle simulations and nonequilibrium equivalent continuum models for reaction propagation at higher length scales to be exercised; it is therefore of strong multiscale character.

Accordingly we assume that the designer's preferences are associated with two functional characteristics—maximizing (1) the strength of the system (container and energetic material), and (2) the reaction propagation of energetic material properties. An indicator of the strength is used in this chapter to simplify the design problem, namely, the overall deformation achieved by the material-container system in a Taylor impact test (details are provided in Section 9.8.3). Similarly, the amount of reaction products (iron, in this material system) accumulated at a specified time after the shock starts is used as a functional indicator for reaction propagation within the energetic material. Utility functions are specified for both responses. These utility functions are ideally generated based on a designer's preferences.

In this specific problem, the utility functions are generated based on assumed performance targets. The actual details of utility values affect the final solution but not the application of the design method.

The cDSPs shown in Table 9.3 are based on robust design formulation where, in addition to achieving target performance, the objective is also to minimize the deviation in performance. Hence, in addition to the preferences for target achievement of the two response values, designers also have preferences for the deviation in these target values. These preferences for deviation in performance are important in the design problem due to the need for making decisions that are robust to variation in performance of the material. The variation in performance can be due to various factors such as inherent randomness in material properties, imprecision in simulation models, and the imprecision introduced due to simplification of design processes.

In the "Find" section of the cDSPs, information about the design variables is captured that designers can control in order to satisfy the design objectives. For the material decision, there are four design variables (size of aluminum particles, size of Fe_2O_3 particles, size of voids, and volume fraction of voids), and for the product decision, there is one design variable (the amount of energetic material stored in the container). This design variable indicates the size of the container. All the other variables associated with the product and the materials are fixed.

The "Satisfy" section in the cDSP captures the information about bounds on design variables, any problem constraints, and the design goals. In this case, the goals are related to achieving targets ($=1$) for utilities. The "Minimize" section captures the overall objective function to be minimized, which in this case is a weighted combination of the deviation of goals from targets. The interactions between the two decisions are modeled using the interaction patterns in Section 9.7.2.

9.7.2. Modeling and Choosing the Interactions Between Decisions Using Interaction Patterns (Step 2)

The decisions regarding product and material are modeled using the interaction patterns: P4 for independent, P5 for sequential, and P6 for coupled. Ideally, the decisions about the material and the container should be made in a coupled fashion because the design variables associated with both product and material affect the overall system performance. Various processes can be used to make decisions in a coupled fashion. One such process is to make decisions individually and iterate until the solution converges to a single design point. The coupled nature of design decisions and the solution using an iterative process is shown in Figure 9.19. The coupled decisions correspond to interaction pattern P6, the highest level of complexity.

The decision patterns can also be simplified into a sequential interaction pattern in which the decisions about container (product)-level parameters are made by first assuming a set of values for container-level design variables. The container-level parameters are then utilized

Table 9.3(a) cDSP representation for the product decision.

Product Decision
Given
Simulation models at three levels (discussed in Sections 9.8.1, 9.8.2, and 9.8.3) Values of material design variables • Size of aluminum particles • Size of Fe_2O_3 particles • Size of voids • Volume fraction of voids Preferences for • Deformation (G1) • Accumulated iron (G2) • Variation in deformation (G3) • Variation in accumulated iron (G4)
Find
Values of product-related design variable • Amount of energetic material stored in the container Values of deviation functions • d_i+, d_i- i = 1..4
Satisfy
Bounds on design variables • Amount of energetic material stored in the container • Utility goals • $G_i + d_i- - d_i+ = 1$ i = 1..4 • These four goals are related to deformation, accumulated iron, variation in deformation, and variation in accumulated iron • Constraints on deviation variables • $d_i+ . d_i- = 0$ i = 1..4 • d_i+, $d_i- \geq 0$ i = 1..4
Minimize
$$Z = \sum_{i=1}^{4} k_i d_i^-, \quad \sum_{i=1}^{4} k_i = 1$$

for making decisions regarding the material. This interaction pattern assumes that the effect of material parameters on the container parameters is insignificant, but the effect of the container level on the material level parameters is significant. This sequential decision making process corresponds to the interaction pattern P5, shown in Figure 9.20. The sequence of decisions can also be reversed by making the material-level decision first and then using the information about material-level parameters to decide upon the product-level parameters. Finally, the two decisions can also be made in an independent fashion. Hence, there are four different configurations in which the decisions about product and material can be made (coupled; sequential with material decision first; sequential with product decision first;

Table 9.3(b) cDSP representations of the materials decision.

Material Decision
Given

Simulation models at three levels (discussed in Sections 9.8.1, 9.8.2, and 9.8.3)
Values of product-related design variable
  • Amount of energetic material stored in the container
Preferences for
  • Deformation (G_1)
  • Accumulated iron (G_2)
  • Variation in deformation (G_3)
  • Variation in accumulated iron (G_4)

Find

Values of materials related design variables
  • Size of aluminum particles
  • Size of Fe_2O_3 particles
  • Size of voids
  • Volume fraction of voids
Values of deviation functions
d_i^+, d_i^- i = 1..4

Satisfy

Bounds on design variables
  Size of aluminum particles
  Size of Fe_2O_3 particles
  Size of voids
  Volume fraction of voids

Utility goals
$G_i + d_i^- - d_i^+ = 1$ i = 1..4 These four goals are related to deformation, accumulated iron, variation in deformation, and variation in accumulated iron Constraints on deviation variables $d_i^+ \cdot d_i^- = 0$ i = 1..4 $d_i^+, d_i^- \geq 0$ i = 1..4
Minimize
$$Z = \sum_{i=1}^{4} k_i d_i^-, \quad \sum_{i=1}^{4} k_i = 1$$

and independent decision). Having formulated the decisions, we next discuss the details of simulation models used for solving the design problem.

9.8. Materials and Systems-level Simulation Models for Energetic Material Design

The third step in the design method is to identify the simulation models and associated interaction patterns. For the energetic material design problem, we rely on three

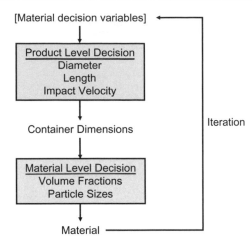

Figure 9.19 Coupled material and product decision making (pattern P6).

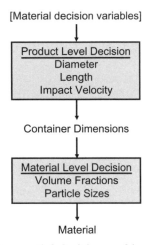

Figure 9.20 Sequential decision making (pattern P5).

models—particle-level shock simulation (micro-level), nonequilibrium mixture theory model (continuum-level), and container-level model (system-level) as shown in Figure 9.21. The particle-level shock simulation model is used to analyze the effect of various material parameters on the energetic material properties. The material properties are captured in the EOS parameters and average reaction initiation temperature at hotspots corresponding to proximal reactant interfaces. The EOS parameters are used in the system-level model to predict the strength characteristics and the reaction initiation temperature is used in the nonequilibrium theory model to predict the energy release characteristics of the material. The flow of information between these models, the design variables, and the responses are shown in Figure 9.21. A nonequilibrium continuum model is used as a coarse grain model

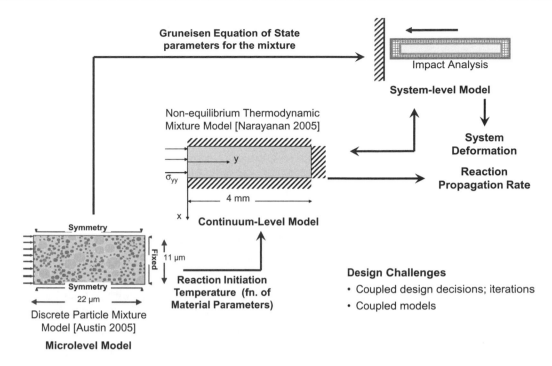

Figure 9.21 Three simulation models used in the energetic material design problem and the information flow between them.

to compute reaction propagation. The details of these three models are discussed in Sections 9.8.1, 9.8.2, and 9.8.3, respectively. The flow of information between these three models is discussed in Section 9.8.4 and is modeled using the interaction patterns in Section 9.8.5.

9.8.1. Particle-level Shock Simulation Model (PSSM)

The particle-level shock simulation model (PSSM) is a microscale finite element simulation that provides spatial resolution of the coupled thermal, mechanical, and chemical responses at the particle level during shock compaction. The model has been developed by Austin and coauthors (Austin 2005, Austin, McDowell et al. 2006). Presentation of details of the model is adapted from the authors' publications. This model is used to understand the effect of changing size of constituent particles (aluminum and iron-oxide), volume fractions, spatial arrangements or correlations, particle locations in space, and different loading conditions on the overall properties of the material, specifically the EOS and the probability of reaction initiation. The model is used to predict the average temperature at hotspots, equation of state properties of the material, the size of high temperature hotspots, and the number of reaction initiation sites as assessed by application of certain criteria at hotspots. The reactions may initiate at hotspots at the interface of aluminum and iron oxide particles (reactants). The

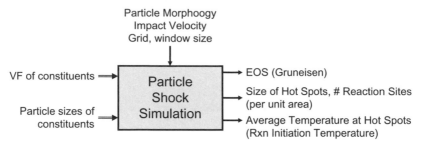

Figure 9.22 Inputs and outputs for the PSSM.

inputs and outputs of the model are shown in Figure 9.22. Some details of the model are discussed next.

The first step in the PSSM is the reconstruction of digital microstructures, based on experimental data. Information obtained from microscopy is used to generate size distributions of the particles and voids, for given volume fractions, which are lognormal distributed. Experimental data are also used to generate nearest-neighborhood distributions of particles. Information regarding size and particle distributions is used to randomly generate discrete sets of micron-scale particles (aluminum particles, iron oxide agglomerates, and voids). The particle size is controlled based on the mean and variance of particle size distributions observed from the microscopic images, and with prescribed volume fractions of the SVE under consideration. The distribution of particles in the SVE is controlled by the nearest neighborhood distributions. The remainder of the SVE is filled with epoxy.

After the particle mesostructure is generated, the next step is to perform numerical simulation using finite element techniques. In this model, shock waves are propagated through the reactive particle system to quantify the thermomechanical conditions that lead to reaction initiation. The simulation is performed using an Eulerian hydrocode Raven developed by Benson (Benson 1995). The boundary conditions on the SVE are shown in Figure 9.21. The shock propagation phenomenon is idealized as a 1D shock wave. A compressive shock wave is propagated through the mixture by applying a Lagrangian velocity boundary condition to the left surface of the SVE. Symmetry planes serve as Lagrangian boundary conditions for the top and bottom surface of the model. A fixed Lagrangian boundary condition is imposed on the right-hand-side surface. The simulation is carried out until the shock wave propagates through 95% of the SVE to avoid wave reflections. The material properties are modeled in terms of the hydrostatic and deviatoric components of stress-strain response. Modeling details are discussed in Refs. (Austin, McDowell 2006, Choi, Austin 2005).

9.8.2. Nonequilibrium Thermodynamics Mixture Model (NTMM)

In the nonequilibrium thermodynamics mixture model (NTMM) employed here, shock-induced chemical reactions in aluminum and iron-oxide mixtures are modeled in the

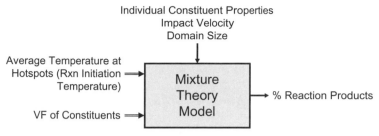

Figure 9.23 Inputs and outputs for the mixture theory model.

context of nonequilibrium continuum thermodynamics, in which both thermochemical and mechano-chemical processes are accommodated (Lu, Narayanan 2003). The discussion in this section is adapted from (Choi 2005). The constitutive model and the conservation equation are formulated by introducing a combination of internal state variables and extended irreversible state variables (additional flux terms). The internal state variables include mass fractions of reactants and products, and void content. The extended irreversible state variables include chemical reaction rate, heat flux, and pore collapse flux. The irreversibility of these processes are implied in the nonnegative entropy production rate (i.e., the second law of thermodynamics) and their contribution to the dissipation. Relaxation times during to the duration of the chemical initiation and sustained reactions are in the range of 100–200 nanoseconds. A uniformly blended mixture theory is used to describe the porous mixture. The chemical reaction of the constituents is $2Al + Fe_2O_3 \rightarrow 2Fe + Al_2O_3$. Conservation equations, constitutive models, and chemical reaction equations are described in detail in (Lu, Narayanan 2003). The simulation model is implemented in MATLAB®. The example is shown in Figure 9.21. Top, bottom, and right boundary conditions are fixed with regard to normal displacement and zero tangential shear traction, and initial loading (σ_{yy}) is applied on the left boundary. The relevant output of this analysis is the degree of chemical reaction in the material system (amount of product phase). In order to assess the extent of chemical reaction, the parameter to be captured is mass fraction of iron, since it is the product of the chemical reaction. In this study, we calculate the sum of the predicted mass fraction of iron at all nodes in the finite difference meshes in the nonequilibrium thermodynamic mixture model at 300 ns after the initial loading. This parameter is termed the accumulated mass fraction of iron (*acFe*).

In summary, the simulation model is a nonequilibrium thermodynamic model incorporating shock-induced chemical reactions. In this model, void collapse flux, chemical reaction flux, heat flux and associated relaxation times in the constitutive models are included, which explains the delayed initiation and sustained chemical reaction. However, reaction initiation conditions in the model are assumed and these reaction initiation criteria must be obtained from the microscale discrete particle mixture model (PSSM). The inputs and outputs for the simulation model are shown in Figure 9.23. The discussion of flow of information between the models is presented in Section 9.8.4. Complete details of the simulation models are beyond the scope of this chapter. The nonequilibrium simulation code is executed for different values of the volume fraction of constituents and average temperature at hotspots. A response surface of accumulated iron is generated as a function of the inputs for the models, and is then used in design exploration for the overall system.

9.8.3. Container-level Simulation Model (CSM)

The first two material models, presented in Sections 9.8.1 and 9.8.2, predict the performance of a shock-loaded energetic material. The models facilitate understanding the effects of changing the composition and morphology of material on the material properties/responses. In contrast to these material-level models, the container model is a system-level model that allows designers to vary the system-level parameters such as impact velocity, container shape, and dimensions, etc. It is noted that the container may either be an external structure encapsulating the energetic material or an integrated reinforcement structure such as a cellular material filled with energetic material. The objective of the container-level model is to simulate the effect of these system parameters on the overall system performance. The overall performance is comprised of the reaction initiation, and propagation, as well as the strength of the system. The reaction initiation behavior of the material is modeled in the particle shock simulation model (see Section 9.8.1) and the reaction propagation behavior is discussed in the context of nonequilibrium mixture model (see Section 9.8.2). In addition to the reaction behavior, another important requirement for the container is the ability to *withstand loads that result from impact during transportation or storage for purposes of safety.* The ability to withstand impact loading is a function of the strength of the overall system, which depends on the dimensions of the container, the amount of energetic material included, impact, parameters, and the material properties. The loads generated during the impact also depend on the properties of the impacting body. Hence, different containers should ideally be designed for different impact scenarios. For purposes of this chapter, a simplified model is developed in LS-DYNA® to incorporate the effect of system-level parameters on the strength of the overall system. This simplified model is discussed in this section.

The simplified impact response indicator used in this case is a simulated Taylor impact test (Taylor 1946, Taylor 1948); this provides an indication of the container strength, albeit somewhat indirectly, by measuring the degree of permanent deformation and damage upon high velocity normal impact on a rigid anvil. It might represent, for example, a crash scenario. In the Taylor impact test, a cylindrical container is given an initial velocity and impacted against a rigid wall. Due to the impact, the container with energetic material deforms into a "mushroom" shape. The amount of deformation depends on the material properties, the initial velocity and the container dimensions, and the configuration. In the simulation model developed for this chapter, the cylindrical container consists of an outer hollow steel shell filled with the energetic material that is to be designed.

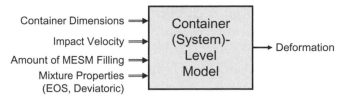

Figure 9.24 Inputs and outputs for container-level model.

The outer diameter of the steel cylinder is specified as 50 mm. The length of the container is fixed at 100 mm. The inner diameter of the steel shell is a design variable and is assumed to vary between [10 46] mm. The impact velocity is fixed at 1000 m/sec, an extreme crash velocity for purposes of illustration. This container is impacted against a rigid wall. The impact is simulated using the explicit Finite Element code LS-DYNA. The deformation of the container is measured after a predefined fixed time (t = 5 μs in this case), since the impact velocity is not varied. The maximum radius of the deformed shape is measured. This maximum radius of the deformed shape is an indicator of the strength of the container. The inputs and outputs for the model are shown in Figure 9.24.

The container-level model is based on elastic-plastic behavior of steel. The behavior of the energetic material is modeled as a combination of the hydrostatic and deviatoric behavior. The hydrostatic behavior is modeled using a Gruneisen EOS (see the description of Gruneisen EOS in Section 9.8.4.1). The devatoric behavior of the material is modeled using the experimental data reported by Patel (Patel 2004). For simplicity, the Taylor impact model employed here assumes that the deviatoric behavior of the model is the same for different values of material microstructure parameters, such as the size of constituent particles, volume fraction of voids; clearly, deviatoric strength behavior is sensitive to microstructure and will be modeled in more detail in later work. The variation in material properties is incorporated in the model by considering the changes in material's hydrostatic behavior, i.e., changes in parameters associated with the EOS.

9.8.4. Linkage Between the Three Material Simulation Models

9.8.4.1. Linkage Between the Particle-level Shock Simulation Model (PSSM) and the Nonequilibrium Thermodynamic Mixture Model (NTMM)

The performance of the reactive particle system is evaluated based on (1) the number of sites per unit volume at which reaction initiates during shock propagation, (2) the average temperature at the hotspots, and (3) the hydrostatic behavior of the overall mixture. Reaction initiation is possible where the reactants are in intimate contact. The initiation of reaction is characterized by unbounded growth of hotspots that develop at reactant interfaces due to the heat liberated by exothermic chemical reactions. The reaction initiation predicted using the shock simulation model is realized at the microscale level, which differs from reaction propagation, which occurs at a macroscopic level. Reaction propagation is not predicted using the particle shock simulation model. The prediction of reaction initiation is based on the Merzhanov criterion (Merzhanov 1966), which determines conditions by which the thermal explosion of hotspots occur when the rate of heat generated by chemical reaction is greater than the rate of heat conduction to the surroundings. The factors affecting the reaction initiation criterion include the temperature at the hotspots, the temperature of the hotspot surroundings, and the size of the hotspots. The maximum number of reaction initiation sites over a statistical volume is calculated with time during shock propagation. Two of the outputs of the particle shock simulation are (1) the temperature at various points in the domain and (2) the size of hotspots. The temperature at various hotspots along with the area of hotspots is used to calculate the area-weighted average of the hotspot temperature, i.e.,

$$T_{ignit} = \frac{\sum_{i=1}^{n} A_i \, T_i}{\sum_{i=1}^{n} A_i} \tag{9.6}$$

where n is the number of hotspots, T is the temperature of a hotspot, and A is the size of a hotspot. It is assumed that hotspot temperature is the sole criterion for initiating chemical reaction initiation; more complex reaction initiation criteria can be assumed. The weighted average of temperature is calculated at the time step when the first reaction starts anywhere in the domain. The critical temperature at which chemical reaction will be initiated is the average of the hotspot temperatures with weighting by the spot sizes; this weighted average temperature is the input parameter in the NTMM (discussed in Section 9.8.2) as the reaction initiation condition. It is necessary to introduce an average rather than a discrete measure to accord with the form of the continuum NTMM description in this case, as the latter does not assume heterogeneity of the initiation sites for reaction.

9.8.4.2. Linkage Between the Particle-level Shock Simulation Model (PSSM) and the Container-level Simulation Model (CSM)

The PSSM is also used to determine the parameters in the constitutive model of the energetic material used in the container-level simulation model (CSM). The constitutive model parameters are functions of the microstructure of the particle system and represent effective (homogenized) material properties. This represents a hierarchical rather than concurrent multiscale modeling strategy. This hierarchical description facilitates assessment of the degree of model coupling among scales that is required for decision support rather than presupposing complete model precision and certainty, as would be implied using a seamless, concurrent multiscale modeling scheme. Moreover, such a concurrent scheme would be computationally intensive. It is emphasized that hierarchical coupling is useful in the process of informing design decisions and does not generally require concurrent, multiresolution and/or multiscale modeling. Hierarchical modeling also facilitates design of experiments strategies for assessing coupling of models at different length and time scales. It is presumed that limited physical prototyping is necessary in any case to evaluate efficacy of the design and to assist in converging on design decisions.

The constitutive model used here for the mixture is the Gruneisen EOS. An EOS describes the relationship between the pressure, mass density, and specific internal energy of a material, e.g., $P = P(\rho, e)$ (Austin 2005). Since the operating conditions of the material lie in the high-pressure range, simple linear elastic relations are unsuitable. In such conditions, the Gruneisen EOS is widely used. The details of the Gruneisen EOS are not presented here (Austin 2005). The parameters for Gruneisen EOS are calculated by performing a linear regression on shock wave, speed-particle speed data. The slope (S) and intercept (C) of this regression line are used in the Gruneisen EOS model for the material. These two parameters are useful in the material model in container-level simulation. As a summary, the inputs and outputs of the simulation code for the sake of design are shown in Figure 9.22. The simulation code is executed at various points in the design space that are selected using design of experiments techniques. The variation in response due to changing material morphology is captured by generating different particle distributions and executing the model multiple times at a given point in the design space (which is specified by the values of design variables).

In summary, the information flow between the models is shown in Figure 9.21. The inputs of the particle shock simulation model include volume fraction of various constituents (Aluminum, Iron Oxide, Epoxy, and Voids) and the size distribution of these particles. The outputs of this model include the number of reaction sites, the average size of hotspots, average temperature at reaction initiation, and the parameters for the Gruneisen EOS of the mixture. The area-weighted average of the hotspot temperature of reaction initiation is used as an input in the nonequilibrium mixture theory model (see Figure 9.23). The average hotspot temperature, in association with the volume fraction of constituents, is used by the NTMM to predict the amount of accumulated reaction products, which is an indicator of the extent of reaction propagation; it is used to account for the changing material parameters and their morphology in the reaction propagation behavior. The Gruneisen EOS, which is an output from the particle shock simulation, is used as an input for the container-level simulation for accounting for the evolving material properties in the system-level simulation. The output of the container-level simulation is the deformation achieved in the Taylor impact experiment, with the energetic material filler described by an EOS that depends on microstructure morphology. Hence, a combination of the three models can be used to predict the strength properties (through deformation from the container-level model) and the reaction properties (through the accumulated reaction products from the mixture theory model).

9.8.5. Simulation Model Interaction Patterns for the Energetic Material Design Problem

The information flows between the three models discussed previously are characterized using the interaction patterns shown in Figure 9.8. We note that the interaction pattern P1 refers to an independent set of models, pattern P2 refers to a sequential flow of information, and pattern P3 refers to a bidirectional flow of information between models.

The flow of information between the simulation models is highlighted in Figure 9.21. Based on this information flow, the interaction between the *particle shock simulation* and the *container-level simulation* is sequential, i.e., pattern P2. Similarly, the interaction between the particle-level shock simulation and the nonequilibrium mixture theory model is also sequential, i.e., pattern P2. The patterns are highlighted for the energetic material design example in Figure 9.25.

A sequential interaction pattern can be simplified into an independent interaction pattern if, instead of utilizing the output of the first model into the second model, the input of the second model is set to a constant value. The constant value may be the average value of the output from the first model. For example, in the case of interaction between the particle shock simulation (the first model) and the container-level model (the second model), the output from the first model is a set of parameters for the EOS of the material. By varying the inputs of the particle shock simulation, the values of the parameters change. These parameters are used in the container-level model to model the effect of varying material properties. If the interaction pattern between the two models is changed to an independent interaction pattern, an average set of values for the EOS parameters are set as the inputs to container-level simulation. Hence, the output of the container-level simulation (deformation) is only

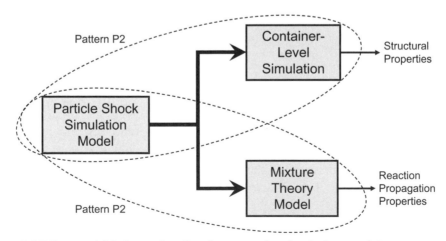

Figure 9.25 Sequential information flow between the simulation models represented as interaction pattern P2.

a function of the container-level parameters. Physically, this means that the changes in material properties are ignored during the calculation of deformation. The scenario where the interaction between particle shock simulation and the container-level simulation is simplified from pattern P2 to pattern P1 is shown in Figure 9.26. In this figure, the sequential interaction between particle shock simulation and the nonequilibrium mixture theory model is preserved. Following the same logic, four different combinations of interaction patterns can be generated between the three models— PSSM, the Nonequilibrium Mixture Theory Model (NTMM), and the Container-level Simulation Model (CSM). These include (1) pattern P2 for both the interactions, (2) pattern P1 for PSSM-NTMM interaction and pattern P2 for PSSM-CSM interaction, (3) pattern P2 for PSSM-NTMM interaction and pattern P1 for PSSM-CSM interaction, and (4) pattern P1 for both interactions.

In summary, four interaction patterns for simulation models are shown, along with four separate interaction patterns for the simulation models. These interaction patterns for simulation models and decisions (discussed in Section 9.7.2) can be combined because the simulation models are used to make decisions. Since there are four different types of model interaction patterns and four different types of decision interaction patterns, we have a total of 16 alternatives for the decision related to the simplification of design process. These 16 design process related alternatives are labeled A through G in a matrix form in Figure 9.27. Each row in the matrix corresponds to a specific model interaction pattern and each column corresponds to a specific decision interaction pattern. Each cell in the matrix corresponds to a unique combination of model and simulation interaction patterns. The model and decision interaction patterns used in these alternatives are listed in Table 9.4. The designers' objective is to determine the appropriate design process alternative that is simple enough to allow efficient design exploration and allows for achieving the desired system performance.

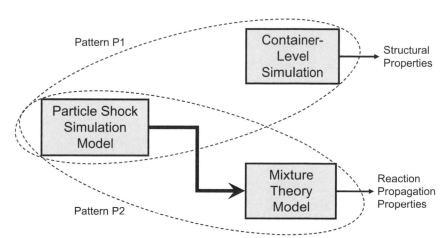

Figure 9.26 Simplification of the interaction between the particle-level shock simulation and the container-level simulation from P2 to P1.

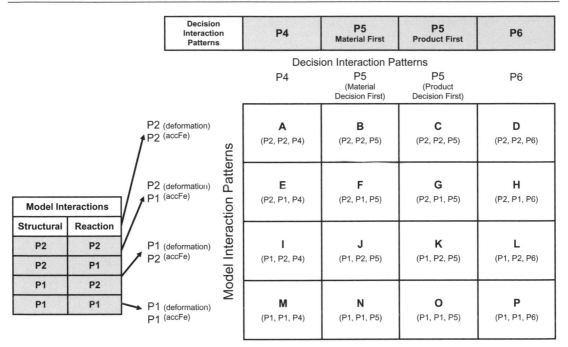

Figure 9.27: Alternatives (A through P) for design process simplifications.

The details and results from the application of Steps 4–9 of the method to the energetic material problem are provided in Section 9.9.

9.9. Results Using the Stepwise Refinement of Interaction Patterns (Steps 4–9)

In Steps 4–9, presented in Section 9.5.2, instead of exploring each design process alternative A through P, a designer starts with a simple design process and sequentially refines it until further refinement does not improve the final design. The simplest design process alternative in Figure 9.27 is alternative M, for which all the interactions between models and decisions are ignored. The most complex alternative is D. The refinement of design process alternatives from M through D can be carried out in different ways. An example of sequential refinement of patterns is provided in Figure 9.28. Initially a designer starts with the simplest set of interaction patterns (P1, P1, P4), which is labeled as process alternative M in Figure 9.27.

The process alternatives are represented by circles, and the refinement steps are shown with arrows. Solid arrows represent the refinement of decision interaction patterns from P4 → P5 or P5 → P6, whereas dashed arrows represent the refinement of model interaction patterns from P1 → P2, or P2 → P3. The patterns are sequentially refined to process alternative

Table 9.4 Patterns of interactions used in the process alternatives A–P.

Scenario	Interaction Patterns		
	Model Interactions		Decision Interactions
	Structural	Reaction	Decision
A	Sequential (P2)	Sequential (P2)	Independent Decisions (P4)
B	Sequential (P2)	Sequential (P2)	Sequential, Materials Decision First (P5)
C	Sequential (P2)	Sequential (P2)	Sequential, Product Decision First (P5)
D	Sequential (P2)	Sequential (P2)	Coupled Decisions (P6)
E	Sequential (P2)	Independent (P1)	Independent Decisions (P4)
F	Sequential (P2)	Independent (P1)	Sequential, Materials Decision First (P5)
G	Sequential (P2)	Independent (P1)	Sequential, Product Decision First (P5)
H	Sequential (P2)	Independent (P1)	Coupled Decisions (P6)
I	Independent (P1)	Sequential (P2)	Independent Decisions (P4)
J	Independent (P1)	Sequential (P2)	Sequential, Materials Decision First (P5)
K	Independent (P1)	Sequential (P2)	Sequential, Product Decision First (P5)
L	Independent (P1)	Sequential (P2)	Coupled Decisions (P6)
M	Independent (P1)	Independent (P1)	Independent Decisions (P4)
N	Independent (P1)	Independent (P1)	Sequential, Materials Decision First (P5)
O	Independent (P1)	Independent (P1)	Sequential, Product Decision First (P5)
P	Independent (P1)	Independent (P1)	Coupled Decisions (P6)

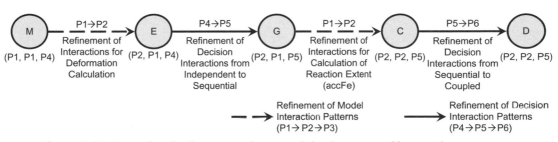

Figure 9.28 Example of refinement of sequential refinement of interaction patterns.

D with interaction patterns (P2, P2, P5). The refinement is carried out in four steps—two steps of model refinement and two steps of decision refinement. In this chapter, we carry out the refinement only one pattern at a time. However, more than one pattern can be refined in a step. The refinement is carried out until a low improvement potential is achieved.

If the designer refines a single pattern in one step, then there are 12 possible refinement paths from alternative M to D. These refinement paths are shown in Figure 9.29. Although a designer follows only one path for refinement from alternative M through D, all other paths are shown for illustrative purposes. For each process alternative, the improvement potential and the expected utility are shown.

9.9.1. Inferences from the Improvement Potential

First, we focus the discussion on the impact of refinement of design processes on the improvement potential (presented in Equation 9.4).

1. *Refinement of Decision Interaction Patterns:* Based on the solid arrows, it is observed that as designers refine the interactions between decisions from P4 through P6 (e.g., $M \rightarrow O, E \rightarrow G, C \rightarrow D$, etc.), the improvement potential decreases. This indicates that the possibility of achieving a benefit by increasing the level of coupling between decisions reduces as we move from independent to sequential and from sequential to coupled decisions. This reduction is due to the reduced uncertainty in the overall utility. This uncertainty is quantified using the lower and upper bounds on overall utility (as shown in Figure 9.12) and arises from lack of knowledge about other designers' decisions. By including more interactions among the design decisions, the lower and upper bounds on the overall utility approach the exact value. The improvement potential for the combination "D" is zero because there is no uncertainty due to simplification in

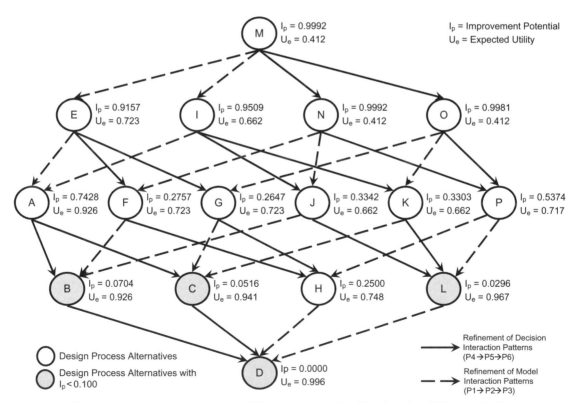

Figure 9.29 Improvement potential and expected utility for the different design process alternatives.

decision interactions or due to simplification of model interactions. The contribution of uncertainty in the overall utility value due to inherent system variability is also zero in this case (which is just a characteristic of the system and may not be true in all cases). The improvement potential for the combination "M" is the maximum ($= 0.9992$) because it results from both decision decoupling and model decoupling. Note that in combination "M," the simplest patterns are used for models and decisions. The trend indicates that more complex interaction patterns result in decisions that are less uncertain, which is in line with the expected trend. The same trend is observed as designers improve the interaction patterns between models (from P1 → P2).

2. *Magnitude of Improvement Potential:* Although the general trend is according to the expectation, more insight into the problem behavior is gained by looking at the relative values of the improvement potential. The independent decision interaction pattern P4 (in alternatives A, E, I, and M) has the highest improvement potential. When model pattern P2 is used for deformation calculations, the reduction in improvement potential from P4 → P5 (i.e., A → B, A → C, E → F, and E → G) is significantly higher compared to the reduction from P5 → P6 (i.e., B → D, C → D, F → H, and G → H). As examples, (1) the reduction in improvement potential from A → B is 0.6724 (0.7428 − 0.070 4 = 0.6724) and (2) the reduction in improvement potential from B → D is 0.0704 (0.0704 − 0.0000 = 0.0704). From a design-process decision-making standpoint, this implies that if pattern P2 is used for deformation calculations, the benefit of moving from an independent interaction pattern P4 to sequential model interaction pattern P5 is much more compared to moving from P5 to a coupled interaction pattern P6. Hence, *if pattern P2 is used as models for deformation calculations, pattern P5 should be used for decisions.* If improvement potential is the only criterion for meta-level decision making, then designers would choose combinations B, C, L, and D. All these patterns result in an improvement potential that is less than 0.100. In other words, the maximum possibility of improvement in the payoff by the addition of information is less than 0.100. These combinations are highlighted in Figure 9.29.

9.9.2. Inferences from Overall Utility

It is important to note that although improvement potential is an important metric to consider while making design process decisions, it is not the only metric. While making design process–related decisions, designers should also consider other factors such as design freedom, robustness of process, complexity of the process, cost of executing the process, etc. Due to the scope of this chapter, these factors are not included. These factors are avenues for further research. To keep this discussion of the results simple, consider only the achieved utility values. The achieved utility values for different process alternatives are also shown in Figure 9.29 and discussed in the following section. The overall utility at a decision point reflects how good a design is; it is a direct reflection of the quality of the design outcome (which in turn depends on the design process followed). Based on the achieved utility values shown in Figure 9.29 that exceed 0.9, we observe the following:

1. The utility values increase as the interaction patterns are improved from independent to sequential to coupled. This indicates that by introducing complete information flows between decisions and models, the quality of the final design is better. For example, in the case of a fixed-model interaction pattern (P2 for deformation and P2 for accumulated iron), the utility of pattern P6 (0.996) in combination D is better than the utility (0.941) for sequential interaction pattern P5 with the product decision made before the material decision (i.e., alternative C), which is in turn higher than the utility achieved using independent interaction pattern P4 (i.e., alternative A). A similar trend is observed by fixing the decision interaction pattern and varying the model interaction pattern. This is again an intuitive result—better design processes should result in better designs.

2. The maximum utility is equal to 0.996 and is achieved by using alternative D, where the decisions are modeled using coupled interaction patterns (P6). The minimum utility of 0.412 is achieved in alternative M, where both decisions and simulation models are modeled using independent interaction pattern (P4 and P1 respectively).

3. The decisions made using sequential decision patterns where the material decision is made before the product decision (i.e, B, F, J, and N) results in the same overall utility as the case where decisions are made independently (i.e., A, E, I, and M). Such a trend is attributed to the fact that there are four design variables related to the material and only one design variable corresponding to the product. If the material decision is made first, then most of the design freedom is locked. In this specific case, the effect of material design variables on overall utility is significantly greater than the effect of product design variables. Hence, there is not a significant difference between making the product decision independently or with knowledge about material parameters. It is important to note that this trend is valid only in the context of the specific design problem and preferences formulated in Section 9.7.1. By changing the preferences, design variables, or their ranges may change this observed trend.

4. Based on the first three points, we conclude that the selection of decision interaction patterns is highly dependent on the model interaction pattern used. This is mainly because the type of model interaction pattern used determines the information about interdependencies captured between parameters. It is the interdependencies between parameters that make the decisions coupled, sequential, or dependent.

Based on the values of overall utility obtained in all the combinations, the combinations A, B, C, D, and L result in an overall utility of 0.900 or above. Hence, these combinations are considered good alternatives for the design process decision. The combinations E, F, G, H, I,

J, K, and P result in designs with overall utility values better than 0.65 and the combinations M, N, and O result in an overall utility value of around 0.41.

9.9.3. Product and Material Decisions Using Different Process Alternatives

9.9.3.1. The Design Process Decision

Combining the results for improvement potential and overall utility for the problem under consideration, we conclude the following:

- Based on the minimization of improvement potential, the best process options (i.e., the combinations of interaction patterns for decisions and models) include B, C, D, and L.

- Based on the maximization of overall utility, the best process options include A, B, C, D, and L. The common set of process options using both criteria are B, C, D, and L. The meanings of these process options are as follows:

  B: Sequential decision with material decision made before the container-level decision; sequential interaction patterns for both deformation and accumulated iron.
  C: Sequential decision with container-level decision made before the material-level decision; sequential interaction patterns for both deformation and accumulated iron.
  D: Coupled material and container-level decisions; sequential interaction patterns for both deformation and accumulated iron.
  L: Coupled material and container-level decisions; sequential interaction pattern for accumulated iron and an independent interaction pattern for deformation.

In this section, it is shown that the most complicated design process pattern (D) is not required for making design decisions in the energetic material design scenario. Simple design processes (B, C, and L) result in equally good designs. This is apparent from the high utility values resulting from these three scenarios, in association with the low potential for improvement. The approach allows designers to identify those simple design process options. Both improvement potential and overall utility indicate the appropriateness of interaction patterns. Hence, both these metrics should be considered for making design process decisions.

9.9.3.2. Product and Material Decisions

The outcomes of decisions corresponding to each combination (A through P) are summarized in individual rows in Table 9.5. The columns in the table list the corresponding overall utility values at the decision point, the values of design variables at the design point, and the corresponding response values. The design variables include four material-level variables: (1) size of aluminum particles (SizeAl), (2) size of iron oxide particles (SizeFe$_2$O$_3$), (3) size

Table 9.5 Utilities, responses, and design variable values for various interaction pattern scenarios.

Scenario	Overall Utilities	Responses		Design Variable				
		Average Deformation (mm)	AccFe_avg	Material				Product
				SizeAl (mm)	SizeFe2O3 (mm)	SizeVoids (mm)	VFVoids	RadFilling (mm)
A	0.926	8.20	12.03	0.0005	0.0006	0.0007	0.05	23
B	0.926	8.20	12.03	0.0005	0.0006	0.0007	0.05	23
C	0.941	8.09	13.41	0.0015	0.0007	0.0002	0.1	23
D	0.996	7.99	12.03	0.0006	0.0006	0.0006	0.05	9
E	0.723	8.08	12.03	0.0005	0.0002	0.0010	0.04	23
F	0.723	8.08	12.03	0.0005	0.0002	0.0010	0.04	23
G	0.723	8.08	12.03	0.0005	0.0002	0.0010	0.04	23
H	0.748	8.00	12.03	0.0011	0.001	0.0010	0.04	9
I	0.662	8.22	12.03	0.0005	0.0006	0.0007	0.05	23
J	0.662	8.22	12.03	0.0005	0.0006	0.0007	0.05	23
K	0.662	8.22	12.03	0.0005	0.0006	0.0007	0.05	23
L	0.967	7.99	12.03	0.0005	0.0006	0.0007	0.05	9
M	0.412	8.22	12.03	0.0005	0.0002	0.0002	0.04	23
N	0.412	8.22	12.03	0.0005	0.0002	0.0002	0.04	23
O	0.412	8.22	12.03	0.0005	0.0002	0.0002	0.04	23
P	0.717	7.99	12.03	0.0005	0.0002	0.0002	0.04	9

of voids (SizeVoids), (4) volume fraction of voids (VF voids) and the product (container)–level design variable—amount of energetic material stored in the container (RadFilling). The response values shown in the table are average deformation in the container at the predefined time, and the average mass fraction of accumulated reaction product (AccFe_avg). The preferences for these responses are modeled using utility functions such that the amounts of deformation and accumulated iron are close to a target value and the deviation in these responses due to uncertainty resulting from simplification of design processes is as low as possible. The table is used to compare the decisions made using each combination of interaction patterns labeled A through P.

Limitations of the approach: Note that these process-level decisions also depend on the time (or expense) consumed to execute the design process. The time for execution of design process has not been included in this study. The results of this section can be extended by include time considerations by including utility functions for time during the calculation of the overall utility value. Further, we note that in this section, the results are presented as if the information about all the interaction patterns is available simultaneously; moreover, the decision is made about the interaction patterns with the knowledge relating to the outcome from all process options. This approach is adopted in this section to illustrate the tradeoffs between simplification and the quality of decisions made. However, in a real design scenario, a designer starts with a simple interaction pattern and calculates the overall utility and the value of additional information for that decision. Based on these two values and the resources available to improve the design process, he or she may decide to use that design process option or to use the current process option. The approach presented in this chapter is intended to assist designers to make conscious decisions about the improvement of the design process. Note that when a designer is utilizing a particular process option and he or she is not aware of the performance of other options, the metrics guide a designer as to whether or not improvement in the process is necessary. They do not provide any guidance in terms of *how much improvement/refinement is necessary*. For example, based on the available information about the simplest interaction pattern combination (M), designers cannot determine whether he or she should choose the combination N, O, J, I, or E. It is only after he or she executes the processes using other combinations that he or she can determine the right level of refinement. This is a limitation of the approach. However, it is suggested that only limited analyses involving highly coupled simulations are necessary to guide decisions.

9.10. Closing Remarks

In this chapter, we deal with the issue of complexity in the integrated design of products and materials. Two of the main aspects discussed in this chapter include the complexity of simulation models and the complexity of design processes. The approach presented in this chapter has a broad range of applications to other concurrent materials and product design problems. For example, it can be used for relatively simple material and product design problems, such as the design of pressure vessels or other structural elements using composite materials, or for more complex scenarios where models at multiple scales are utilized. In the pressure vessel example, value-of-information can be used to determine whether the material model needs to be refined further by more accurately modeling manufacturing variability or other physical phenomena. Similarly, for multilevel design problems involving multiple material length scales, value-of-information can be used to determine whether concurrent multiscale modeling is required or hierarchical modeling is appropriate. A summary of both these aspects is presented next.

9.10.1. Part 1: Model Refinement

The question posed in this chapter is "How much refinement of a simulation model is adequate for design?" The qualitative answer to this question is "Simulation models should be refined to the extent that they help designers in making good decisions efficiently." The "goodness" of decisions is dependent on the satisfaction of design requirements.

Although a qualitative answer to this question is useful, a quantitative answer is required to make decisions.

To provide a quantitative answer to the question, we have introduced a value-of-information-based metric for determining the appropriate level of refinement of simulation models, largely distinct from issues related to refinement associated with numerical convergence. Value-of-information is the improvement in the overall utility value when an exact model is used as compared to a simplified model. However, due to the unavailability of exact models in most design scenarios, designers cannot directly evaluate the improvement in the overall utility. Instead of the improvement, we use an upper bound on the value-of-information as a metric for refinement of simulation models. The metric, called the improvement potential, is used to determine the level of refinement beyond which the payoff from refinement is insignificant. This metric is evaluated as the difference between the maximum expected payoff (utility) that can be achieved at any point in the design space, and the lowest expected payoff (utility) achieved at the current decision point. The value-of-information metric presented in this chapter is based on the assumption that information about error bounds (i.e., the lower and upper bounds on the imprecise variables) is available.

The improvement potential has the various properties that are important for making model-refinement-related decisions. These properties differentiate the metric presented in this chapter from existing value-of-information based metrics. These properties are highlighted as follows:

- The metric *is a quantification of the impact of imprecision* in simulation models on the design decisions via the lower and upper bounds of payoff. This allows designers to account for imprecision in addition to the statistical variability. Existing metrics do not include the effect of imprecision.

- The metric is based on the difference between maximum and minimum payoffs achieved. Hence, it also quantifies the possible variation in payoff due to imprecision. If a designer refines the simulation model in stages, the deviation in payoff reduces, which is reflected in the reduced improvement potential. This allows designers to measure the improved confidence in decisions as the imprecision range reduces.

- The metric allows designers to *quantify the opportunity for improving the design solution* because the metric is based on the upper bound of payoff throughout the design space. This allows designers to assess the maximum possible improvement in the design solution by addition of information via model refinement.

9.10.2. Part 2: Design Processes

The primary motivation for this chapter is that multiscale material systems are inherently complex due to the interactions between scales, nonlinearities, and irreversibilities, functional requirements, and resulting design decisions. These interactions result in increased complexity of the associated design processes. To successfully design these systems at multiple levels, it is important to simplify the design processes consciously in a manner that the overall system performance is not substantially affected. Stated differently, an important task in multilevel design is to find the design processes that are robust to simplification.

The approach presented in this chapter consists of the following key aspects: (1) use of interaction patterns to model design processes at various levels of abstraction in terms of simulation models, design decisions, and their interactions, (2) use of a value-of-information based metric improvement potential, and (3) stepwise refinement of interaction patterns instead of relying on completely coupled interactions.

The unique aspect of this approach is the systematic consideration of design process decisions such as determining which couplings are important from a decision-making standpoint, what level of refinement is necessary in the simulation models, etc. These decisions are critical to the effective and efficient utilization of computational resources and the information generated by simulation models at multiple scales. In this chapter, it is shown that by eliminating noncritical couplings between decisions and simulation models, complex systems design can be carried out more efficiently. The systems-based approach presented in this chapter is distinct from the multiscale modeling approach commonly espoused, wherein a single coupled (concurrent) simulation model is developed for accurate prediction of the system behavior. Not only are fully coupled multiscale models computationally intensive, they are costly to develop, requiring accurate models at each scale/level; the coupled model chain is only as strong as its weakest link. This allows designers to design complex multiscale systems efficiently by managing their complexity and permitting application of simplified models.

The approach presented in this chapter is an extension of the robust design philosophy proposed by Taguchi (Taguchi 1986). Taguchi's approach involved making the system performance insensitive to the variation in manufacturing processes, whereas in this chapter, we make the system performance insensitive to the simplifications in the design process.

The approach presented in this chapter has been implemented using various commercially available tools and some in-house software. Most of the algorithms for stepwise refinement and the evaluation of improvement potential are implemented in MATLAB®[1]. Minitab®[2] is used for statistical analysis and response surface modeling. iSIGHT® (Engineous Inc. 2004) is used for the design of experiments. The ModelCenter® (Phoenix Integration Inc. 2004) application, from Phoenix Integration Inc., is used for the remote execution of models. Some of these tools for distributed model integration and execution are discussed in the following chapter.

---

[1] http://www.mathworks.com/products/matlab/
[2] http://www.minitab.com/

References

Agogino, A.M., 1997. Information in the design process. In: Frontiers of Engineering: Reports on Leading Edge Engineering from the 1996 NAE Symposium on Frontiers of Engineering. National Academy of Engineering, pp. 13–16.

Antonsson, E.K. (Ed.), 2001. Imprecision in Engineering Design. Engineering Design Research Laboratory, Division of Engineering and Applied Science: California Institute of Technology, Pasadena, CA. (http://design.caltech.cdu/Research/Imprecise/Reading_List/Imprecision_Book.pdf).

Antonsson, E.K., Otto, K.N., 1995. Imprecision in engineering design. J. Mech. Des.: Spec. Comb. Issue Trans. ASME Commemorat. 50th Anniv. Des. Eng. Div. ASME, 117 (B), 25–32.

Arrow, K., Hurwicz, L., 1972. An optimality criterion for decision-making under ignorance. In: Carter, C.F., Ford, J.L. (Eds.), Uncertainty and expectation in economics: Essays in honour of G. L. S. Shackle. Blackwell, Oxford, UK, pp. 1–11.

Ashby, M.F., 1999. Materials Selection in Mechanical Design. Butterworth-Heinemann, Oxford, UK.

Aughenbaugh, J.M., Ling, J., Paredis, C.J.J., 2005a. Applying information economics and imprecise probabilities to data collection in design. ASME IMECE 2005, Computers in Engineering Conference, Orlando, FL. Paper number: IMECE2005-81181.

Aughenbaugh, J.M., Ling, J.M., Paredis, C.J.J., 2005b. The use of imprecise probabilities to estimate the value of information in design. ASME International Mechanical Engineering Congress and Exposition, Orlando, FL. Paper number: IMECE2005-81181.

Austin, R.A., 2005. Numerical Simulation of the Shock Compaction of Microscale Reactive Particle Systems. M.S. thesis, GW Woodruff School of Mechanical Engineering, Georgia Institute of Technology, Atlanta, GA.

Austin, R.A., McDowell, D.L., Benson, D.J., 2006. Numerical simulation of shock wave propagation in spatially-resolved microscale particle systems. Model. Simul. Mater. Sci. Eng. 14 (4), 537–561.

Benson, D.J., 1995. A multi-material Eulerian formulation for the efficient solution of impact and penetration problems. Comput. Mech. 15 (6), 558–571.

Box, G.E.P., 1979. Robustness in the strategy of scientific model building. In: Launer, R.L., Wilkinson, G.N. (Eds.), Robustness in Statistics. Academic Press, New York, pp. 201–235.

Bradley, S.R., Agogino, A.M., 1994. An intelligent real time design methodology for component selection: an approach to managing uncertainty. J. Mech. Des. 116 (4), 980–988.

Bradley, S.R., Agogino, A.M., Wood, W.H., 1994. Intelligent engineering component catalogs. In: Artificial Intelligence in Design '94. Kluwer, Dordrecht, the Netherlands, pp. 641–658.

Chen, W., 1995. A Robust Concept Exploration Method for Configuring Complex Systems. Ph.D. dissertation, Mechanical Engineering department, GW Woodruff School of Mechanical Engineering, Georgia Institute of Technology, Atlanta, GA.

Choi, H.-J., 2005. A Robust Design Method for Model and Propagated Uncertainty. Ph.D. dissertation, Mechanical Engineering department, GW Woodruff School of Mechanical Engineering, Georgia Institute of Technology, Atlanta, GA.

Choi, H.-J., Allen, J.K., Rosen, D., McDowell, D.L., Mistree, F., 2005. An inductive design exploration method for the integrated design of multi-scale materials and products. Design Automation Conference, Long Beach, CA. Paper number: DETC2005-85335.

Choi, H.-J., Austin, R., Allen, J.K., McDowell, D.L., Mistree, F., 2004. An approach for robust micro-scale materials design under unparameterizable variability. 10th AIAA/ISSMO Multidisciplinary Analysis and Optimization Conference, Albany, NY. Paper number: AIAA-2004-4331.

Choi, H.-J., Austin, R., Allen, J.K., McDowell, D.L., Mistree, F., 2005. An approach for robust design of reactive powder metal mixtures based on non-deterministic micro-scale shock simulation. J. Comput-Aided Mater. Des. 12 (1), 57–85.

Engineous Inc., 2004. iSIGHT, Version 8.0. http://www.engineous.com/product_iSIGHT.htm.

Fernández, M.G., Seepersad, C.C., Rosen, D., Allen, J.K., Mistree, F., 2005. Decision support in concurrent engineering—The utility-based selection decision support problem. Concurrent Eng. Res. Appl. 13 (1), 13–28.

Ferson, S., Donald, S., 1998. Probability bounds analysis. In: Mosleh, A., Bari, R.A. (Eds.), Probabilistic Safety Assessment and Management. Springer-Verlag, New York, pp. 1203–1208.

Howard, R., 1966. Information value theory. IEEE Trans. Syst. Sci. Cybern. SSC-2 (1), 779–783.

Keeney, R.L., Raiffa, H., 1976. Decisions with Multiple Objectives: Preferences and Value Tradeoffs. John Wiley and Sons, New York.

Lawrence, D.B., 1999. The Economic Value of Information. Springer, New York.

Lu, X., Narayanan, V., Hanagud, S. 2003. Shock-induced chemical reactions in energetic structural materials. 13th American Physical Society Topical Conference on Shock Compaction of Condensed Matter, Portland, OR, July 20–25, 2003.

Merzhanov, A.G., 1966. On critical conditions for thermal explosion of a hot spot. Combust. Flame. 10 (4), 341–348.

Messer, M., 2008. Systematic Approach for Integrated Product, Materials, and Design-Process Design. Ph.D. dissertation, Mechanical Engineering department, GW Woodruff School of Mechanical Engineering, Georgia Institute of Technology, Atlanta, GA.

Mistree, F., Hughes, O.F., Bras, B.A., 1993. The compromise decision support problem and the adaptive linear programming algorithm. In: Kamat, M.P. (Ed.), Structural Optimization: Status and Promise. AIAA, Washington, DC, pp. 247–286.

Mistree, F., Seepersad, C.C., Dempsey, B.M., McDowell, D.L., 2002. Robust concept exploration methods in materials design. 9th AIAA/ISSMO Symposium on Multidisciplinary Analysis and Optimization, Atlanta, GA. Paper number: AIAA-2002-5568.

Moore, R.E., 1966. Interval Analysis. Prentice-Hall, Englewood Cliffs, NJ.

Pahl, G., Beitz, W., 1996. Engineering Design: A Systematic Approach, second ed. Springer-Verlag, New York.

Panchal, J.H., 2005. A Framework for Simulation-Based Integrated Design of Multiscale Products and Design Processes. Ph.D. dissertation, Mechanical Engineering department, the GW Woodruff School of Mechanical Engineering, Georgia Institute of Technology, Atlanta, GA.

Panchal, J.H., Choi, H.-J., Allen, J.K., McDowell, D.L., Mistree, F., 2007. A systems based approach for integrated design of materials, products, and design process chains. J. Comput-Aided Mater. Des. 14 (Suppl. 1), 265–293.

Panchal, J.H., Choi, H.-J., Shephard, J., Allen, J.K., McDowell, D.L., Mistree, F., 2005. A strategy for simulation-based multiscale, multifunctional design of products and design processes. ASME Design Automation Conference, Long Beach, CA. Paper number: DETC2005-85316.

Patel, N., 2004. Intermediate Strain Rate Behavior of Two Structural Energetic Materials. MS thesis, the GW Woodruff School of Mechanical Engineering, Georgia Institute of Technology, Atlanta, GA.

Phoenix Integration Inc., 2004. ModelCenter®, Version 5.0. <http://www.phoenix-int.com/products/ModelCenter.html>.

Poh, K.L., Horvitz, E., 1993. Reasoning about the value of decision model refinement: methods and application. Proceedings of ninth conference on uncertainty in artificial intelligence. San Francisco: Morgan Kaufmann, 174–182.

Seepersad, C.C., 2001. The Utility-Based Compromise Decision Support Problem with Applications in Product Platform Design. M.S. thesis, Mechanical Engineering department, the G.W. Woodruff School of Mechanical Engineering, Georgia Institute of Technology, Atlanta, GA.

Seepersad, C.C., Allen, J.K., McDowell, D.L., Mistree, F., 2008. Multifunctional topology design of cellular structures. ASME J. Mech. Des. 130 (3) 031404(1-13).

Seepersad, C.C., Mistree, F., Allen, J.K., 2005. Designing evolving families of products using the utility-based compromise decision support problem. Int. J. Mass Customisat. 1 (1), 37–64.

Simon, H.A., 1996. The Sciences of the Artificial. The MIT Press, Cambridge, MA.

Simpson, T.W., Rosen, D., Allen, J.K., Mistree, F., 1998. Metrics for assessing design freedom and information certainty in the early stages of design. J. Mech. Des. 120 (4), 628–635.

Taguchi, G., 1986. Introduction to Quality Engineering. UNIPUB, New York.

Taylor, G.I., 1946. The testing of materials at high rates of loading. J. Inst. Civil Engrs. 26, 486–519.

Taylor, G.I., 1948. The use of flat ended projectiles for determining yield stress, I: Theoretical consideration. Proc. R. Soc. Lond. A194, 289–299.

Von Neumann, J., Morgenstern, O., 1947. The Theory of Games and Economic Behavior. Princeton University Press, Princeton, NJ.

Ward, A.C., Lozano-Perez, T., Seering, W.P., 1990. Extending the constraint propagation of intervals. Artif. Intell. Eng. Des. Manuf. 4 (1), 47–54.

Wood, K.L., Antonsson, E.K., 1989. Computations with imprecise parameters in engineering design: Background and theory. ASME J. Mech., Transm., Autom. Des. 111 (4), 616–625.

Wood, W.H. 2000. Quantifying design freedom in decision based conceptual design. ASME Design Engineering Technical Conferences, Baltimore, MD. Paper number: DETC2000/DTM-14577.

Wood, W.H., Agogino, A.M., 2005. Decision-based conceptual design: Modeling and navigating heterogeneous design spaces. J. Mech. Des. 127 (1), 2–11.

CHAPTER 10

Distributed Collaborative Design Frameworks

The emphasis in the previous chapters is on methods for integrated design of materials and products based on computational modeling and simulation. Chapter 5 centers on multi-objective decision support, Chapters 6 through 8 focus on uncertainty management, and Chapter 9 emphasizes the management of complexity. In this chapter, we discuss the fourth and important aspect for enabling integrated products and materials design—frameworks for distributed collaborative design.

An overview of the chapter and its relationship to other chapters is given in Figure 10.1. In Section 10.1, an overview of the framework for integrated materials and products design framework is presented. In Section 10.2, we review the capabilities of existing distributed design frameworks. A motivational design example used to illustrate the concepts presented

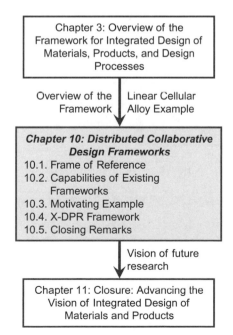

Figure 10.1 Overview of Chapter 10 and its relationship with other chapters in the book.

DOI: 10.1016/B978-1-85617-662-0.00010-7

in this chapter is discussed in Section 10.3. The details of a computational framework called X-DPR is presented in Section 10.4. Closing remarks are presented in Section 10.5.

10.1. Frame of Reference—Framework for the Distributed Concurrent Design of Materials and Products

Increasingly, design efforts are being carried out in a globally distributed manner. Computer-assisted design of materials and products employs information, models, and expertise that are generally distributed in nature. On the one hand, online materials databases such as MatWeb (MatWeb 2008) are increasingly becoming available for obtaining materials property information over the Internet. On the other hand, sophisticated computational models are being developed by materials researchers around the world to simulate the behavior of a variety of materials for a wide range of applications. Integration of these resources in a design process enables designers to implement the design processes in a faster and resource-efficient manner. To integrate computational resources such as databases and models over the Internet, a *computing framework* is required. In the present context, a computing framework is defined as "a computational backbone that facilitates deployment and utilization of software components that can be executed remotely to perform tasks in a design process."

Aspects of a framework for integrated, concurrent materials and product design are shown in Figure 10.2. Components of the framework include databases of material structure and

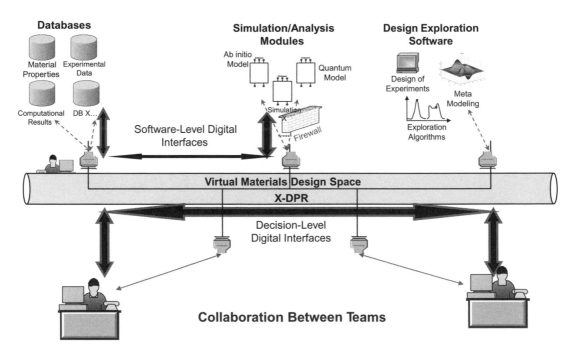

Figure 10.2 A Distributed environment for integrated materials and product design.

properties, experimental data, results from model execution, simulation models at various scales (such as continuum scale finite element analyses, molecular dynamics models, material synthesis models, etc.), and various design exploration tools (such as statistical analysis software, design of experiments, meta-modeling, decision-making tools, and optimization routines). A computational backbone is responsible for the seamless integration and information exchange between the software tools. The computational backbone is needed to automate the details of executing and linking various models, enabling a designer to reuse existing models and to concentrate on higher-level design issues. The computing framework should be easily extensible so that new models and databases can be easily integrated. The computing framework should also be platform-independent so that models on diverse platforms can be integrated with each other. It also needs to support various data-related features such as archival organization of large amounts of data. It should facilitate real-time data sharing with other designers, allow for systematic communication, and search-based retrieval of complex design information. Tools are also needed for online collaboration, communication, and project management, including real-time data sharing.

There are two distinct ways in which the framework can be implemented—as *middleware* and as *end-user* software. Both of these have certain advantages and disadvantages as discussed here.

- *Middleware* is generic software that can be used to connect software applications running on one or more computers across a network. The software is general and can be utilized to integrate various types of applications. Middleware tools free users from having to write their own routines to handle reliable data transfer between applications or from having to worry about complexities when multiple systems are integrated. However, users still must write codes to integrate application functionalities. Examples of middleware toolkits include OMG's Common Object Request Broker Architecture (CORBA) (Schmidt 2002) and Microsoft's Distributed Component Object Model (DCOM) (Microsoft 1996). If the framework is implemented as middleware, it may be flexible enough to be customized for different design processes, but it requires a significant effort to particularize it for specific types of software applications. Hence, to customize the framework for different types of problems, and for different design processes, a significant effort is required from the designers.

- Alternatively, the entire framework (including databases, analysis codes, etc.) may be developed by a single software vendor and provided to the users as *end-user software*. For example, in the product development arena, various Product Lifecycle Management software vendors provide complete suite of tools to perform various types of tasks such as CAD modeling, finite element analysis, manufacturing process planning, etc. The amount of effort involved in deploying and using the framework is low so long as new tools are not integrated into the framework. Significant costs

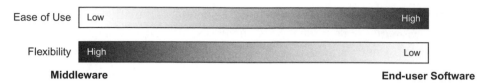

Figure 10.3 Comparing middleware and end-user software in terms of flexibility and ease of use.

are associated with integrating new tools into the framework or extending the framework's functionality because of the proprietary nature of information exchange standards used by the software vendor in developing the framework. Hence, if an engineering framework is developed as end-user software, the user must put forth only minimal effort but, in general, these frameworks are inflexible and cannot be modified easily when new situations arise, such as when the software tools or the design processes change.

Middleware tools provide standardization of communication protocols and leave a lot of integration work to users, whereas engineering frameworks (end-user software) provide easier integration capabilities but are inflexible. The tradeoff between middleware and end-user frameworks is illustrated in Figure 10.3. Choosing between the flexibility and ease-of-use of engineering frameworks is one of the key decisions faced by an organization involved in distributed design of products and materials.

Using a simple example, we present the general steps involved in developing a model to be used by remote users. Imagine a materials designer developing a finite element model for predicting the behavior of a material mixture under shock, as discussed in Chapters 8 and 9. The materials designer wants to deploy it to a network so that it is available for remote execution by other users. To do this, the designer needs to do the following:

1. Determine the parameters that serve as inputs and outputs for the program. For example, in the shock simulation model, the inputs are parameters that determine the morphology of the material mixture (particle sizes, nearest neighborhood distances, and associated uncertainties), the material properties of the constituents (e.g., equation of state parameters), the boundary conditions (symmetry, shock wave speed), the domain size, and the size of the mesh. The outputs include the pressure and temperature distribution at various time steps, and the number and location of hot spots.

2. Having determined the inputs and outputs to the model, the designer needs to specify the manner in which these inputs will be provided to the model. Generally, different model developers use diverse means to provide the input information to their models; two common means are (1) simple text files and (2) graphical user interfaces. Graphical user interfaces are useful when the model is used as a stand-alone computer program. Text

files are more common when a model's intended use is in a batch mode or in a distributed setting. Hence, the model developer needs to define clear, simple-to-parse structures of files used as inputs and outputs. The input and output files serve as the interfaces to this model that the remote users will use. These interfaces are also useful when the model is used in conjunction with other models and there is information flow between multiple models. If the model already exists, a separate wrapper (a small main procedure to be invoked from other software) may be needed to extract the input parameters from the input file and feed it into the model, because the original model may not be developed so that remote users or other models can directly access the simulation program. This step is carried out on the *server side* where the program will reside.

3. After the input and output files of the models are specified, the designer needs to develop a description file (or documents) containing the *specifications* of input and output and related information regarding the scope of application of the simulation program to let other users know how to use the program. The description file may contain information such as the title of the program and a description of what it does. The programming language and platform on which the model is developed, the scientific theories on which the model is based, theoretical assumptions, ranges of input values within which the model is applicable, approximate execution time, description of outputs, and sample input and output files. This description file is essential to ensure the correct use of the model by other users.

4. Notify the users that a new simulation model is available by registering the model on a registry server that maintains the list of models/databases that are available.

In the problem described previously, we assume that only one model is being developed and will be executed by remote users independently. However in a realistic design scenario, a designer would like to integrate these models and databases "on the fly" for the problem at hand. This typically involves in-the-loop decisions made by designers and hence the framework is not fully automated. If we imagine a scenario where simulation models are being developed by many different experts across the globe, and users are interested in using different models in a design process characterized by many information flows between the models, the effort involved in *integrating* these models may be enormous due to the fact that each model has a different type of input and output. The file formats are different, the ranges of applicability of models are different, the models reside on different computers (hence, bandwidth issues need to be accounted for), corporate firewalls may prevent access to models, and so on.

The problem of integrating models and databases is further complicated by the number and variety of applications (e.g., analysis, simulation, optimization, decision support, etc.) required in a complex integrated product and materials design scenario. Additional complications arise if (1) the models need to be changed and updated frequently, (2) model precision, accuracy, and convergence issues must be dealt with, and/or (3) the interactions between models change frequently (for example, when deciding which interaction pattern is the best, as discussed in Chapter 9). These are some of the challenges in developing distributed engineering frameworks for distributed collaborative materials and products design. The details of requirements associated with developing such framework are discussed by Panchal and coauthors (Panchal 2002, Panchal, Choi 2007). In this chapter, we present the current state-of-the-art technology in developing frameworks that can be used for integrated materials and product design. The emphasis here is on developing a framework by accounting for both flexibility and usability as discussed in this section.

The outline of this chapter is as follows. A literature review of distributed computing frameworks is presented in Section 10.2. In Section 10.3, we provide an illustrative design scenario of linear cellular alloy (LCA) design. In Section 10.4, we discuss the development of an open, adaptable framework, the eXtensible Distributed Product Realization (X-DPR) environment. Finally, in Section 10.5, we close the chapter with suggestions for future developments.

10.2. Review of Existing Frameworks

The approaches for distributed collaborative design can be broadly classified into two categories: *Web-based systems* and *agent-based systems* (Wang, Shen 2001). These categories are discussed in Sections 10.2.1 and 10.2.2, respectively.

10.2.1. Web-based Systems

Web-based systems use the client server architecture with the Internet as a backbone for communication. The Web-based architecture supports multiple teams to access the central information base and to communicate through a central Web server. The collaboration between designers is generally through tools like chat tools, whiteboards, webcams, etc. Most of the currently available Web-based tools are developed using Java technologies. These Web-based systems are used for remote usage of distributed software applications through applet-servlet pairs (Xiao, Kulkarni 2001) or through other means like LISP (Chalfan 1986), PROLOG (Rodgers, Huxor 1999), ActiveX (Huang, Lee 1999), etc. The Web-based systems can be categorized further into domain-specific software integration tools and general distributed computing tools.

10.2.1.1. Domain-specific Integration Tools

Domain-specific integration tools are generally CAD-CAE integration tools. Cramer and coauthors (Cramer, Jayaram 2002) developed a collaboration architecture to allow distributed designers to work on the same CAD model concurrently. The architecture incorporates three components: a server, a controller, and multiple members. The communication between members is through the controller and server by exchanging data packets. The system is developed specifically for exchanging information between CAD and CAE applications. The system uses CORBA (Schmidt 2002) for sharing information between CAD tools and CAE applications. In this system, the objects and the communication between objects are clearly defined. The system can be used with other CAE applications only if the applications provide application programming interfaces (APIs) for interactions. A number of collaborative CAD tools are developed using Virtual Reality Modeling Language (VRML) files for remote viewing of CAD models. The Virtual Web Plant (Ebbesmeyer, Gausemeier 2001) is developed for distributed access to engineering data at a central location. The tool integrates 3D models from various CAD plant design tools and displays them interactively. It uses VRML for displaying CAD information remotely. The central data repository is an object-oriented database. The system also uses Java applets as clients for accessing the central data repository. Wang and coauthors (Wang, Bjarnemo 2005) developed a Web-based environment for mobile phone customization named Virtual Mobile Phone Design Space (VMPDS), which allows users to collaborate on conceptual design of mobile phones. VMPDS is developed using VRML, Java, and JavaScript. Lin and Afjeh (Lin and Afjeh 2004) present an Extensible Markup Language (XML)–based framework for Web-based aircraft engine simulation. The framework allows easier data flow across different simulation components. Simpson and coauthors (Simpson, Umapathy 2003) present an interactive Web-based product platform customization framework for enhancing customer interaction and reducing design and manufacturing lead time for custom orders. Other similar applications for integrating CAD tools over the Internet are discussed in (Gupta, Lin 2002, Kim, Kim 1998, Shyamsundar, Dani 1998, Wang and Wright 1998).

10.2.1.2. General Distributed Computing Applications

Rezayat (Rezayat 2000) introduced the notion of an e-Web portal to illustrate how Web-based standards and distributed object technologies can be integrated to provide controlled access to any type of information and resource within an enterprise. The author has argued that out of a number of systems that provide client-server services (including CORBA, DCOM, HTTP/CGI, RPC etc.), only CORBA and DCOM provide the degree of sophistication needed to implement practical object-based client server systems at an enterprise level. The authors also recognized a need for using standards such as XML for formalizing the semantics of the information. TeleDM (Jiang, Fukuda 2002) is an e-service test bed for verifying Web-based product design developed with the aid of a prototype-manufacturing environment. A client-server infrastructure with Web-based technologies is used in this test bed. The clients are Java applets with corresponding servlets on the server side. The clients can also be CAD tools (AutoCAD, Pro/E, etc) or Rapid Prototyping (RP) planner tools (ACIS viewer, etc.). The process of "design for X" is modeled as a design-coordinate-redesign process. Wang and coauthors (Wang, Ma 2005) developed a Web-based generic distributed mechanical system simulation platform based on a gluing algorithm. The platform allows the integration of distributed simulations into a system-level simulation. An XML description of individual simulation models is provided, which is a key element that links together different parts of the system-level simulation model. The individual simulation models are integrated using a gluing algorithm. The benefits of such an approach include independence of subsystem models and support for collaborative design in a supply chain. Other examples of similar distributed computing applications include Ansys AI Workbench™ (ANSYS 2006), MSC. Acumen (MSC Software 2006), EDS Teamcenter (UGS 2006), PTC Windchill (PTC 2003), and Alibre Design (Alibre 2006).

10.2.2. Agent-based Systems

An *agent* is defined as a software component that can be invoked remotely to perform computational tasks. Similar to the Web-based design systems, agent-based systems also provide a collaborative environment for the sharing of design information, data, and knowledge among distributed design teams. However, unlike the Web-based design systems using the client/server architecture, an agent-based system is a loosely coupled network of models that work together to solve design problems. The agent-based systems are generally based on direct communication between agents instead of a client-server type communication that is common to the Web-based systems. The Web-based systems are easier to develop using the available technologies. An agent-based system is desired when the system is rapidly changing and the process is too complex. Agents are suited for ill-structured and modular systems.

Agent-based technologies date back to early 1990s when the Web was not as widely used. Agent-based systems are based on simplified architecture. The basic aim of the agent-based systems is software reusability and flexibility in using the same software programs for different scenarios. The agents are dynamically linked. This dynamic linking between agents can be achieved by having common information exchange protocols, syntax, and semantics for communication. Most agent-based systems (Cutkosky, Engelmore 1993, Gupta, Lin 2002, Olsen, Cutkosky 1995, Ray 2002, Rezayat 2000) adhere to knowledge-based standards for achieving interoperability between agents. Knowledge-based standards involve defining common ontologies and/or definitions that the agents agree upon. Whenever there is communication between different agents, they use common ontologies. However, these agents may internally use different software-level standards for processing data. This provides flexibility in the development of agents. One such knowledge-based agent framework is PACT (Cutkosky, Engelmore 1993), which is one of the earliest agent-based systems for engineering design applications. The PACT framework is developed with a focus on integrating legacy software tools using knowledge interchange languages like Knowledge Query Modeling Language (KQML), Knowledge Interchange Format (KIF), Agent Communication Language (ACL) etc. The system uses wrappers based on knowledge contained in various systems. A common ontology is defined for knowledge interoperability between agents. The PACT system provides flexibility in terms of the fact that the agents can use their own data models and the tools need to be committed to a single standard for defining data models. The SHARE project (Toye, Cutkosky 1994) was concerned with developing open systems for concurrent engineering, particularly for design information and data capturing and sharing. The system provided collaboration services including multimedia mail, desktop conferencing, online catalog ordering, and fabrication services. Rajagopalan and coauthors (Rajagopalan, Pinilla 1998) propose an agent-based infrastructure to provide designers with access to multiple layered manufacturing services, including design, process planning, and manufacturing services. Madhusudan (Madhusudan 2004) presents a Web service-based framework to expose intraorganizational information sources. In this framework, processes are dynamically composed using artificial intelligence planning mechanisms. Wu and coauthors (Wu, Xie 2004) integrated Web services and the agent technology and developed an information framework for collaborative product development. One of the key features of this framework is its flexible client-side product development environment, which is especially useful for addressing the need for negotiation while managing conflicts in engineering design processes.

Four of the recent frameworks that describe different types of distributed frameworks are DOME (Pahng, Senin 1998), NetBuilder (Dabke, Cox 1998), Web-DPR (Xiao, Kulkarni 2001), and Fiper (Engineous Inc. 2005), (Simulia 2009). These four frameworks are selected because they respectively represent product-centric, process-centric, Web-based, and agent-based frameworks. DOME and NetBuilder represent agent-based systems. While DOME is a product-centric framework where each agent models a subsystem of the artifact, NetBuilder is a process-centric framework where each agent models an activity in the design process. Web-DPR is selected because it represents the Web-based systems and is a foundation for the X-DPR framework presented in this chapter. Fiper represents the current state of commercial distributed design frameworks and is to date the most advanced commercial engineering framework. In this remaining part of this section, we review these frameworks. For a detailed feature-by-feature review of these frameworks in the context of open engineering systems, please refer to (Panchal, Choi 2007). DOME, NetBuilder, Web-DPR, and Fiper are discussed in more detail in Sections 10.2.3, 10.2.4, 10.2.5, and 10.2.6, respectively.

10.2.3. Distributed Object-based Modeling and Evaluation (DOME)

The Distributed Object-based Modeling and Evaluation (DOME) framework (Pahng, Senin 1998) is intended to integrate designer-specified mathematical models for multidisciplinary and multi-objective design problems. The focus of the DOME framework is to create a modeling scheme that handles the different variable types needed in engineering design, integrate multi-objective evaluation and optimization with design models, and provide an object-based methodology to facilitate the integration of design models. In this framework, a product design problem is modeled in terms of interacting objects, called *modules,* each representing a specific aspect of the problem. One of the key assumptions of the framework is that product design problems can be decomposed into subproblems. The decomposition reflects both the physical subdivision of the product into components or subassemblies and the division of analysis expertise. Each object represents a subset of an aspect of the problem and acts as a stand-alone model, managing the data and services that it can provide. An integrated design model is realized by objects representing the different parts of the problem. These objects are executed simultaneously.

In summary, the DOME environment is focused on simulation-based design and breaking down the design artifact into subsystems that can be represented mathematically and may be distributed over the network. The framework is not designed with an open system paradigm but with a product-dependent distributed objects framework, which is more intuitive from a designer's point of view. It is platform-dependent and, because it uses a CORBA protocol, requires lots of effort to create wrappers and the appropriate graphical user interfaces. DOME does not have a supporting tool for the management of objects in the framework and real-time information handling.

10.2.4. NetBuilder

NetBuilder (Dabke, Cox 1998) provides a mechanism for coordinating collaborative activities in a distributed environment. There are two key aspects to the NetBuilder approach. First, NetBuilder provides a compositional framework that allows designers to combine individual tools into *meta-programs* that capture the simulation process. These meta-programs can be executed and stored for future use. Second, NetBuilder supports wrapping individual modules so that they can be invoked as part of meta-programs in a uniform way. NetBuilder leverages mechanisms of distributed computing such as CORBA to provide a seamless integration of networked resources. NetBuilder provides the capability of capturing dependencies among simulation subtasks in terms of links connecting meta-program modules. When a meta-program is running, the NetBuilder scheduler determines which modules may be executed by checking to see whether the appropriate input data is ready. Each analysis tool is wrapped which allows it to accept input and produce output in a standard format. NetBuilder also contains a module-wrapping toolkit to support the encapsulation of existing tools as CORBA-compliant modules.

NetBuilder has most of the features that are needed for an adaptable framework. Real-time management of process information is a valuable feature, as is the mapping communication protocol. However, there are some features that are only partially implemented, which limits NetBuilder's usage as an adaptable framework. CORBA itself requires that separate wrappers must be developed for all modules being integrated. The framework enables interfaces between modules on heterogeneous platforms, but components of the framework (such as meta-programs) cannot run on heterogeneous platforms. The descriptions of service assets are clearly defined in the Resource Catalog; however, in this catalog, there is not enough information for a user to find an appropriate service asset, and the format of the Resource Catalog is not an industry standard. In summary, NetBuilder enables the rapid and dynamic assembly of systems distributed on a large scale, but it has limitations in serving as an adaptable framework. However, it represents valuable progress toward an adaptable engineering framework.

10.2.5. Web-DPR

Web-DPR (Xiao, Kulkarni 2001) has been developed based on the communication framework of PRE-RMI (Gerhard, Rosen 2001). The major objective of Web-DPR is to make agent services accessible with a simple Java-enabled Web browser. The essential components of the Web-DPR framework are a Web server, Framework Database, coordinator, and Agent Template (Choi 2001). The architecture of the WebDPR framework is shown in Figure 10.4.

The Web-DPR framework database stores information about available agents, the event channels to which they are registered, and other information regarding the design process. A client sends a request to the Web server, and this request is then transferred to event channels. The event channels then forward the request to agents. Information is transferred between various agents either as messages or as engineering data. A message is a short note or a command to other engineers, which is independent of product design domains.

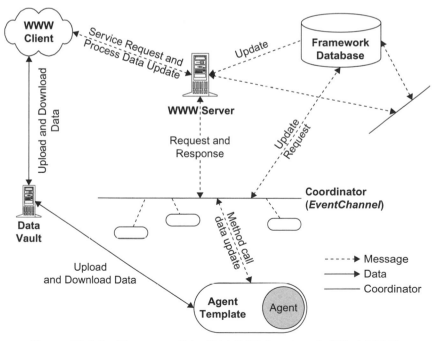

Figure 10.4 Architecture of the WebDPR framework (Choi 2001).

Engineering data include data files, CAD models, etc. These engineering data are archived in a central data vault. In Web-DPR, the event is split into message and engineering data in order to ensure that an agent's functions are completely independent of the functions of other agents.

A Java-based application, *Agent Template,* is used to create and deploy agents easily into the framework. With the Agent Template, users do not need to have much knowledge about the internal implementation details about the framework. Web-DPR has features including a general message construct based on Java-RMI, dynamic Web-browser user interface (UI), standardized wrapper (Agent Template), etc. However, it cannot support the detailed access to remote objects since it wraps distributed modules using an Agent Template, which only provides abstract access to these remote objects. The dynamic Web browser UI can neither take range values nor select alternatives. The Web-DPR framework does not support parameter mapping between tasks or task decomposition. This framework uses a Web server to publish services to the Web so there can be a bottleneck on the Web server.

10.2.6. Federated Intelligent Product Environment (Fiper)

Federated Intelligent Product Environment (Fiper) (Engineous Inc. 2005) is composed of three different layers—Core Infrastructure, Core Extensions, and Application Components. The Core Infrastructure provides the foundation for the environment and is comprised of a collection of services for handling process management and data communication and storage (Wujek, Koch 2002). The Core Extension contains modules that can be plugged into the Core Infrastructure and allow organizations to use the existing IT infrastructure. The Application Components provide the functionality desired by the users and can be published to the environment. Fiper uses a standard Java-based wrapping mechanism to allow easy creation of components for the environment. The Fiper Standard Development Kit (SDK) is provided to help write necessary Java code and execute it. The Fiper library is a virtually centralized and physically distributed repository for publishing, searching for, and retrieving components. It facilitates collaboration by sharing the services offered by the Application Components. It also allows an engineer to assemble components into a workflow model of his or her design process. Kao and coauthors (Kao, Seeley 2004) present the use of the Fiper framework for aircraft engine combustor design. Fiper enables real-time business-to-business collaboration at GE and Parker.

In addition to Fiper, two other similar frameworks are commercially available: iSIGHT® (Engineous Inc. 2004) and ModelCenter® (Phoenix Integration Inc. 2004). iSIGHT is a software application developed by Engineous Software Inc. (now a subsidiary of Dassault Systèmes) (Engineous 2001) for integrating various simulation codes that are generally used in a design process. iSIGHT also provides built-in functionality to carry out design of experiments, approximations, and optimization. iSIGHT is mainly developed to integrate multiple design tools, coordinate the design process, and automate information flow and execution of computational tasks. It enables engineering designers to explore more design alternatives and reach optimal designs faster, leading to significantly lower product costs and increased overall product performance (Engineous 2001). iSIGHT employs a variety of optimization methods, including quality engineering, approximations, and design of experiment methods. It is capable of integrating software programs that accept inputs in the form of simple text files and can provide outputs in the form of simple text files. iSIGHT uses its inbuilt text parser to parse these text-based input and output files to extract and write parameter values.

Phoenix Integration Inc. provides two modules, ModelCenter and Analysis Server. ModelCenter operates on client-server-based architecture and all the applications are installed on Analysis Server and the user can use ModelCenter to define a design process. Those two modules provide integration strategy for engineering organizations (Phoenix 2001). These tools are lightweight, portable, and configurable. It provides Web-based integration of engineering process. ModelCenter automates the process of executing multiple analysis codes in a design process. Data for a design is automatically passed from one program to another using ModelCenter, enabling the designer to concentrate on the results of the design rather than on the tedious job of manually executing individual/heterogeneous programs. ModelCenter enables users to link input data parameter to individual output data parameter with drag and drop operation. Analysis Server offers four different data interfaces, which are File wrapper to Macro/Script APIs, Java-based command-line interaction, COM/OLE integration through Visual Basic Scripting, and C/C++ API interface.

Although the processes modeled in existing frameworks such as Fiper can be stored as templates and reused for designing the same product with different specifications, the main restriction is the reusability of processes for designing *different* products even if the tasks and distributed applications involved remain the same. Currently, the processes in Fiper and other similar commercial frameworks (such as iSIGHT and ModelCenter) are inherently defined as a series of tasks with flow of product parameters between these tasks. Hence, the processes defined at a computational level in frameworks such as Fiper cannot be used to design different products, whose parameter sets are different. The reusability of processes for different products is addressed by Panchal and coauthors (Panchal, Fernández 2004). Further, Fiper does not support product information modeling.

10.3. Motivating Example: Design of Linear Cellular Alloys (LCAs)

As discussed in Chapter 3, linear cellular alloys (LCAs) are prismatic 2D cellular materials that are processed through compounding of a slurry (binder phase mixed with metal powder oxides), which is then extruded under pressure through a multistage die and subjected to

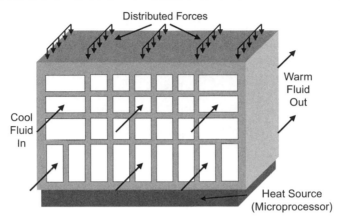

Figure 10.5 LCAs with rectangular cells (Seepersad 2004).

drying and reduction into the metallic phase in a hydrogen-rich environment, followed by sintering to achieve nearly fully dense metal composites (Hayes, Wang 2001, Seepersad, Dempsey 2002). A wide range of cell sizes and shapes can be achieved, including functionally graded structures, which provides multifunctional structural and thermal performance. Cell sizes on the order of half a millimeter and up and cell wall thicknesses on the order of 50 to 100 microns can be achieved, resulting in very fine as well as very coarse structures. These metallic structures can be produced with any arbitrary 2D cross-sections and are suitable for multifunctional applications involving both structural and thermal functions (Seepersad, Dempsey 2002) (see Figure 10.5). Applications of these materials include heat sinks for microprocessors, combustor liners, and so forth.

The process of design of LCAs is shown in Figure 10.6. It involves six steps starting from the requirement-gathering phase to the final geometry of LCAs. The first step is to capture the requirements for designing these LCAs. These requirements expressed are in terms of the expected behavior of the LCA in a specific target application. These requirements are used to create geometries in a CAD modeling tool. Based on experience, the designer starts with a cell geometry, which is subsequently iteratively modified to match the expected behavior with simulated behavior. Since this is a multifunctional application, the analysis is carried out for both structural and thermal requirements. This simulated behavior is compared against the expected behavior. If these two do not match, appropriate changes are made to the geometric parameters to achieve desired performance. Some of the steps in this example design process, namely thermal and structural analyses, are carried out using automated software applications, whereas other steps like capturing requirements require manual inputs.

In the next section, we discuss our effort toward an adaptable framework, X-DPR, and explore the features and capabilities of X-DPR in the example shown in Figure 10.6.

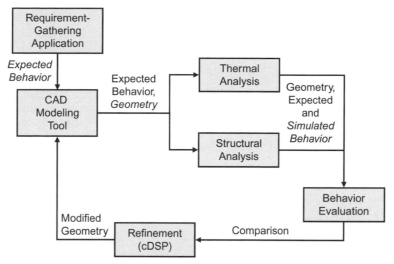

Figure 10.6 Design process for multifunctional LCAs.

10.4. X-DPR (eXtensible Distributed Product Realization) Environment

In this section, we provide an overview of the eXtensible Distributed Product Realization (X-DPR) framework (Section 10.4.1), discuss the main features of the framework (Section 10.4.2), use the framework for LCA design (Section 10.4.3), and show how the framework can be characterized as an adaptable framework (Section 10.4.4). The framework is based on industry standards. The models for capturing and passing information are also based on various standards.

10.4.1. Overview of X-DPR

The X-DPR framework is designed with peer-to-peer communication between agents, where each agent is an independent entity communicating with other agents. A peer-to-peer communication framework enables independent communication between different agents. X-DPR is an open system in which different modules can be easily integrated into the system for enhancing the functionality of the overall system. Engineers can integrate their own applications residing on their machines with X-DPR, which will help create a global library of engineering tools over the Internet. This library can then be integrated with tools from other areas such as marketing, sales, or other business services to realize a global enterprise. The X-DPR framework uses the Decision Support Problem (DSP) technique (Bras and Mistree 1991, Muster and Mistree 1988) to support meta-design, a process of designing systems that includes partitioning the system for function, partitioning the design process into a set of decisions, and planning the sequence in which these decisions will be made.

The system is designed so that a designer can easily model his or her design process using visual tools. This capability for meta-design is unique in X-DPR. Engineers can then connect process models with services available in the global library using the Internet and execute complete design processes online. X-DPR provides flexibility at a design process level. It enables designers to design a process and replace entities of process with other entities later. The framework allows engineers to develop and execute process models collaboratively. Thus multiple designers distributed around the globe can work together as a team on product realization projects. A detailed discussion about each element of the framework is presented in Section 10.4.2.

10.4.2. Elements of the X-DPR Framework

In this section, we describe the elements of the X-DPR framework (Figure 10.7) in further detail. Two agents are shown in Figure 10.7, along with the client application. The exchange of information between various elements of the framework is also shown. In the X-DPR

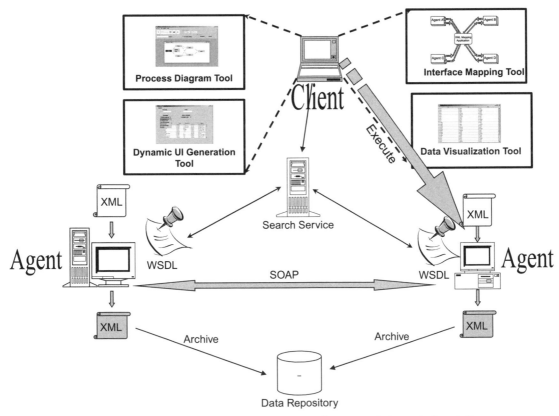

Figure 10.7 Block diagram of the X-DPR framework (Panchal 2002).

framework, an agent is defined as a software component that can be invoked remotely to perform tasks in a product realization process. Agents can be invoked by the client application or by other agents by sending XML messages. On receiving these messages, the agent processes the input information and replies back with XML messages. Agents in X-DPR are associated with (1) an associated input message template in XML format, (2) a processing mechanism in the form of software, (3) output message template in XML format, (4) a WSDL description file that provides information about the location of service and the way to invoke this agent, and (5) an XML-based UI description file (optional) that is used by the client application to generate a custom UI. The details of these elements are discussed in Sections 10.4.2.1 through 10.4.2.7.

The client application used to create process diagrams, manage the flow of information, and access agents remotely has four elements—a process diagram tool, a dynamic UI generation tool, an XML data viewer tool, and an XML mapping application. With the process diagram tool, users can create their own networks of tasks. These tasks can be assigned to agents available over the Internet. The user can search for available agents with a search service that is essentially a database containing the location and description of the agents. Once a task in the process diagram is assigned to an agent, it can be executed from the process diagram tool. The dynamic UI generation tool extracts information from the Web service description (WSDL) of an agent and creates a UI for the client based on the inputs taken by the agent. The XML mapping tool maps the XML-based input-output UIs between different agents. It facilitates the smooth and seamless flow of information from one agent to another. The information generated throughout a process is archived in a data repository. One of the most important capabilities of the client application is its flexibility to execute any agent remotely by dynamically creating Simple Object Access Protocol (SOAP) messages from the WSDL file and the XML-based agent input template. The Java .class files in the client application responsible for dynamic agent execution are packaged as a Java .jar file and deployed with the remote agents so that these agents can directly invoke other agents.

10.4.2.1. Data Repository

The data repository is a database of all the information computed, exchanged, or stored during the design process. The data repository in the X-DPR framework is developed using STandard for Exchange of Product Data (STEP) (Nell 2003) and the XML standards. STEP is an international standard for engineering information models. STEP standards have various predefined schemas that can be reused directly for application specific information models. STEP standards are used in the X-DPR framework in order to make the data repository standards-based. The Express language (Schenck and Wilson 1994), which is part of the ISO standard (ISO 10303-11), is used for developing the information model for the product that is being designed. In the LCA design example presented in Section 10.3, the information model consists of:

1. Expected behavior (i.e., the requirements)

2. Form, which consists of topology and material used to manufacture the LCAs

3. Simulated behavior, which consists of thermal and structural behaviors.

The details of these entities are not presented in this chapter. While developing the information model supporting LCA design, the following STEP parts are used (see Figure 10.7 for context): Part 42 (for geometry and topology representation), Part 104 (for finite element analysis information), Part 45 (for materials information), Part 50 (for mathematical constructs), and Part 47 (for tolerances). The schema is written in an Express file, and an instance is a Part 21 file. The data access from the Part 21 file is carried out using the JSDAI™ toolkit developed by LKSoft (LKSoft 2006) The JSDAI Express Compiler creates Java APIs from STEP Express schemas. These Java APIs are used to extract information required by different agents from Part 21 files. This extracted information is formatted as XML files and sent as inputs to agents. A sample XML file for the cDSP template is shown in Figure 10.8.

This method facilitates capturing the engineering information in the object-oriented STEP database and also allows information transfer through a standardized, platform-independent XML standard. Although the data repository is an integral component of the X-DPR framework, its functionality is similar to that of other agents. It accepts XML messages from agents that are stored in the repository and sends back XML messages as requested by other agents. A user can also implement custom data repositories as agents in the X-DPR framework. However, these custom data repositories have to be explicitly instantiated as tasks in design processes using the process diagram tool discussed next in Section 10.4.2.2.

10.4.2.2. Process Diagram Tool

The process diagram tool, shown in Figure 10.9, is used to model a product realization process, and then it can be used to invoke the available agents integrated into the framework. The tool is coded in Java and hence is platform-independent. This tool contains a whiteboard on which the process diagram can be created by simple drag-and-drop operations. The process diagram construction toolbar aids in this process of creating flow diagrams with blocks and connecting lines. These blocks represent various tasks in a design process and the connecting lines indicate the flow of information from task to task. The tasks can be assigned

```
<?xml version="1.0" encoding="UTF-8" ?>
<!-- edited with XML Spy v3.5 NT (http://www.xmlspy.com) by Jitesh (SRL)  -->
<dsides:DSIDES xmlns:dsides="urn:test:dsides" title1="jietsh12" title2="dfhisdhl312ih">
    <GIVEN nSysVars="4" nRealVars="4" nIntegerVars="0" nBooleanVars="0" nLinearCons="2" nEqualityCons="0" nInEqualityCons="1"
      nLinearGoals="0" nNonLinearGoals="1" nPriorityLevels="1" PerformCalculations="1" PrintFinalSolutions="0" nSynthesisCycles="12000.0"
      StationarityDeviationFunction="0.05" StationaritySystemVariables="0.05" />
    <FIND>
    - <SYSTEM_VARIABLES>
          <PARAMETER name="R" min="25.0" max="150.0" guess="50.0" type="real" />
          <PARAMETER name="L" min="25.0" max="240.0" guess="50.0" type="real" />
          <PARAMETER name="Ts" min="1.0" max="1.375" guess="1.2" type="real" />
          <PARAMETER name="Th" min="0.5" max="1.0" guess="0.75" type="real" />
      </SYSTEM_VARIABLES>
    - <DEVIATION_VARIABLES>
          <PARAMETER name="d_1+" />
      </DEVIATION_VARIABLES>
    </FIND>
    - <SATISFY>
    - <SYSTEM_CONSTRAINTS>
          <CONSTRAINT name="const" type="L" relation="(1,-0.0193) (2,0.0) (3,1.0) (4,0.0)" sign="GE" value="0.0" />
          <CONSTRAINT name="consra2" type="L" relation="(1,-0.00954) (2,0.0) (3,0.0) (4,1.0)" sign="GE" value="0.0" />
          <CONSTRAINT name="cons14568" type="NL" relation="3.14*R*R*L + (4/3)*3.14*R*R*R - 1296000" sign="" value="" />
      </SYSTEM_CONSTRAINTS>
    - <SYSTEM_GOALS>
          <GOAL name="goal1" relation="4000/( 0.6224*Ts*R*L + 1.7781*Th*R*R + 3.1661*Ts*Ts*L + 19.84*Ts*Ts*R ) - 1"
            value="" type="NL" />
      </SYSTEM_GOALS>
    </SATISFY>
    - <MINIMIZE>
    - <DEVIATION_VARIABLES>
          <LEVEL name="Level1" relation="(+1,1.0) (-1,0.0)" />
      </DEVIATION_VARIABLES>
    </MINIMIZE>
</dsides:DSIDES>
```

Given

Find

Satisfy

Minimize

Figure 10.8 A sample XML file representing a cDSP template.

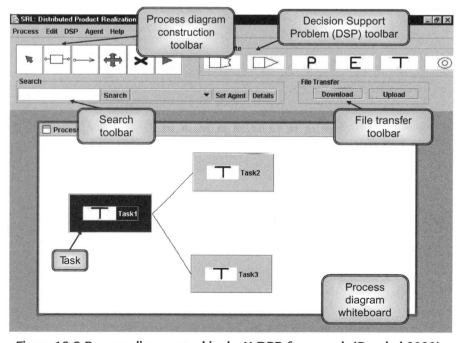

Figure 10.9 Process diagram tool in the X-DPR framework (Panchal 2002).

to any of the Web services available over the network. Using the process diagram tool, we can define process templates that can be edited for specific design problems.

The search toolbar is used to search for available services. The Decision Support Problem (DSP) technique (Bras and Mistree 1991) toolbar is used to model a design process in terms of phases, events, and tasks and it also contains links to decision support tools for the design process. The file transfer tool is used for sending and receiving files (for example, CAD files) to various agents. The process diagram tool supports a hierarchical process development decomposing a task into subtasks. This means that a designer can move from a higher level in the process and then design a particular task as a network of subtasks. These processes are then saved in a central database that can be accessed by distributed team members and software agents. This process database contains information regarding the tasks in the process, flow of information between these tasks, agents assigned to these tasks, the tasks that are currently completed, in progress, or uninitiated. In the current implementation of X-DPR, the agents are executed automatically in a sequential fashion. All the tasks that require the outputs from finished tasks as their inputs are activated as soon as these inputs are available. Currently, complex and conditional sequencing is not available in the X-DPR framework. The process diagram tool is implemented using the Swing library in Java. The Java application extends the JFrame class in javax.swing package. The file transfer toolbar contains two buttons to upload and download files. These files can be any file that the agent needs for execution or any file as a result of the execution of the agent. This toolbar is especially useful when the agents require processing of binary files like CAD files.

10.4.2.3. Dynamic UI Generation

If an agent requires user input, a graphical UI must be developed for this purpose. The kind of interaction of an agent with the user varies from case to case and different graphical UIs are required for different agents. Since it is not possible to create a separate UI and code it into the client, a dynamic graphical UI is created based on the number and types of input that the agent requires.

Implementation of the dynamic UI generation—Two types of dynamic generation of UI generation are developed in X-DPR. The first type corresponds to a situation in which the inputs required from the user are very simple—for example, a few different parameters must be specified in a function. In this case, the description of the required inputs to the agent can be extracted directly from the WSDL file. Inputs from the user are generally taken with simple text boxes. The process of customized UI generation can be accomplished as follows: (1) the client looks for the WSDL document published by the service, (2) from the WSDL document, the client extracts inputs and the corresponding data types, and (3) the client generates a graphical UI for the user inputs. Based on the data entered by the user, the agent is executed. The process is shown in Figure 10.10.

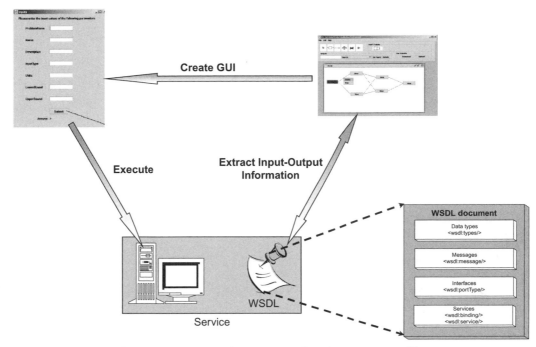

Figure 10.10 Dynamic UI generation from a WSDL file.

The second type of UI generation corresponds to a situation where the inputs of an agent are complex XML tree structures. For example, the input to a Design of Experiments (DOE) agent implemented in iSIGHT (Engineous Inc. 2001) is in the form of an XML file, which requires complex interactions with the user. The user must enter all the DOE parameters and their ranges. The user also needs to enter the type of DOE to be performed. In this case, the complete description of the inputs and how the user inputs will be configured are not available in the WSDL file. A single XML file describing the user interface must be created at the agent and deployed with the agent itself. This XML file contains nodes for individual entities in the UI to be created such as text box, label, combo box, table, checkbox, radio button, etc. For each element, an XML tree representation is provided that contains the information required to generate the UI component. For example, for a label, the information required to generate is its location on the form, its size, and the text of the label. The XML schemas for some of these elements are standardized in X-DPR. The client application accesses this description file remotely, and a UI is created automatically at the client for that agent. The process is shown in Figure 10.11.

In the LCA design example, inputs to various agents are complex XML tree structures containing geometry information, boundary conditions, analysis results, etc. The second type of the UI generation technique is used where a separate XML file describing the UI is created. This UI description XML file is used by the client application to generate the UI remotely.

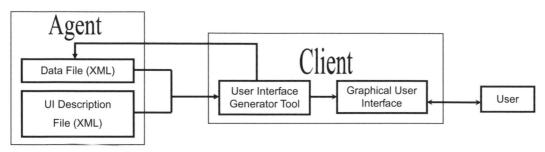

Figure 10.11 Dynamic UI generation using the UI description XML file.

The dynamic UI generation tool helps in maintaining consistency between agent service descriptions and the client's UI.

10.4.2.4. Interface Mapping Tool

In a generic framework in which different applications provide vastly different functionalities, it is very likely that the output of one agent will not be exactly that which is required as the input to another agent. To achieve a seamless flow of information between agents, the outputs need to be converted into a format compatible with the inputs to other agent. In general, if there are n agents, the number of conversions required will be $n*(n-1)$. In the LCA design example, a structural designer creates a response surface to investigate the effect of design variables (thickness of ribs, overall height of LCAs, etc.) on the overall strength. The first step in the process is to carry out a design of experiments in the design space. The design of experiments is performed using a commercial application—iSIGHT. The result of the DOE is a set of points in the design space at which the analysis is carried out. The analysis of the component at these points is carried out in an in-house code or an Finite Element Method (FEM) program such as ANSYS (Ansys Inc. 2003). The output of the design of experiments from iSIGHT is in iSIGHT's own ASCII file format and the input to the ANSYS FEM solver requires the ANSYS ASCII file format.

For automatic transfer of information from iSIGHT to ANSYS, a designer must write a parser to convert one file format into another. To overcome this problem of developing separate converter applications, an interface-mapping tool is created that can be used to map information output from one agent to the inputs of another. Here, the term *interface* refers to the structure of information input and output by agents. This tool has the capability of mapping the XML structure from the output of one agent to the XML input of another agent (see Figure 10.12). The mapping tool shows the tree structures of the output XML file of one agent and the input XML file of another agent on two sides of a window. The user selects corresponding information entities (i.e., XML nodes) on the input and output sides and maps them. Once the mapping is established between two agents, the mapping rules are saved in a separate XML file for future use. Hence, the users need to establish the mapping between

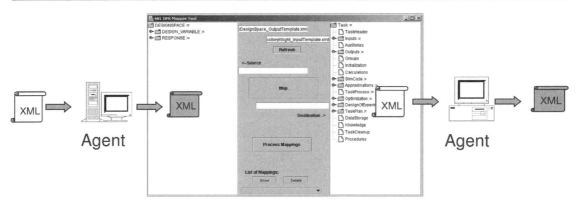

Figure 10.12 Mapping of interfaces between two agents.

two file formats only once. It is important to note that the mapping tool implemented in X-DPR can be used only with XML inputs and outputs. In the case of applications such as iSIGHT and ANSYS, where the inputs/outputs are simple text files, the applications need to be wrapped (i.e., text files converted into XML files) such that the agents have XML inputs and outputs. This is also illustrated in the block diagram of X-DPR (see Figure 10.7), where agents are shown with XML inputs and outputs.

Implementation of interface mapping tool—The implementation of this tool is done using XPath, which is a language for specifying the path of an element in an XML document. The tool consists of the following two components encoded in Java: (1) mapping definition UI and (2) mapping execution class. The XML mapping definition UI is shown in Figure 10.12. The form has two tree elements corresponding to the XML output of the first agent and the XML input of the second agent. The XML structures are shown in the tree elements. The user selects an element from either side and creates the mapping between these elements using the Map button. Two text boxes show the source element path and the destination element path, respectively. These paths correspond to the XPath of the selected elements. The information regarding mapping between all the elements is stored in an XML *mapping information file* that can later be used to extract information from the outputs of the first agent and populate the XML input file of the second agent. The mapping execution class reads the *mapping information file* and extracts information from the XML output file of the first agent and fills up the XML input of the second agent.

10.4.2.5. Messaging and Agent Description in X-DPR

The transfer of information between different software applications in X-DPR is through XML-based standards such as SOAP (Box, Ehnebuske 2000). The SOAP standard for interfacing different software applications is programming language–independent. XML is a platform- and language-neutral standard for representing information. The benefit of

XML is that it separates data from meta-data (i.e., information about the data). XML is also being adopted as a universal standard for representing information in distributed computing frameworks. SOAP is a communication mechanism based on XML. It is also a platform- and language-independent standard. Previous communication protocols, such as CORBA, DCOM, Enterprise Java Beans (EJB), and Java-RMI, share the common problem that they are incompatible with each other and that the applications deployed with these protocols cannot be accessed through a firewall. The SOAP protocol addresses this problem. SOAP uses a simple Hypertext Transfer Protocol (HTTP) request/response-based communication, allowing it to pass through corporate firewalls (Marcato 2002). A SOAP message typically contains an XML message along with an HTTP header.

In X-DPR, we use XML to define interfaces between different design activities, SOAP for message transfer between distributed applications over different platforms, and WSDL to describe different Web services. Since we are using standards common to all Web service–based frameworks, X-DPR is compatible with other similar frameworks.

10.4.2.6. Publishing a Service

The agents are published in the X-DPR framework simply by creating a description file based on the WSDL standard. The client retrieves the information from the WSDL description and creates a UI for the agent. WSDL documents can either be created manually or can be created automatically using commercially available toolkits. The Microsoft SOAP toolkit (Microsoft 2003) can be used to create WSDL for COM objects, and Systinet Server for Java (Systinet Corporation 2004) can be used for creating WSDL for Java classes.

10.4.2.7. Asset Search Service

The task of searching for agents appropriate for a particular task is implemented as a Web service in itself. This Web service is called the Search Service. The Search Service maintains a database of links to WSDL files with a description of the service. Currently, the new agents in the database are populated manually and the database is created in Microsoft Access. However, it is planned that the Search Service will perform a running search on the Web for WSDL description files.

The agent search service also maintains information about whether the service is currently in use or not. In the X-DPR framework, an agent's lifecycle is described by three states— *available*, *busy, and unavailable*. In the *available* state, the agent can receive requests for execution from clients or other agents. When the agent is being executed by the client or another agent, it shifts into the *busy* state. The agent shifts between the available and busy state automatically. An agent is *unavailable* when it is registered in the database but cannot execute. This may happen when the agent is physically disconnected from the network. In the current implementation, the agent developer has to manually set the state to *unavailable*.

Whenever a user searches for an agent, the search service automatically gives a list of *available* and *busy* agents. Maintaining the Search Service as a separate module is helpful because it can be developed independently of the framework and thus replaced with a different service at a later date. This also leaves open the possibility that commercial Web service search engines developed in the future may be integrated into the framework. The agent search service can be extended in future by using the standard Universal Description, Discovery, and Integration (UDDI) protocol. UDDI describes a standard method for publishing, managing, discovering Web-based services (OASIS 2004). UDDI is also based on other XML-based standards such as SOAP, WSDL, and XML Schema.

The search service is implemented as a Java class. The Java class performs SQL queries on the database that contains the following information: (1) agent name, (2) agent description, (3) location of the description file (WSDL file), (4) input file template (if any), (5) output template (if any), (6) location of the user interface description file (if any), and (7) an entry that specifies the current state of the agent. The interface-mapping tool described in Section 10.4.2.4 uses the input and output template files to map the outputs and inputs of different agents. From the process diagram tool, the search toolbar can be used to perform a keyword search on agents available for use. The process diagram tool sends a request to the search service and the search service returns a list of available agents matching the keywords. The user can select any of the agents from the list and assign it to the blocks in the process diagram.

10.4.3. Using the X-DPR Framework for LCA Design

In the LCA design example, seven distributed software applications are involved. These include applications for:

- Requirements capturing (in-house Java application),
- Problem definition,
- Design of experiments (iSIGHT),
- Thermal analysis (in-house MATLAB® code),
- Structural analysis (in-house MATLAB code),
- Response Surface Model (iSIGHT),
- Updating the geometric parameters (in-house Java code).

These applications are deployed as agents in the overall robust design process, as shown in Figure 10.13.

To explain the use of the X-DPR framework in the context of LCA design, we revisit the five tasks listed in Section 10.1 that are performed by an agent developer. The first task is

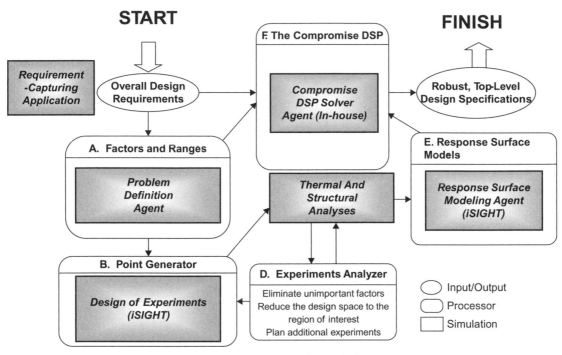

Figure 10.13 Agents in the robust design of LCAs.

Figure 10.14 XML interface for iSIGHT.

to specify input and output data constructs for each application. The inputs and outputs for all seven applications are described as individual XML files. The second step is to develop a wrapper for each of these applications. The wrapper development involves converting the input XML file into the application's native input format and converting the output from the native format into the XML format described in the previous step. The wrapper for iSIGHT has been developed in Visual Basic, and for other applications, it has been developed in Java (see Figure 10.14). The hierarchical schema representation of the input of the iSIGHT XML files is shown in Figure 10.15.

The third step is to develop a service description file for each application to be published as an agent. This step is performed automatically by the use of applications such as the Microsoft SOAP toolkit for Visual Basic–based wrappers and using the Systinet Server for Java–based wrappers.

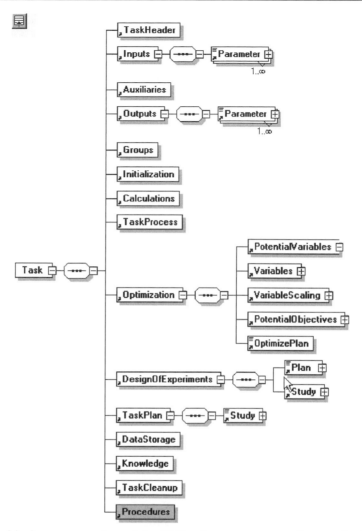

Figure 10.15 Graphical representation of the XML schema corresponding to the iSIGHT input file (generated in XMLSpy®).

The fourth step is to notify the framework of the newly available service. This step is performed by adding individual entries to the agent database described in Section 10.4.2.7. The fifth step is to create a UI for the user interaction with agents if required. The UI for each agent is described for each agent using the UI description file (WSDL file). These UIs are created at the client side only if the user wants to interact with the agent. Although all of the steps described in Section 10.1 are required in the X-DPR framework, the prime advantage of using an adaptable framework such as X-DPR is its openness, which reduces rework if there is a change in either the design process or the agents themselves. The features that provide this flexibility to the framework are discussed in Section 10.4.4.

Having discussed the activities performed by agent developers in deploying the applications, we now move on to the steps that the user follows for executing the LCAs design process remotely. Using the process diagram tool, the user creates a design process as a network of tasks where each task corresponds to an agent to be executed remotely. For the LCA design process, these tasks are:

- Capturing customer goals in terms of target heat transfer and stiffness of the LCAs, and

- Defining design variables, responses, and associated ranges, which are mapped to a design of experiments task.

The design of experiments task outputs a list of points in the design space where the analyses (thermal and structural) are executed. The output of the design of experiments task is mapped to the inputs of both thermal and structural analysis tasks. The outputs of analysis tasks are inputs for the response surface modeling task, where an approximate surface is fit between the input variables and output variables. The output of the response surface modeling task is input to a task that updates the LCA's geometry. At this time, these tasks are not tied to any software agent. The user then performs keyword searches corresponding to each agent using the search toolbar, which results in a list of available agents. For example, when the user searches for "Design of Experiments," two agents are shown in the list—iSIGHT and Minitab. Both of these agents have similar functionality. The user then selects the suitable agents and assigns them to tasks defined in the process diagram using the search toolbar. Through this assignment, the task associated with an agent is linked with the corresponding WSDL file and also sets up a link with its input and output file template. The user then maps the outputs and inputs of agents that are linked in the process diagram using the process diagram tool. For example, the output XML template file of the design of experiments agent is mapped to the input XML template files for structural and thermal analyses. After the inputs and outputs are mapped, the process is executed from the process diagram tool. This initiates an execution chain of agents wherein the agents are executed sequentially until all the agents have executed once. Loops of execution are not supported in the current implementation of X-DPR.

10.4.4. X-DPR as an Adaptable Framework

We have discussed the elements of X-DPR, implemented as an adaptable framework. The main features of X-DPR as a standardized yet flexible framework are discussed in the following subsections 10.4.4.1-10.4.4.4.

10.4.4.1. Flexible Mapping of Information Interfaces Between Agents Using the XML Mapping Tool

In X-DPR, the Web is used as a backbone along with the associated technologies (Java, Web browsers, etc.) and standards (XML, SOAP, WSDL, etc.) for communication. In X-DPR, XML is used because it formalizes the semantics of the contents of information and facilitates electronic data exchange. As we have seen previously, most of the earlier frameworks focused on standardizing the structure of data models and information exchange between agents. This caused problems while integrating new tools into the framework. Any new tool to be integrated to the framework must abide by the standardized schemas in which information is communicated between agents, which limit the flexibility of agents to implement their own input/output schemas.

The interface-mapping tool helps in achieving flexibility in defining data structures for storing and passing information. Hence, the agents have flexibility in defining the structure of information flowing in and out and by using the XML standard in the X-DPR framework. This provides the required adaptability and user transparency when the inputs/outputs of the agents change or when there is a process change resulting in changes in the agents that interface with each other. The use of interface mapping tool can also be used for mapping different domain ontologies. Information schemas from various domains can be mapped to each other to accomplish an enterprise-level transfer of information rather than just information transfer between software applications.

10.4.4.2. Standardized Means of Describing UIs

Another important issue while developing distributed agent-based systems is the way in which users interact with remote agents. Most of the existing frameworks are based on the assumption that a fixed set of agents are deployed into the system and fixed UIs have been created for each type of agent. However, in an open engineering system in which it is not known what kinds of agents will be deployed on the framework and what kinds of interaction will be required by the system, it is difficult to create individual UIs for each agent. This problem is amplified when each agent is configured differently for different processes. In X-DPR, dynamic user interface generation at the client side is used to overcome this problem. This feature ensures that the UIs can be reused on *heterogeneous platforms* and for *different programming languages*. It provides both the adaptability to changes in UIs without changing the framework, and the ease of use because changing the UIs only requires changing the associated XML file.

10.4.4.3. Standardized Means of Representing Process Information (Using the Decision Support Problem Technique)

In the X-DPR framework, the capability for meta-design is provided using the DSP technique (see Section 10.4.2.2). It helps designers rapidly configure design processes and use distributed resources execute the processes. This capability supports rapid configuration of the product realization environment.

10.4.4.4. Standardized Description of Assets using WSDL

In the X-DPR framework, we use the XML-based standard WSDL to describe the capabilities of agents and ways that they can be invoked. This provides the adaptability to changes in agents and their services. Whenever there is a change in the services provided by an agent, these changes are automatically reflected in the corresponding WSDL files.

10.4.5. Summary of X-DPR Framework

Having reviewed various candidate approaches, we have designed and implemented an adaptable engineering framework for distributed product realization, X-DPR. Such a framework is not unique, but the key elements are quite relevant to our domain of concurrent design of materials and products using computational modeling and simulation for decision support. X-DPR balances the need for flexibility and standardization. Features that balance the requirements include flexible mapping of information between agents, standardized means for describing user interfaces, a standardized means for representing process information, a standardized message protocols based on XML, and a standardized description of assets using WSDL. In X-DPR, the existing communication infrastructure is used as a backbone along with the technologies (Java, Web browsers etc.) and standards (XML, SOAP, WSDL, etc.) for communication. These provide the flexibility and enable the future expansion of the framework. The interface-mapping tool also helps to achieve this flexibility, by easily allowing agents to reconfigure information flow. The client application supports rapid reconfiguration of engineering task and decision-making activities and also task decomposition to formulate hierarchical product realization processes. Standardized service description files (represented in WSDL) are used for creating graphical UIs and interacting with agents. In the X-DPR framework, an emerging industry standard Remote Procedure Call (RPC) protocol, the SOAP-based agent wrapper provides much more flexibility and ease of implementation than is available in the other frameworks. Workflow information is shared among distributed users by the Process Diagram Whiteboard in real time. The most important advantage of the X-DPR framework is that it is compatible with other frameworks. We envision the X-DPR framework as a link between an engineering framework (that manages design chains) and business framework (that manages supply chains).

Coordination and conflict management play an important role in distributed engineering design frameworks. Brazier and coauthors (Brazier, Langen 1995) have categorized these conflicts into two broad categories: (1) conflicts during management of information content and (2) coordination between activities. The first category includes conflicting design requirements and conflicts while updating the design description, whereas the latter category includes design process and agent coordination conflicts. In the X-DPR framework, coordination between agents is achieved by capturing the sequence of agent execution and maintaining their status in the database (see Section 10.4.2.7). The X-DPR framework takes advantage of the STEP data repository and its Java interface discussed in Section 10.4.2.1 for conflict management during updates of the design description.

10.5. Closing Remarks: Future Needs

In this chapter, we discuss various aspects of collaborative, distributed design and the nature of an information management infrastructure that can enable integrated design of materials and products. There are two possible directions an enterprise can take in deploying such a framework—implementing a custom framework for the organization using middleware, or purchasing an end-user software framework. If the organization chooses the first option, the flexibility in integrating specialized in-house software is high. But at the same time, the cost of implementing the framework is also high. The organization would require a dedicated team of software developers who can maintain the framework and help in its growth by integrating new tools. If the second option is selected and the organization decides to purchase end-user frameworks such as Fiper or ModelCenter, the cost of using the framework in the organization is low. At the same time, the flexibility of integrating new types of tools is also limited.

In the context of materials design, the flexibility of commercial frameworks is limited by: (1) the *types of information flows* allowed by these frameworks and (2) the *types of design support tools* available (such as optimization routines, statistical analysis routines, etc.). For example, the information flow between different analysis models in ModelCenter is defined as a set of parameters. If the information flow is more complex that a simple set of parameters, then it is not possible to define information flow between models. This is a significant challenge in materials design problems because the information flow between models is typically complex; examples include information regarding microstructure morphology, grain boundary character, uncertainty in material structure, composition, etc.

Similarly, the types of design support tools available with these applications are limited from the standpoint of materials design. The design tools are mainly centered on geometric modeling and parametric optimization. There is a need to develop and implement algorithms for robust design in these frameworks that give full recognition to the myriad of experimental and modeling issues in material process-structure and structure-property relations discussed elsewhere in this book. Moreover, the frameworks do not support reconfiguration of design processes. A detailed discussion on the limitations and lack of flexibility of commercially available frameworks is provided in (Panchal, Fernández 2005, Panchal, Fernández 2004, Panchal, Vrinat 2004).

One of the key prerequisites for design in a distributed simulation-based design environment is to develop a consistent and standardized way of representing information that can be used by all stakeholders involved in the process. Currently, there is no comprehensive information model for representing the complex information of multifunctional materials. The various complexities in the required information arise from the need to represent and integrate structures and properties at the macro, meso, micro, and quantum levels. Another source of complexity is the various types of information associated with multifunctional, multiscale materials design, such as scalars, matrices, tensors, images, graphs, mathematical equations, computer programs, etc.

Currently, a wide range of materials modeling and simulation efforts are being carried out. Some efforts are focused on specific types of materials, some are focused on modeling specific physical phenomena, some are focused on simulating the material behavior at specific scales, while other efforts are focused on specific application areas. In the future, there is a need for a common platform that can be used to classify, manage, and integrate these diverse modeling efforts in a holistic manner. In order to achieve these capabilities, we envision an open collaborative framework for integrated modeling and design of materials and products. As more and more modeling and simulation tools become widely available to support materials design, there will be a need for establishing standard formats for exchanging information between different tools. Standards such as STEP (Nell 2003, Peak, Lubell 2004) are available in the product design domain, but similar standards are currently lacking in the materials design domain. We envision a need for new standards based on existing International Organization for Standardization (ISO) standards developed at the National Institute of Standards and Technology (NIST); specifically, extending the STEP standard for modeling and exchange of complex materials design related information. Although STEP standards are well developed for representing data in the latter stages of design, these should be extended to represent material data with robust design considerations throughout the design process. Various existing tools can be extended for this purpose. For example, Express (Schenck and Wilson 1994) modeling language supports object-oriented modeling and is developed specifically for complex engineering data. Through the formalization of the underlying information model, standardized digital interfaces at the computing level can be identified between teams of designers. The information model is the key for capturing, archiving, translating, and retrieving information from the corresponding database. Such a material information model is a fundamental and essential starting point for moving toward a future standard that will accelerate development of commercial software applications for materials modeling. A further advantage of developing an information model based on existing standards is the possibility of extending existing CAD applications to include material design considerations.

There is also a growing need for integrating comprehensive materials databases into the frameworks to support integrated products and materials design. MatWeb (MatWeb 2008) allows exporting materials data to various CAD/ FEA software tools such as ANSYS, SolidWorks, Algor, and Comsol Multiphysics. However, the kinds of material properties that can be exported are limited. The link between the MatWeb materials database and CAD/FEA applications is only to support the design of products using *existing* materials; support for *materials design* is severely limited. One of the reasons is that the database only captures material properties. The relationships between material (micro)structure and material properties, which are essential for materials design, are not captured. In the future, we envision new materials databases that capture the relationships between all the four aspects—processing, structure, properties, and performance—with more focus being placed on the material structure (either in digital format or in terms of correlation functions) as the medium of communication for design information rather than properties.

Finally, we emphasize the need to pursue open, adaptable XML-based frameworks, such as X-DPR in Section 10.4, to enable implementation of the gamut of algorithms discussed earlier in this volume for materials design, including top-down design based on bottom-up simulations (Inductive Design Exploration Method) and management of complexity based on successive refinement of models and patterns of coupling using value-of-information approaches. The need for adaptive, reconfigurable design processes is apparent.

References

Alibre, 2006, Alibre Design 9.1. <http://www.alibre.com/products/>.

ANSYS, 2006. ANSYS Workbench™. <http://www.ansys.com/products/workbench.asp/>.

Ansys Inc, 2003. Ansys Products. <http://www.ansys.com/ansys/index.htm/>.

Box, D., Ehnebuske, D., Kakivaya, G., Layman, A., Mendelsohn, N., Nielsen, H. F., Thatte, S., and Winer, D. 2000. Simple Object Access Protocol (SOAP) 1.1, <http://www.w3.org/TR/SOAP/>.

Bras, B.A., Mistree, F., 1991. Designing design process in decision-based concurrent engineering. SAE Trans. J. Mater. Manuf. 100 (3), 451–458.

Brazier, F.M.T., Langen, P.H.G.v., Treur, J., 1995. Modeling conflict management in design: an explicit approach. Artif. Intell. Eng., Anal. Manuf. 9 (4), 353–366.

Chalfan, K.M., 1986. A knowledge system that integrates heterogeneous software for a design application. AI Mag. 7 (2), 80–84.

Choi, H.-J., 2001. A framework for distributed product realization, M.S. thesis, G.W. Woodruff School of Mechanical Engineering, Georgia Institute of Technology, Atlanta, GA.

Cramer, D., Jayaram, U., and Jayaram, S., 2002. A collaborative architecture for multiple computer aided engineering applications. ASME 2002 Design Engineering Technical Conferences. Montreal, Canada, September 30–October 4. Paper Number: DETC2002/CIE–34498.

Cutkosky, M., Engelmore, R., Fikes, R., Genesereth, M., Gruber, T., Mark, W., Tenenbaum, J., Weber, J., 1993. PACT: An experiment in integrating concurrent engineering systems. IEEE Comput. 26 (1), 28–37.

Dabke, P., Cox, A., Johnson, D., 1998. NetBuilder: An environment for integrating tools and people. Comput. Aided. Des. 30 (6), 465–472.

Ebbesmeyer, P., Gausemeier, J., Krumm, H., Molt, T., 2001. Virtual Web plant: An Internet-based plant engineering information system. J. Comput. Inf. Sci. Eng. 1 (2), 257–260.

Engineous, Inc. 2001. Rapid Design Exploration and Optimization. <http://www.engineous.com/toc_ht.html/>.

Engineous Inc., 2001. Product Overview: iSIGHT. <http://www.engineous.com/overview.html/>.

Engineous Inc., 2004. iSIGHT, Version 8.0, <http://www.engineous.com/product_iSIGHT.htm/>.

Gerhard, F.J., Rosen, D.W., Allen, J., Mistree, F., 2001. A distributed product realization environment for design and manufacturing. J. Comput. Inf. Sci. Eng. 1 (3), 235–244.

Gupta, S.K., Lin, E., Lo, A.J., and Xu, C., 2002. Web-based innovation alert services to support product design evolution, *ASME 2002* Design Engineering Technical Conferences. Montreal, Canada, September 30–October 4. Paper Number: DETC2002/CIE-34462.

Hayes, A.M., Wang, A., Dempsey, B.M., and McDowell, D.L., 2001. Mechanics of linear cellular alloys. Proceedings of IMECE, International Mechanical Engineering Congress and Exposition. New York, November 11–16.

Huang, G.Q., Lee, S.W., Mak, K.L., 1999. Web-based product and process data modeling in concurrent 'design for x'. Rob. Comput. Integr. Manuf. 15 (1), 53–63.

Jiang, P., Fukuda, S., Raper, S.A., 2002. TeleDM: An Internet Web service testbed for fast product design supported by prototype manufacturing. J. Comput. Inf. Sci. Eng. 2 (2), 125–131.

Kao, K.J., Seeley, C.E., Yin, S., Kolonay, R.M., Rus, T., Paradis, M., 2004. Business-to-business virtual collaboration of aircraft engine combustor design. J. Comput. Inf. Sci. Eng. 4 (4), 365–371.

Kim, C.S., Kim, N., Kim, Y., Kang, S., and O'Grady, P., 1998. Internet-based concurrent engineering: An interactive 3D system with markup. Proceedings of the 18th ASME Computer in Engineering Conference. Atlanta, GA, September 13–16.

Lin, R., Afjeh, A.A., 2004. Development of XML databinding integration for Web-enabled aircraft engine simulation. J. Comput. Inf. Sci. Eng. 4 (3), 186–196.

LKSoft. 2006. JSDAI, <http://www.jsdai.net/>.

Madhusudan, T., 2004. An intelligent mediator-based framework for enterprise application integration. J. Comput. Inf. Sci. Eng. 4 (4), 294–304.

Marcato, D., 2002. Distributed Computing with SOAP. <http://www.devx.com/upload/free/features/vcdj/2000/04apr00/dm0400/dm0400.asp/>.

MatWeb, 2008. Online Materials Information Resource. <http://www.matweb.com/>.

Microsoft, 1996. DCOM Technical Overview. <http://msdn.microsoft.com/library/default.asp?url=/library/en-us/dndcom/html/msdn_dcomtec.asp/>.

Microsoft, 2003. SOAP Toolkit 3.0. <http://msdn.microsoft.com/downloads/default.asp?URL=/downloads/sample.asp?url=/msdn-files/027/001/948/msdncompositedoc.xml/>.

MSC Software, 2006. MSC.Acumen. <http://www.mscsoftware.com.my/products/software/msc/acumen/index.htm/>.

Muster, D., Mistree, F., 1988. The decision support problem technique in engineering design. Int. J. Appl. Eng. Educ. 4 (1), 23–33.

Nell, J., 2003. STEP on a Page (ISO 10303). <http://www.nist.gov/sc5/soap/>.

OASIS, 2004. Introduction to UDDI: Important Features and Functional Concepts. <http://uddi.org/pubs/uddi-tech-wp.pdf/>.

Olsen, G.R., Cutkosky, M., Tenenbaum, J., Gruber, T.R., 1995. Collaborative engineering based on knowledge sharing agreements. Concurrent Eng. Res. Appl. 3 (2), 145–159.

Pahng, F., Senin, N., Wallace, D., 1998. Distribution modeling and evaluation of product design problems. Comput. Aided Des. 30 (6), 411–423.

Panchal, J.H., 2002. Towards a design support system for collaborative product realization. M.S. thesis, G.W. Woodruff School of Mechanical Engineering, Georgia Institute of Technology, Atlanta, GA.

Panchal, J.H., Choi, H.-J., Allen, J.K., Rosen, D., Mistree, F., 2007. An adaptable service-based framework for distributed product realization. In: Li, W. (Ed.), Collaborative Product Design and Manufacturing Methodologies and Applications. Springer-Verlag, UK, pp. 1–37.

Panchal, J.H., Fernández, M. G., Allen, J.K., Paredis, C.J.J., and Mistree, F., 2005. Facilitating meta-design via separation of problem, product, and process information. 2005 ASME International Mechanical Engineering Congress and Exposition. Orlando, FL November 5–11. Paper number: IMECE2005-80013.

Panchal, J.H., Fernández, M.G., Paredis, C.J.J., and Mistree, F., 2004. Reusable design processes via modular, executable, decision-centric templates. 10th AIAA/ISSMO Multidisciplinary Analysis and Optimization Conference. Albany, NY, August 30–September 1. Paper number: AIAA-2004-4601.

Panchal, J. H., Vrinat, M., and Brown, D. H., 2004. Design and Simulation Frameworks, Collaborative Product Development Associates, LLC <http://cpd-associates.com/>. Report Number: 050302.

Peak, R., Lubell, J., Srinivasan, V., Waterbury, S., 2004. STEP, XML, and UML: Complementary technologies. J. Comput. Inf. Sci. Eng.: Spec. Issue on Eng. Inf. Manage. Prod. Lifecycle Manag. 4 (4), 379–390.

Phoenix Integration Inc., 2004. ModelCenter®, Version 5.0. <http://www.phoenix-int.com/products/ModelCenter.html/>.

PTC. 2003. PTC Windchill. http://www.ptc.com/products/windchill/.

Rajagopalan, S., Pinilla, J. M., Losleben, P., Tian, Q., and Gupta, S. K., 1998. Integrated design and rapid manufacturing over the Internet. ASME 1998 Design Engineering Technical Conferences. Atlanta, GA, September 13–16. Paper number: DETC98/CIE-5519.

Ray, S.R., 2002. Interoperability standards in the semantic Web. J. Comput. Inf. Sci. Eng. 2 (1), 65–71.

Rezayat, M., 2000. The enterprise-Web portal for life-cycle support. Comput. Aided Des. 32 (2), 85–96.

Rezayat, M., 2000. Knowledge-based product development using XML and KCs. Comput. Aided Des. 32 (5–6), 299–309.

Rodgers, P.A., Huxor, A.P., Caldwell, N.H.M., 1999. Design support using distributed Web-based AI tools. Res. Eng. Des. 11 (1), 31–44.

Schenck, D.A., Wilson, P.R., 1994. Information Modeling: The EXPRESS Way. Oxford University Press, New York.

Schmidt, D.C., 2002. Distributed object computing with CORBA middleware. <http://www.cs.wustl.edu/~schmidt/corba.html/>.

Seepersad, C.C., 2004. A robust topological preliminary design exploration method with materials design applications. Ph.D. dissertation, G.W. Woodruff School of Mechanical Engineering, Georgia Institute of Technology, Atlanta, GA.

Seepersad, C.C., Dempsey, B.M., Allen, J.K., Mistree, F., and McDowell, D. L., 2002. Design of multifunctional honeycomb materials. 9th AIAA/ISSMO Symposium on Multidisciplinary Analysis and Optimization. Atlanta, GA, September 4–6. Paper number: AIAA-2002-5626.

Shyamsundar, N., Dani, T., Sonthi, R., and Gadh, R., July 1998. Shape abstractions and representations to enable Internet-based collaborative CAD. Proceedings of the *Japan-USA Symposium on Flexible Automation*, Otsu, Japan, 229–236.

Simpson, T.W., Umapathy, K., Nanda, J., Halbe, S., Hodge, B., 2003. Development of a framework for Web-based product platform customization. J. Comput. Inf. Sci. Eng. 3 (2), 119–129.

Simulia, 2009. FIPER Infrastructure. <http://www.simulia.com/products/fiper.html/>.

Systinet Corporation. 2004. Systinet Server for Java. <http://www.systinet.com/products/ssj/overview/>.

Toye, G., Cutkosky, M., Leifer, L., Tenenbaum, J., Glicksman, J., 1994. SHARE: A methodology and environment for collaborative product development. Int. J. Intell. Coop. Inf. Syst. 3 (2), 129–153.

UGS, 2006. Teamcenter 2005. <http://www.ugs.com/products/teamcenter/>.

Wang, F., Wright, P., July 1998. Internet-based design and manufacturing on an open architecture machine center. Proceedings of the *Japan-USA Symposium on Flexible Automation*. Otsu, Japan, 221–228.

Wang, J., Ma, Z.-D., Hulbert, G.M., 2005. A distributed mechanical system simulation platform based on a gluing algorithm. J. Comput. Inf. Sci. Eng. 5 (1), 71–76.

Wang, L., Shen, W., Xie, H., Neelamkavil, J., Pardasani, A., 2001. Collaborative conceptual design—State of the art and future trends. Comput. Aided Des. 34 (13), 981–996.

Wang, P., Bjarnemo, R., Motte, D., 2005. A Web-based interactive virtual environment for mobile phone customization. J. Comput. Inf. Sci. Eng. 5 (1), 67–70.

Wu, T., Xie, N., Blackhurst, J., 2004. Design and implementation of a distributed information system for collaborative product development. J. Comput. Inf. Sci. Eng. 4 (4), 281–293.

Wujek, B.A., Koch, P.N., McMillan, M., and Chiang, W.-S., 2002. A distributed component-based integration environment for multidisciplinary optimal and quality design. 9th AIAA/ISSMO Symposium on Multidisciplinary Analysis and Optimization. Atlanta, GA, September 4–6. Paper number: AIAA2002-5476.

Xiao, A.H., Kulkarni, R., Allen, J., Rosen, D.W., Mistree, F., and Feng, S.C., 2001. A Web-based distributed product realization environment. ASME Computers in Engineering. Pittsburgh, PA, September 4–6. Paper number: DETC2001/CIE-21766.

CHAPTER 11

Closure—Advancing the Vision of Integrated Design of Materials and Products

It should be abundantly clear to the reader that concurrent design of materials and products is an emerging multidiscipline ripe for creative contributions that are essential to realizing its full potential. It is inherently a collaborative activity at the intersection of disciplines of computational materials science, mechanics of materials, materials characterization and in situ experiments, information technology, design theory, and engineering systems design, among others.

A thread of visionary initiatives and research and development programs from the mid-1990s onward has advanced this idea of developing frameworks for concurrent material and product design. A vision for systems-based materials design and virtual manufacturing was outlined a decade ago for the academic and research communities at a 1998 workshop (McDowell and Story 1998) sponsored by the National Science Foundation (NSF) and entitled "New Directions in Materials Design Science and Engineering (MDS&E)." That workshop report concluded that a change of culture is necessary in U.S. universities and industries to cultivate and develop the concepts of simulation-based design of materials to support concurrent design of material and products. It also suggested that globalization will naturally drive a specialty (tailored) materials supply/development industry and distributed virtual design and manufacturing. Recommendations included establishing a U.S. roadmap addressing:

- Development and maintenance of databases to enable distributed materials design,

- Development of principles of systems design that integrate hierarchical materials systems, and

- Identification of opportunities and deficiencies in science-based modeling, simulation, and characterization "tools" to support concurrent design of materials and products.

The United States Council for Automotive Research (USCAR) program from 1995 to 2000, sponsored by the U.S. Department of Energy, provided an early indication of the feasibility of obtaining substantial improvements by coupling structure-property relations with component-level design (Fan, McDowell et al. 2003, McDowell, Gall et al. 2003) in design of lightweight cast automotive components. The 2000–2005 DARPA/AFRL Accelerated Insertion of

Materials (AIM) program (Apelian, Alleyne et al. 2004, Backman and Dutton 2006) offered insight into how computational materials science and engineering can be harnessed in the future to assist in developing and certifying materials in a shorter timeframe that more closely matches the duration of the product design cycle. AIM was a bold initiative that assembled materials suppliers, original equipment manufacturers (OEMs), and government and academic researchers in a collaborative, distributed effort to build designer knowledge bases for metallic systems (nickel-base superalloys for gas turbine engine disks) and composite airframe materials. The program was founded upon three systemic changes in materials development: (1) revolutionizing the way product designers and materials engineers interact, (2) advancing the integration of computational materials science/mechanics with engineering design tools, and (3) creating a designer knowledge database in which designers can embed experience and build design system templates.

Building on demonstrations from these programs, it is clear that a systems approach is needed to integrate modeling and simulation in a targeted, prioritized manner to effectively support integration of materials development with product development. Indeed, other recent federal initiatives emphasize the interdisciplinary collaboration of bottom-up materials modeling and simulation, high performance computing, networking, and information sciences to accelerate the discovery of new materials. The recent NSF vision for Materials Cyber-Models for Engineering (Billinge, Rajan et al. 2006; Rajan 2009), for example, provides a science-based perspective on using quantum and molecular modeling tools to explore potential new materials and compounds, making the link to properties. To support improvement or tailoring of existing materials, as well as assessing feasibility of synthesis or processing of new materials, we contend that a systems approach is essential that embeds material processing/supply, manufacturing, computational materials science, experimental characterization, and systems engineering and design. This idea is similar to the conceptualization of Integrated Computational Materials Engineering (ICME) outlined in a National Academy of Engineering (NAE) National Materials Advisory Board report (Pollock, Allison et al. 2008), an approach to concurrent design of products and the processes and materials that comprise them. This is achieved by linking materials process-structure and structure-property models at multiple length and time scales with elements of the design system for specific products and applications.

There are important educational challenges in advancing the concepts underlying ICME. The report of the aforementioned 1998 NSF MDS&E workshop (McDowell and Story 1998) stated that "The systems integration that is necessary to conduct materials design must be recognized as part of the materials education enterprise. This has not generally been the case." In spite of fairly widespread materials design courses offered in undergraduate curricula in materials science in research universities, there is still much progress to be made in offering instruction in supportive areas of computational materials science and related aspects of engineering systems design. It is likely necessary for capstone courses in larger engineering disciplines such as mechanical and civil/structural to address elements of materials design in collaboration with materials science and engineering departments. In this regard, the present volume is intended as a means for introducing early concepts for algorithms and computing infrastructures to bridge these disciplines. Augmented with modeling principles and methods, the basic ideas presented here are intended to advance the philosophy underpinning materials design. They serve as a starting point to integrate the present state-of-the-art in systems design with the objective of integrating computational modeling and simulation of material process-structure-property relations. One by-product of this kind of systematic approach to materials design is that it can be applied to intriguing problem sets that excite undergraduate and introductory graduate students and is relevant to real products. It emphasizes the role of the engineer as a designer, regardless of how a student's particular expertise fits into the overall picture of materials and product development.

Of course, the educational and developmental aspects of concurrent design of materials and products have global dimensions. Friedman (2006) argues in *The World Is Flat* that the Internet has ushered in a new era of global communication and collaboration that has fundamentally changed the nature of product development and manufacturing. Competition for providing goods and services is now global, and modeling and simulation can be conducted in "just-in-time" fashion from qualified vendors. This change of paradigm offers both opportunities and challenges that will place a premium on flexibility of design processes and knowledge management, engaging distributed project engineers, materials experts, model developers and scientists, information specialists, and materials suppliers in a much more concurrent manner. It should inexorably result in improved products, decreased time to market, and increased customization of products. Moreover, exciting vistas and potential business models such as mass collaboration over the Web are unveiled to creative and entrepreneurial minds.

Indeed, we have seen fit to compile this book to flesh out the relevant issues and challenges that face more intimate integration of materials development with product design. We expect activity and progress in this field to accelerate with increased reliance on modeling and simulation to support decisions in materials design. Supplemented by supporting curricula in materials science, existing infrastructure for materials selection in the properties/performance linkage (e.g., Granta Design Ltd. 2007), computational methods, applied mechanics, and systems theory, it is anticipated that this book can play an integrative role in advanced undergraduate and introductory graduate courses in engineering design.

The process of preparing this book has invited the identification of enabling technologies that require further development in order to advance the vision of integrated design of materials and products. These enablers might serve as a basis for research and educational initiatives. We have already discussed the need for more complete integration of engineering systems design curricula at the undergraduate level with materials science and engineering (and even materials physics and chemistry). The flat-world paradigm of Friedman and the distributed nature of research and development efforts in both industry and academia strongly suggest that development of distributed collaborative networking is a key element of realizing the vision of integrated materials and product design. There are important scientific and engineering research challenges, some of which are summarized in what follows.

The linkage between integrated material and product design and the information sciences is perhaps rather obvious in view of the foregoing discussion. There is ample room for creative contributions in this regard. The emerging field of materials informatics (Rajan 2008) embodies elements such as knowledge discovery extracted from databases via data mining in interdisciplinary areas such as statistics, materials databases, and results of material simulations to assist in discovery of new materials concepts. These ideas are particularly attractive for cases in which well-established theories and models do not exist, i.e., high uncertainty and little intuitive guidance. In terms of the integration of design methods with the computing and information sciences, we list the following areas of inquiry:

- Development of adaptively reconfigurable digital interfaces between models and databases.

- Methods and protocols for Web-based modeling and simulation environments delivering situation- and application-specific capabilities.

- Mass collaboration in Web-based integrated material and product design.

- Storage of simulation results for subsequent visualization and interrogation, including bandwidth and transmission issues, parallelization, and distributed networking.

- Exploitation of massively parallel computing for bottom-up simulations to support design exploration, reallocating processing to focus on top-down design and detail design in later stages.

- Distinguishing design exploration from detail design.

- Protocols for assigning uncertainty metrics and tracking propagation through networks.

Another point to emphasize in terms of a fundamental shift in paradigm from a focus on materials selection to concurrent materials and product design is the importance of quantitative description of microstructure. Indeed, microstructure attributes ultimately serve as the most effective interface for communication between the materials developer and the systems designer. In the conventional view of materials selection, as discussed in Chapter 1, material properties are the means of communicating information between materials developers, modelers, engineers, and systems designers. In other words, the set of material properties is assumed to form a "clean" interface between two groups of individuals that by and large do not iterate together within the design process to achieve system solutions. This perspective of the material properties as the interface between materials engineers and product designers has perhaps been fueled by a traditional decomposition of effort and expertise in both industry and academia. Materials selection based on properties is a "fence" that demarcates organizations and departments in many cases. It has also led to a focus by the materials development and characterization communities on the goal of populating property databases and mining them, largely avoiding more complete integration of tailorable aspects of material systems with mechanics- and system-level issues. For example, how many universities integrate computational materials science, finite element methods, controls, manufacturing, and systems design topics in their materials science and engineering curriculum? Similarly, how many programs in mechanical, aerospace, and civil engineering provide coursework for undergraduates beyond strength of materials and basic principles of material structure and properties? And yet, materials design is an integrative activity across disciplines. Its elements can be taught starting in the freshman year, as evidenced by the curriculum in the Design of Materials Science and Engineering (MSE) course at Northwestern University (McDowell and Olson 2008). This effort at Northwestern has also developed in parallel with a noteworthy, successful spinoff startup company, (QuesTek LLC 2009), which specializes in design and development of advanced alloy systems for ground vehicle and aerospace applications.

Perhaps more than anything else, the need to address materials design and development in a manner that is concurrent with product design requires acknowledgement of the pivotal role of microstructure as the interface between materials engineers and systems modelers/designers. Unlike *properties*, a much richer palette of *responses* can be associated with microstructure, many of which we do not regularly measure or estimate but play an important role in a specific design. Many such responses are evolutionary in nature and cannot be done justice by expressing in terms of single point properties. For example, how does one express the full range of implications of crystallographic texture of polycrystalline materials using only the notion of anisotropic elastic constants or anisotropy of yield and ultimate strength? Much is omitted from these kinds of single point properties—for example, work-hardening behavior in these and intermediate orientations of loading, multiaxial stress state dependence of strain to localization and failure, variability in properties due to variability in microstructure, influence of grain boundary networks, and so on. Whereas coarse, low-order descriptors such as mean grain size and classification of mode of texture can be linked to a small set of point properties, the richness of information contained in a digital instantiation (or representation based on n-point statistics (Adams et al. 2004; Fullwood et al. 2007; Li et al. 2005; Torquato 2002)) of the microstructure represented by a large set of degrees of freedom can serve many purposes, and offers a quantitative basis for judging relative improvements in a material, comparing one material versus another material, or exploring responses in multiphysics, multifunctional domains using computational simulations that operate on these instantiations.

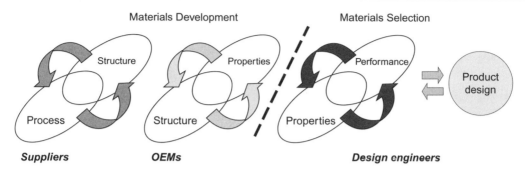

Figure 11.1 Typical time scale and organizational separation of materials development and materials selection components of the integrated materials and product design process.

Furthermore, in practice, the duration and number of iterative cycles, mix of computation and experiments, and timeframe of assessing process-structure relations in materials development do not match those of evaluating structure-property relations. Another way of saying this is that processing of materials is typically not efficiently coordinated with structure-property evaluations for a number of reasons. Moreover, product design is often conducted with system-level considerations by distinct groups separated by a "fence" (delineated by the bold dashed line on the right side of Figure 11.1), representing the step of materials selection. As such, the property requirements may be specified, but there is a lack of integration of the degrees of freedom of material microstructure into the overall systems design problem. For example, we might be interested in how grain size or fine-scale-strengthening precipitates affect performance of a turbine disk or crankshaft. Using properties at the point of materials selection does not confer transparency (or even accessibility) to such relationships. In addition, the cooperative materials development process between materials suppliers and OEMs typically involves substantially different timeframes for materials processing and assessment (experiments and simulations) of structure-property relations, often carried out in different organizations. In an analogy to multiscale modeling, both time (cycle duration for processing and certification experiments) and length (organizational distribution) scales serve as limiting factors in coordinating the design process.

This is essentially just a somewhat more detailed decomposition of the issues addressed by the AIM program discussed earlier in this chapter. On the other hand, the approach suggested here is to effectively remove the rightmost bold dashed line in Figure 11.1, effectively integrating structure-property relations with materials selection by coupling computational materials simulation with the systems product design process. If this is realized, then the primary interface for exchange of disparate information is between materials suppliers and the OEMs, as shown in the short dashed line on the left. The mediatory "language" for communicating design variables then becomes the material microstructure. The focus for accelerating the insertion of new and improved materials into products is then on balancing the timeframe for materials processing with the structure-properties-performance sequence; the latter is closely integrated, while the former must consider details of microstructure attributes as the targets of processing rather than just properties. Within the ICME rubric, there is evidence that this is happening in OEMs (Pollock et al. 2008).

Systems engineering approaches for concurrent design of hierarchical materials and structures are made feasible by the confluence of several fields:

- Ubiquitous computing.

- Computational materials science.

- Advances in high resolution materials characterization and in situ measurements.

- Advances in computational micromechanics, including stochastic and probabilistic methods.

- Advances in decision theory, networking, and computer-assisted design of products.

There are several particularly noteworthy future directions for concurrent systems design of materials and products that intersect with current trends in computational materials science and mechanics but have distinct objectives and methods:

- Synthesis of multiscale and multiphysics modeling with decision-based design strategies for multilevel materials design.

- Algorithms to conduct feasibility studies for potential design solutions in robust concept exploration—early stage exploration of ranges of potential solutions to meet specific design requirements, extending beyond experience or intuition.

In materials design problems, one often finds that models are either nonexistent or insufficiently developed to support decision making. This includes both models for process-structure relations and 3D microstructure, as well as associated 3D models for structure-property relations. Of particular need is the coordination of model repositories for rapid availability in design exploration. A complicating factor that is rarely addressed is quantifying uncertainty of model parameters and model structure that is necessary in robust design of materials. Another very important consideration is that availability of mechanistic models is often the limiting factor in applying decision-based design frameworks in view of the need for reliability and confidence estimates; however, guidance is required to decide how to best invest in model development that will maximize payoff or utility in the design process. Not all models are equally important in terms of their role in design, and this depends heavily on the design objectives and requirements.

On the other hand, one can readily identify gaps in multiscale modeling methods without regard to utility in design. One example is the gap in reliable, robust models between the level of atomistics and polycrystal plasticity for metallic systems. This gap is closing each year with advances in discrete dislocation plasticity and other methods, but progress in predictive methods for dislocation patterning at mesoscales has been slow, in part due to the lack of top-down calibration compared to polycrystal plasticity. From the perspective of decision support in materials design, much can be done using models at lower and higher scales of the hierarchy without a requirement to form a bridge to accurately predict these substructures. The relative need to bridge this gap is problem-dependent.

Where are the opportunities for improvement of algorithms and tools in materials design? Several can be identified:

- Rapid methods for feasibility study and robust concept exploration. Early stage exploration of ranges of potential solutions to specific requirements, beyond experience or intuition. This requires assessment of value-of-information metrics (utility theory), and identification where models are needed, establishing model/ database priorities.

- Methods for quantifying and managing uncertainty in material process-structure and structure-property models and in model to support decision-based robust design.

- Microstructure-mediated design. Improved portable file format representations of material microstructure. Balancing iterations in process development with structure-property-performance iterations—managing assets and deciding on the nature of information interfaces between processing and structure-property relations (cf. Figure 11.1); distinguishing design exploration from detail design.

- Parallel processing algorithms for robust concept exploration. Materials design is an ideal candidate for parallelization in the initial design exploration process (cf. the Inductive Design Exploration Method (IDEM) outlined in Chapter 8). Such searching is normally mentioned in connection with data mining, but we believe the wider exploration of potential design space is a daunting task worthy of massively parallel computing.

- In addition to improving product affordability, innovation, and performance, there are other important capabilities that accrue to this broad technology, including but not limited to:
  - Prioritizing models and computational methods in terms of measures of utility in supporting design decisions.
  - Prioritizing mechanisms and materials science phenomena to be modeled for a given design problem.
  - Conducting feasibility studies to establish the probable return on investment of candidate new or improved material systems.
  - Assessing to what degree the gaps (time, cost) between processing-related materials development and incorporation into products can be closed based on application of available technologies.

While by no means exhaustive, investment in basic research programs and emerging curricula in these topics should pay dividends in terms of establishing useful protocols for integrated design of materials and products. We look forward to an accelerating trend of new and profound advances in this emergent multidiscipline, fully acknowledging that some cannot be anticipated at the time of this writing.

References

Adams, B.L., Lyon, M., Henrie, B., 2004. Microstructures by design: linear problems in elastic–plastic design. Int. J. Plast. 20 (8–9), 1577–1602.

Apelian, D., Alleyne, A., Handwerker, C.A., Hopkins, D., Isaacs, J.A., Olson, G.B., Vidyanathan, R., Wolf, S.D., 2004. Accelerating Technology Transition: Bridging the Valley of Death for Materials and Processes in Defense Systems Report Number: ISBN-10: 0-309-09317-1. National Materials Advisory Board, NAE, National Academies Press.

Backman, D.G., Dutton, R., 2006. Integrated materials modeling for aerospace components. In: Symp. on the Role of Computational Methods in Materials Research and Development: Applications of Materials Modeling and Simulation, MS&T '06, Cincinnati, OH, October 15–19.

Billinge, S.J.E., Rajan, K., Sinnot, S.B., 2006. From cyberinfrastructure to cyberdiscovery in materials science: Enhancing outcomes in materials research. Report of NSF-sponsored Cyberinfrastructure in Materials Research workshop held in Arlington, VA, August 3–5, <http://www.mcc.uiuc.edu/nsf/ciw_2006/ > .

Fan, J., McDowell, D.L., Horstemeyer, M.F., Gall, K., 2003. Cyclic plasticity at pores and inclusions in cast al-si alloys. Eng. Fract. Mech. 70 (10), 1281–1302.

Friedman, T.L., 2006. The world is flat [updated and expanded]: a brief history of the twenty-first century. Farrar, Straus, & Giroux, New York.

Fullwood, D.T., Adams, B.L., Kalidindi, S.R., 2007. Generalized pareto front methods applied to second-order material property closures. Comput. Mater. Sci. 38 (4), 788–799.

Granta Design Ltd., 2007. Granta Material Intelligence Software. <http://www.grantadesign .com/ > .

Li, D.S., Bouhattate, J., Garmestani, H., 2005. Processing path model to describe texture evolution during mechanical processing. Materials Science Forum, v 495–497. In PART 2, Textures of Materials, ICOTOM 14—Proceedings of the 14th International Conference on Textures of Materials, 977–982.

McDowell, D.L., Gall, K., Horstemeyer, M.F., Fan, J., 2003. Microstructure-based fatigue modeling of cast A356-T6 alloy. Eng. Fract. Mech. 70 (1), 49–80.

McDowell, D.L., Story, T.L., 1998. New directions in materials design science and engineering (MDS&E). Report of a NSF DMR-sponsored Materials Design Science and Engineering workshop, October 19–21.

McDowell, D.L., Olson, G.B., 2008. Concurrent design of hierarchical materials and structures.. Sci. Model. Simul. (CMNS) 15 (1), 207.

Pollock, T.M., Allison, J.E., Backman, D.G., Boyce, M.C., Gersh, M., Holm, E.A., LeSar, R., Long, M., Powell IV, A.C., Schirra, J.J., Whitis, D.D., Woodward, C., 2008. Integrated computational materials engineering: a transformational discipline for improved competitiveness and national security Report Number: ISBN-10: 0-309-11999-5. National Materials Advisory Board, NAE, National Academies Press.

Rajan, K., 2008. Materials informatics, part I: a diversity of issues. JOM 60 (3), 50.

Rajan, K., 2009. Informatics and integrated computational materials engineering: part II. JOM 61 (1), 47.

QuesTek Innovations LLC., 2009. <http://www.questek.com/ > .

Torquato, S., 2002. Random Heterogeneous Materials: Microstructure and Macroscopic Properties. Springer-Verlag, New York.

Index